Pete

Oxford Applied Mathematics and Computing Science Series

General Editors

R. F. Churchhouse, W. F. MacColl and A. B. Tayler

OXFORD APPLIED MATHEMATICS AND COMPUTING SCIENCE SERIES

I. Anderson: *A First Course in Combinatorial Mathematics*

D. W. Jordan and P. Smith: *Nonlinear Ordinary Differential Equations (Second Edition)*

D. S. Jones: *Elementary Information Theory*

B. Carré: *Graphs and Networks*

A. J. Davies: *The Finite Element Method*

W. E. Williams: *Partial Differential Equations*

R. G. Garside: *The Architecture of Digital Computers*

J. C. Newby: *Mathematics for the Biological Sciences*

G. D. Smith: *Numerical Solution of Partial Differential Equations (Third Edition)*

J. R. Ullmann: *A Pascal Database Book*

S. Barnett and R. G. Cameron: *Introduction to Mathematical Control Theory (Second Edition)*

A. B. Tayler: *Mathematical Models in Applied Mechanics*

R. Hill: *A First Course in Coding Theory*

P. Baxandall and H. Liebeck: *Vector Calculus*

D. C. Ince: *An Introduction to Discrete Mathematics and Formal System Specification*

P. Thomas, H. Robinson, and J. Emms: *Abstract Data Types: Their Specification, Representation, and Use*

D. R. Bland: *Wave Theory and Applications*

J. Kondo: *Integral Equations*

J. Kondo

Science Council of Japan

Integral Equations

KODANSHA
TOKYO 1991

CLARENDON PRESS
OXFORD 1991

Exclusive sales rights for Japan:
Kodansha Ltd., 12-21, Otowa 2-chome, Bunkyo-Ku, Tokyo 112 (Japan)

Distribution for all other countries:
Oxford University Press, Walton Street, Oxford OX2 6DP
Oxford New York Toronto
Delhi Bombay Calcutta Madras Karachi
Petaling Jaya Singapore Hong Kong
Nairobi Dar es Salaam Cape Town
Melbourne Auckland

and associated companies in
Berlin Ibadan

Oxford is a trade mark of Oxford University Press

British Library Cataloguing in Publication Data
Kondo, J.
Oxford applied mathematics and computing series
Integral Equations
Integral Equations
I. Title
ISBN 0 19 859691 X (pbk.)
ISBN 0 19 859681 2

Library of Congress Cataloging-in-Publication Data
Kondo. J. (Jiro)
Integral Equations/J. Kondo
—(Oxford applied mathematics and computing science series)
Bibliography: P.
Includes index.
Integral Equations
I. Title. II. Series.
ISBN 0 19 859691 X (pbk.)
ISBN 0 19 859681 2
ISBN 4–06–204031–X (Japan)

Printed in Japan

Preface

The integral equation is a kind of functional equation that includes the integral of an unknown function; its fundamental study has become active only since the beginning of this century. This is strange since differential equations have been a focus of interest to mathematicians for no less than two hundred years.

In the field of the natural sciences, especially in physics and engineering, laws are given, as in Newtonian dynamics and so on, in the form of differential equations. Such an equation denotes an instantaneous balance or local equilibrium. By solving it we obtain results associated with suitable initial or boundary conditions, taken from the observed conditions or experiments. By this means we can understand natural phenomena in comparison with experimental results and complete a design in engineering to agree with the calculated values. When, however, natural law is not established sufficiently as an instantaneous balance or local equilibrium and therefore cannot be expressed in terms of a differential equation, or when the experimental method or observed condition is too closely related to the object of research to be considered separately, and therefore cannot be expressed in the form of the initial or boundary conditions of a differential equation, it is frequently treated within the sphere of integral equations. In addition, cases with singular boundary conditions, or certain kinds of differential equations can be treated more easily when transformed into integral-equation problems.

For some observed values, a problem can be more suitably represented by an integral equation than by a differential equation. In short, we may conjecture that as an expression of natural law, the differential equation describes microscopic balance, while the integral equation shows what we call macroscopic balance and for certain sorts of problems yields more natural expressions. Moreover, the fact that the history of the integral equation is quite young compared with that of the differential equation signifies that there awaits a vast, as yet unopened, treasure house in the field of applications of integral equations. Recently, attention has been drawn to the fact that various problems of applied mathematics—such as potential problems, physical optics, hysteresis, time-series analysis, etc.—can be expressed and solved in terms of integral equations.

It goes without saying that the theory of integral equations is important as a part of pure mathematics, and since the beginning of the twentieth century it has been the central problem of analysis for a number of mathematicians. Today we have a completed theory of linear integral equations. This theory is fluent and deep enough to entice junior researchers into the world of pure mathematics but, on the other hand, its ideas are so new and original that a deliberate attitude of mind is required on entering the field, or else there lies ahead no small danger of missing the coherence of the theory while engrossed in running after the achievements of earlier researchers. This can be seen by the fact that these studies are crowned with the names of just a few scholars: principally Volterra, Fredholm, Hilbert, and Picard.

In this book we intend to unite the theories of integral equations at a little higher level, and to pursue the achievements of the geniuses who paved the way for us. We shall arrange various theories into a unified system, clarifying the associations found among them and showing that a unique idea runs through the methods of solution. Thus we shall try to facilitate the synthetic understanding of these theories. At the same time we have collected applications in as many varied fields as possible in order to understand the use, by defining characteristics and qualities, and to indicate the applicability of integral equations as a means of understanding the nature around us.

Also in this book, emphasis is laid on the method of solution of integral equations, and so not only ordinary forms but also various kinds of special integral equations are treated, while care is taken to lose no precision in the development of the theory. For this purpose, chapters are composed with a special design and the whole volume is divided into two parts. The first considers basic theories and applications, and the second deals with the theories of special integral equations and their applications. The former covers the fundamental theories of the most ordinary cases of linear integral equations, while the special cases treated in the latter part are solved using the fundamental theories of Part I. Therefore, in the fundamental theories, the kinds of integral equations are restricted to the irreducible minimum required for the explanation of the fundamental theories, and more general cases are treated in the special theories or in Part II.

As the author has adopted the specially improved symbolic method originated by Volterra for the convolution of functions, expressions are very simply denoted, and we get a clear perspective of the development of theories. Solutions by differential equations are included here and there so that they may be compared with those by integral equations, and, as will be seen, they are all calculated using operational calculus since the methods of solution of these classic problems are available in many references. Therefore, it would not always to repeat them in the present volume.

For FEM (finite element method) which is applied extensively in engineering, integral equations are more suitable than differential equations to obtain fairly precise solution numerically.

Certainly the theory and applications of integral equations are developing in recent years, however, the fundamental theories were completed at the beginning of this century and are included in this volume.

Tokyo
March 1991

J. Kondo

Contents

General Notation

Numbers refer to the pages on which the explanations first appear.

$\Delta(\lambda)$	Fredholm's determinant	52
$\Delta(\xi, \lambda)$	Fredholm's determinant of the first order	52
$f(t)$	disturbance function	12
∇^2	Laplacian	202
$\|\|f\|\|$	norm	104
$\Phi_i,\ \Psi_i$	biorthoganal system	251
$F[x(\xi)]$	functional	4
$G(t, \xi)$	kernel of solution of Volterra's integral equation	38
$\Gamma(t, \xi)$	kernel of solution of Fredholm's integral equation	35
$K(t, \xi)$	kernel	12
$\mathscr{R}, \mathscr{D}$	functional space	20, 339
$\mathscr{L}$	Laplace transform	336
λ	parameter to Fredholm's integral	13
	eigenvalue	83
$\bar{\lambda}$	conjugate of λ	254
t	variable (real)	12
$\varphi(t)$	eigenfunction	76, 83
$\varphi(x, y)$	velocity potential	205
$\psi(x, y)$	stream function	206
$\tilde{\varphi}$	eigenfunction of conjugate function	83
$\bar{\varphi}$	conjugate function	254
$\psi_i(t)$	normalized orthoganal system	244
$x(t)$	unknown function (one independent valiable — t)	12
$x^{(n)}(t)$	$d^n x/dt^n$	5
$'$	in some special case prime is used to denote first derivative	199
ξ	variable of integration	12

$u(x, y)$	unknown function (two independent variable) —x, y)	292
$V_0V_1V_2V_3$	Volterra type or V type integral equation	12
$\tilde{V}_0\tilde{V}_1\tilde{V}_2\tilde{V}_3$	associate equation of Volterra type integral equation	14
$F_0F_1F_2F_3$	Fredholm type or F type integral equation	12
$\tilde{F}_0\tilde{F}_1\tilde{F}_2\tilde{F}_3$	associate equation of Fredholm type or F type integral equation	14
$\cdot$	product $f \cdot g$	104
$*$	convolution of the first kind ($f * g$)	20
$\circ$	convolution of the second kind ($f \circ g$)	20
D	determinant of coefficients	47
E	unit element	25
S	$a \leqq t \leqq b$, $a \leqq \xi \leqq b$ closed square domain	19
T	$a \leqq \xi \leqq t \leqq b$ closed triangular domain	19
I	$a \leqq t \leqq b$, $t(a, b)$ Closed interval of t	19
s	parameter of Laplace transform	27
	complex number	123
	length of arc	199
v	index of eigenvalue	76

List of theorems with regard to integration and lemmas

Part I

Fundamental theories of integral equations

Introduction of fundamental theories

Part I 'Fundamental theories' is concerned with methods of solution of typical integral equations. In this part special limitation is placed on the form and classification of integral equations in order to clarify the general method of solution. Especially, the theories of Fredholm's integral equation are systematically developed and extensive applications are selected so as to distinguish the essential characteristics of integral equations from those of differential equations.

'Fundamental theories' consists of five chapters. In Chapter 1, various kinds of integral equations are introduced from the viewpoint of the elementary continuation of a linear functional, and the reason why these kinds of integral equations are limited to a few is set down. This introductory chapter contains the theorems and axioms that will be required in the later chapters. In Chapter 2, the exact theories concerning the existence and uniqueness of solutions are explained, and Volterra's iterated kernel and Fredholm's determinant are systematically developed. The theory of eigenvalues and eigenfunctions treated in this chapter is closely linked with the study of D. Hilbert in Chapter 6 of Part II. From the theoretical point of view Chapter 2 is the most important chapter in Part I, and several pages have been devoted to minute explanations of theory; but, as the theory is a little too complicated for use in practice two chapters are reserved to introduce those methods in detail. Chapter 3 treats the analytical method of calculation in rather special cases, and in Chapter 4, the method of numerical solution is given. Of these two methods, the former is neater and easily gives a technically exact solution when the kernel takes a special form, but it lacks generality because it is not applicable in its original form to the cases where kernels are of general forms. The latter is a method of approximation which chiefly employs numerical integral formulae and, although theoretically less interesting, the well-contrived methods of calculation are worthy of attention and are effective when the kernel takes a complicated form. The final chapter of Part I, Chapter 5, deals with some fundamental practical cases, and several pages have been set aside for the explanation of the relation between integral equations and boundary-value problems of differential equations and, especially, for the explanation of the classical potential theory.

1 Introduction and classification of integral equations

Integral equations are named after the mathematicians who first studied them, and they are classified as equations of the first or second kind according to their types. Fredholm's integral equation of the first kind and Volterra's integral equation of the second kind are examples. The reason why the kinds of integral equations are restricted to just a few in our ordinary studies, and the means by which these integral equations are introduced in more general cases, will become clear when the order and linear characteristics of an elementary continuation in a functional space are considered. Essential characteristics and principles of the method of solution of integral equations become clear when the equation is moved from the functional space of infinite dimensions to that of finite dimensions. The author believes that using these essentials to elevate the study of integral equations to the high standpoint of a linear functional and functional space will lead to a clearer idea of the essence of integral equations and a better perception of their development, and thus ultimately will add greatly to the better understanding of integral equations even in a book designed for elementary lectures, such as this.

The idea of the convolution of functions, originated by V. Volterra, is an important means for studying integral equations. Using this idea, we can arrive at a solution by an algebraic operation on symbols, and so not only does the expression become concise but also the method of calculation turns out to be simpler, allowing an easier perspective view to the solution. For these reasons, symbolic calculation is generally adopted throughout the volume.

1.1. Linear functionals

Dirichlet's definition of a function $y = f(x)$ is as follows. Suppose that there are two variables x and y; if x take a value in a certain interval, the value of y is thereby determined. He called y a function of x and x the argument, and the set of values of x for which y is determined is called the interval of definition of the function f.

Volterra's definition of a functional is an extension of Dirichlet's definition of a function. By his definition, a functional $y = F[x(\xi)_a^b]$ is a function of a function; that is, given a function $x(\xi)$ defined in an interval ξ (a, b) there exists a corresponding value y. The correspondence between the function $x(\xi)$ and the variable y is called a functional. $x(\xi)$ is called an *argument function*, and the set of $x(\xi)$ which determines y is the *functional space* which defines the functional F.

When a functional F contains an intermediate argument t, in addition to an argument function $x(\xi)$, y depends on both $x(\xi)$ and t; but when $x(\xi)$ is a constant, y is a common function of t. We express this as

$$y(t) = F[x(\overset{b}{\underset{a}{\xi}}), t].$$

A function $f(x)$ is called *linear* if, for the arbitrary two values x_1, x_2 in the interval of definition we have

$$f(x_1 + x_2) = f(x_1) + f(x_2),$$

and for an arbitrary constant c we have

$$f(cx) = cf(x).$$

Hence we have, for a linear function $f(x)$,

$$f(c_1 x_1 + c_2 x_2) = c_1 f(x_1) + c_2 f(x_2).$$

J. Hadamard (1903) extended this notion and defined the *linear functional*, with the following two characteristics.

Suppose that there are two arbitrary functions $x_1(\xi)$, $x_2(\xi)$, which belong to the functional space that defines a linear functional $F[x(\xi)]$; we have

$$F[x_1(\xi) + x_2(\xi)] = F[x_1(\xi)] + F[x_2(\xi)].$$

And for an arbitrary constant c we have

$$F[cx(\xi)] = cF[x(\xi)].$$

Hence we have, for a linear functional F,

$$F[c_1x_1(\xi) + c_2x_2(\xi)] = c_1F[x_1(\xi)] + c_2F[x_2(\xi)].$$

A linear functional $f(x)$ is expressed in the form $y = kx + b$, but J. Hadamard (1903) studied the general expression of a linear functional and F. Riesz (1909) obtained the following result.

The linear functional $F[x(\xi)_a^b]$ is generally expressed in the following form:

$$F[x(\xi)] = \int_a^b K(\xi)x(\xi)d\xi + \sum \alpha_i x(\xi_i) + \sum \beta_i x^{(1)}(\xi_i) + \ldots$$
$$+ \sum \nu_i x^{(n)}(\xi_i). \tag{1.1}$$

Here $x^{(n)}(\xi_i) = (d^n x/d\xi^n)_{\xi=\xi_i}$, $n = 1, 2, \ldots$, † the α_i etc. are all constants, and $a \leqq \xi_i \leqq b$, $\sum$ is the sum of finite or infinite terms over, and we assume that $\sum|\alpha_i|$ etc. are all convergent. In (1.1) we call the first term on the right-hand side the *regular part*, the second term and the rest on the right form the *exceptional part*, and we say that the functional $F[x(\xi)]$ *depends exceptionally* on the value at $\xi = \xi_i$. We call $K(\xi)$ a kernel.

When a linear functional includes a parameter t, the part that depends exceptionally on the value at $\xi = t$ has the following form:

$$F[x(\xi), t] = \int_a^b K(t, \xi)x(\xi)d\xi + \sum \alpha_i(t)x(t) + \sum \beta_i(t)x^{(1)}(t)$$
$$+ \ldots + \sum \nu_i(t)x^{(n)}(t).$$

This time $K(t, \xi)$ is the kernel. From now on we are chiefly going to treat

$$F[x(\xi), t] = \int_a^b K(t, \xi)x(\xi)d\xi + a_0(t)x(t) + a_1(t)x^{(1)}(t)$$
$$+ \ldots + a_n(t)x^{(n)}(t). \tag{1.2}$$

1.1.1. *Functions defined by lines*

The area $y = \int_a^b x(\xi)d\xi$, which is defined by the curve $x(\xi)$ in $\xi(a, b)$, is a functional because it is determined by the function $x(\xi)$, and the functional space of this functional is the set of integrable functions. If the electric current through the circuit $x(\xi)$ is t, its effect y on the magnetic needle in the magnetic field of this circuit is fixed by the form of the circuit and the strength of the current t, so the effect is a functional of $x(\xi)$ with parameter t.

The notion of functional first appeared in a series of Volterra's theses presented in *Rendiconti della R. Accademia dei Lincei* from 1887, where he at first used the term *function of a line*. It was Hadamard (1903) who used the term *functional* first.

† In this volume the number in parentheses on the upper right-hand side of a symbol indicates the order of differentiation. Thus $x^{(n)}(t) \equiv d^n x/dt^n$ in which (n) gives the degree of the derived function of the function $x(t)$. In particular we express $x^n(0) \equiv x_0^{(n)}$.

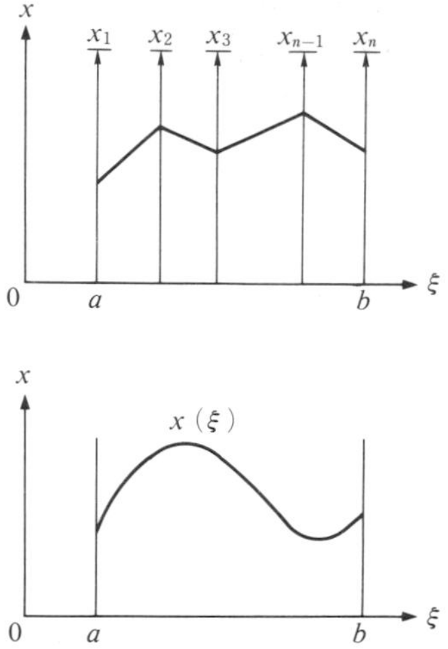

Fig. 1.1. The expression of n-dimensional space and infinite-dimensional space

A functional can also be introduced as the function of an infinite number of variables by way of an extension of the function of a finite number of variables. For example, a function of three variables x_1, x_2, x_3, namely $y = f(x_1, x_2, x_3)$, is a function defined for the point set in three-dimensional space, where each point in this space can be expressed as a set of points in the (ξ, x)-plane (x_1, ξ_1), (x_2, ξ_2) and (x_3, ξ_3), each on one of three vertical lines drawn at ξ_1, ξ_2 and ξ_3. In general, a function of n variables, namely $y = f(x_1, x_2, \ldots, x_n)$, can be considered as a function defined for the point set on n vertical lines drawn in the (ξ, x)-plane, these n points or the broken lines that link these points serve to mark one point in n-dimensional space, as is indicated in Fig. 1.1.

Generalizing this, and when the independent variables are infinite, a function $y = f(x_1, x_2, \ldots)$ is a function defined on the point set of infinite-dimensional space, where points in the infinite-dimensional space are to be expressed by a curve $x = x(\xi)$ as the limit of the above broken lines. Therefore the function $y = f(x_1, x_2, \ldots)$ can be taken as a functional $F[x(\xi)_a^b]$ of $x(\xi)$.

Conversely, if we suppose a variable function $x(\xi)$ in $F[x(\xi)]$ to be infinitely differentiable and to be expansible in Maclaurin series,

$$x(\xi) = x_0 + \frac{x_0^{(1)}}{1!}\xi + \frac{x_0^{(2)}}{2!}\xi^2 + \ldots + \frac{x_0^{(n)}}{n!}\xi^n + \ldots,$$

and if we give values to $x_0, x_0^{(1)}, x_0^{(2)}, \ldots, x_0^{(n)}, \ldots$, then $x(\xi)$ is defined, therefore $F[x(\xi)]$ can be considered as a function of an infinite number of arguments, $x_0, x_0^{(1)}, \ldots, x_0^{(n)}, \ldots$. Thus, to consider a functional as the extension of a common function with many variables often provides some important clues for the study of the functional.

If, for example, in a linear function of n arguments

$$y = \sum_i k_i x_i, \qquad (i = 1, 2, \ldots, n),$$

we substitute a continuous parameter ξ for the discontinuous parameter i, a variable function $x(\xi)$ for the variable x_i, $K(\xi)$ for k_i and the continuous sum (integral) of ξ, namely $\int_a^b d\xi$, for the discrete sum $\sum_i$, then we get

$$y = \int_a^b K(\xi)x(\xi)d\xi.$$

If, in a linear transformation of n-dimensional space

$$y_j = \sum_i k_{ji} x_i \qquad (i, j = 1, 2, \ldots, n),$$

we substitute t for j, and replace k_{ji} by $K(t, \xi)$, we get

$$y(t) = \int_a^b K(t, \xi)x(\xi)d\xi.$$

Volterra called the principle of this sort of process the *principle of passage from discontinuity to continuity*. Strictly, this sort of transformation is not permissible without conditions, and we shall investigate its mathematical properties as a linear functional on another occasion, yet this notion is a useful means to study linear functionals as the extension of functions of a finite number of variables.

1.1.2. *An example of linear functional*

Suppose that there is a curved surface generated by the revolution around the t-axis of a curve

$$x = x(t)$$

which links two fixed points P,Q in the (t, x)-plane. Let us calculate the the resistance R which the surface encounters as it moves with constant speed along the direction of the axis in an ideal fluid.

According to Newton's law, resistance is proportional to the second power of normal speed $v\, dx/ds$ to the curved surface. Let a line element

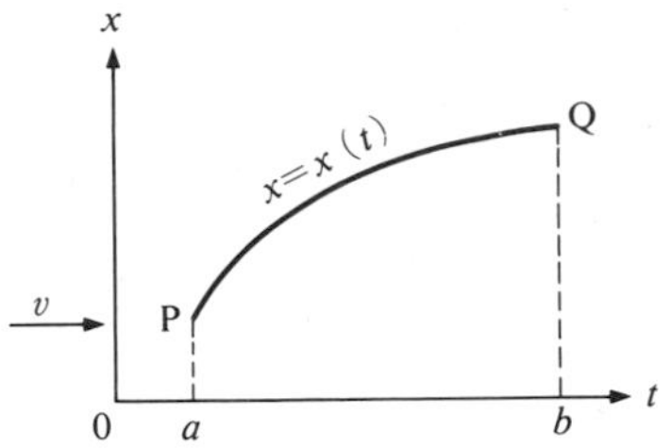

Fig. 1.2. Generation line of the surface of revolution

of the curve be ds; then the t-component dR of the resistance acting on the circular part $2\pi x ds$ is

$$dR = 2\pi x ds \cdot \left(v\frac{dx}{ds}\right)^2 \cdot \frac{dx}{ds}.$$

As

$$ds = \left(1 + \left(\frac{dx}{dt}\right)^2\right)^{\frac{1}{2}}, \qquad dx = \frac{dx}{dt}dt,$$

we have

$$R = 2\pi v^2 \int_a^b \frac{x\, x^{(1)\,3}}{1 + x^{(1)\,2}}\, dt;$$

this is undoubtedly a functional of $x(t)$, but it is not linear.

1.2. Elementary continuity of a functional

The definition of continuity of a functional changes with the change of definition of distance in the functional space. When we have arbitrary elements A, B, C of an abstract set, the distance between A and B, denoted by $\widehat{AB}$, must satisfy the following three postulates.

1. When A coincides with B the distance is *zero*: $\widehat{AA} = 0$.
2. The distance between A and B is equal to the distance between B and A: $\widehat{AB} = \widehat{BA}$.
3. The distance between A and C does not exceed the sum of the distances between A and B, and B and C: $\widehat{AC} \leqq \widehat{AB} + \widehat{BC}$.

Here if two elements $x_1(\xi)$, $x_2(\xi)$ of functional space are continuous then we take

$$\text{Max}\,|x_1(\xi) - x_2(\xi)|,\ \xi(a, \text{b})$$

as the distance between x_1 and x_2. It is obvious that this definition satisfies the above three postulates.

Now we define continuity of a functional in the same way that we define continuity of the usual function $f(z)$. With regard to the function $f(z)$, the domain of definition of the function is the total set of all the values that the independent variable z is able to take. In this domain, the distance between two points z_1, z_2 is expressed by $|z_1 - z_2|$. The η-neighbourhood of z_0 is the set of z in a domain which satisfies

$$|z - z_0| < \eta$$

for an arbitrary small positive number η. That $f(z)$ is continuous at $z = z_0$ means that, to any arbitrary small positive number ε, however small, there corresponds a positive number η such that we have

$$|f(z) - f(z_0)| < \varepsilon \quad \text{for all } z \text{ in } |z - z_0| < \eta.$$

The definition of continuity of the functional $F[x(\xi)]$ can be extended in a similar manner to this.

Definition. Elementary continuity of a functional

The set of all the elements of a variable function $x(\xi)$ which defines a functional $F[x(\xi)_a^b]$ constitutes a functional space which defines a functional $F[x(\xi)]$. In this space the distance between two elements $x_1(\xi)$, $x_2(\xi)$ is defined by $\text{Max}\,|x_1(\xi) - x_2(\xi)|$. The η-neighbourhood of the element $x_0(\xi)$ is the set of functions $x(\xi)$ which satisfy

$$\text{Max}\,|x(\xi) - x_0(\xi)| < \eta \qquad \text{in } c \leqq \xi \leqq b.$$

If $F[x(\xi)]$ is *continuous* at $x_0(\xi)$ (*elementary continuous of the zeroth order*) then a non-zero positive number η exists corresponding to each arbitrary small positive number ε, and in the defined space of the functional $F[x(\xi)]$, we have

$$|F[x(\xi)] - F[x_0(\xi)]| < \varepsilon$$

for all $x(\xi)$ in $\text{Max}\,|x(\xi) - x_0(\xi)| < \eta$. Further, *elementary continuous of the first order* means that to any given positive number ε there corresponds a positive number η such that the inequality $|F[x(\xi)] - F[x_0(\xi)]| < \varepsilon$ holds for any functions x and x_0 in the domain of definition of F, when $\text{Max}\,|x(\xi) - x_0(\xi)| < \eta$ and $\text{Max}\,|x^{(1)}(\xi) - x_0{}^{(1)}(\xi)| < \eta$.

A similar notion can be applied to the elementary continuity of higher orders.

For example, a functional

$$F[x(\xi)] = \int_0^1 (1 - t + t\xi)x(\xi)d\xi, \qquad 0 \leqq t, \xi \leqq 1$$

has the following elementary continuity of the zeroth order at $x_0(\xi) = \xi$:

$$\begin{aligned} |F[x(\xi)] - F[x_0(\xi)]| &\leqq \int_0^1 |1 - t + t\xi|\,|x(\xi) - \xi|\,d\xi \\ &< \int_0^1 |x(\xi) - \xi|\,d\xi \\ &\leqq \operatorname{Max}|x(\xi) - \xi|. \end{aligned}$$

However, denoting the upper limit in a section $\xi(0, 1)$ by $\overline{\lim}$ in a functional:

$$F[x(\xi)_0^1] = \overline{\lim}\, x^{(1)}(\xi),$$

if we take two elements

$$x_0(\xi) = k, \qquad x(\xi) = k + \varepsilon \sin\frac{2\pi\xi}{\varepsilon},$$

we have

$$\operatorname{Max}|x(\xi) - x_0(\xi)| < \varepsilon;$$

hence $x(\xi)$ is quite close to $x_0(\xi)$. Yet, since

$$x_0^{(1)}(\xi) = 0, \qquad x^{(1)}(\xi) = 2\pi \cos\frac{2\pi\xi}{\varepsilon},$$

we have

$$F[x_0^{(1)}(\xi)] = 0, \qquad F[x^{(1)}(\xi)] = 2\pi.$$

Therefore, it does not have elementary continuity of the zeroth order, although it has elementary continuity of the first order.

The example of a functional in section 1.1.2, namely

$$R[x(t)] = 2\pi v^2 \int_a^b \frac{x\, x^{(1)2}}{1 + x^{(1)2}} dt,$$

also has elementary continuity of the first order.

1.2.1. Degree of elementary continuity of the linear functional

A linear functional

$$F[x(\xi), t] = \int_a^b K(t, \xi)x(\xi)d\xi + a_0(t)x(t) + a_1(t)x^{(1)}(t) + \ldots$$
$$+ a_n(t)x^{(n)}(t)$$

has elementary continuity of the nth order.

In particular, the linear functional which has elementary continuity of the zeroth order is written

$$\int_a^b K(t, \xi)x(\xi)d\xi,$$

when it lacks an exceptional part and

$$\int_a^b K(t, \xi)x(\xi)d\xi + a_0(t)x(t)$$

when it has an exceptional part.

1.3. Integral equations

An equation which contains an unknown quantity x, say

$$f(x) = y,$$

is considered as an equation in one-dimensional space. Here we take y on the right-hand side to be known, and at this stage we do not restrict the degree of $f(x)$. Extending this notion of an equation, the equations which contain n unknown quantities are the simultaneous equations which can be written as

$$f_i(x_1, x_2, \ldots, x_n) = y_i \qquad (i = 1, 2, \ldots, n).$$

This is an equation in n-dimensional space. In another form as this equation can be expressed as

$$f_i[x_j] = y_i \qquad (i, j = 1, 2, \ldots, n).$$

Equations in the functional space are infinite-dimensional simultaneous equations and they introduce the notion of a functional in

$$f_i[x_j] = y_i \qquad (i, j = 1, 2, \ldots, n, \ldots).$$

Substituting F for f, t for i, ξ for j and f for y, this equation yields

$$F[x(\xi), t] = f(t). \tag{1.3}$$

Now if in (1.3) we consider $x(\xi)$ to be an unkown function, as an extension to infinite dimensions of unknowns x_j, and $f(t)$ to be a known function which is an extension of a known quantity, then (1.3) is a functional equation with regard to the unknown $x(\xi)$.

If, especially, the left-hand side is a linear functional, it takes the form

$$\int_a^b K(t, \xi)x(\xi)d\xi + a_0(t)x(t) + a_1(t)x^{(1)}(t) + \dots$$
$$+ a_n(t)x^{(n)}(t) = f(t), \tag{1.4}$$

and this contains an integral and derivatives of unknown functions. We call this a *linear integro-differential equation.*

When the degree of elementary continuity is of the zeroth order, we obtain

$$\int_a^b K(t, \xi)x(\xi)d\xi + a_0(t)x(t) = f(t), \qquad [F_3]$$

and we call such an equation which contains an unknown function under the integral sign an *integral equation.* This the most typical type of linear integral equation and we call it an *integral equation of the third kind* or *Picard's integral equation.* We call $K(t, \xi)$ a *kernel* of the integral equation. It is a known function in $a < t, \xi < b$, and the first term is called the regular part and the second term the exceptional part. Now $f(t)$ is known function in the interval $t(a, b)$, and it is sometimes called a *disturbance function.* We treat ξ as a variable of integration and $x(t)$ as an unknown function. To solve an integral equation is to express an unknown function in terms of known functions $f(t)$, $a_0(t)$, $K(t, \xi)$ in $t(a, b)$ and we call the $x(t)$ which satisfies $[F_3]$ a *solution* of the integral equation.

Now $[F_3]$ contains a definite integral of $x(t)$, but when $K(t, \xi) = 0$ in $\xi > t$, we have

$$\int_a^t K(t, \xi)x(\xi)d\xi + a_0(t)x(t) = f(t), \qquad [V_3]$$

where the regular part takes the form of an indefinite integral. We say that the former type is of *Fredholm type* or simply *F* type and the latter *Volterra type* or *V* type. Of course there is no essential difference between the two, and we may consider *V* type to be a special case of

F type; the difference in their treatment will become clear in the course of the work.

In $[F_3]$, when $a\,(t) \neq 0$ in $t(a, b)$, dividing both sides by $a_0(t)$ and putting

$$\frac{K(t, \xi)}{a_0(t)} a_0(\xi) \equiv -\lambda K(t, \xi), \qquad \frac{x(t)}{\alpha_0(t)} \equiv x(t), \qquad \frac{f(t)}{\alpha_0(t)} \equiv f(t)$$

we obtain

$$x(t) - \lambda \int_a^b K(t, \xi) x(\xi) d\xi = f(t). \qquad [F_2]$$

Here λ is a parameter introduced in order to simplify the argument. This equation is called an *integral equation of the second kind.*

If in $[F_2]$ we have $f(t) = 0$, then

$$x(t) = \lambda \int_a^b K(t, \xi) x(\xi) d\xi, \qquad [F_0]$$

and this is called a *homogeneous integral equation.*

If $[F_3]$ lacks an exceptional part or $a_0(t) \equiv 0$, it becomes

$$\int_a^b K(t, \xi) x(\xi) d\xi = f(t), \qquad [F_1]$$

which is called an integral equation of the first kind.

In the case of V type,

$$x(t) - \int_a^t K(t, \xi) x(\xi) d\xi = f(t) \qquad [V_2]$$

is an integral equation of the second kind,

$$\int_a^t K(t, \xi) x(\xi) d\xi = f(t) \qquad [V_1]$$

is an integral equation of the first kind, and

$$x(t) = \int_a^t K(t, \xi) x(\xi) d\xi \qquad [V_0]$$

is a homogeneous integral equation. For the V type equation there is no essential difference if we substitute 0 for the lower limit a, and so we shall usually do this.

The equations $[F_0]$, $[F_1]$, $[F_2]$, and $[V_0]$, $[V_1]$, $[V_2]$ are the most common linear integral equations, and will be our chief study from now on in Part I.

In Part I, the kernel $K(t, \xi)$ is supposed to be bounded. However, when $\int K(t, \xi)x(\xi)d\xi$ exists, even though the kernel attains infinity somewhere in the domain, the kernel is called a *singular kernel*, and the integral equation is called an *integral equation with singular kernel.* It happens that $[F_2]$, $[F_1]$, $[F_0]$ have singular kernels when $a_0(t) = 0$ in $t(a, b)$ for $[F_3]$ (see Chapter 8).

When one or both of the upper and lower limits of the integral of the regular part become infinite; for instance, when

$$\int_0^\infty K(t, \xi)x(\xi)d\xi = f(t) \quad \text{or} \quad x(t) - \lambda \int_{-\infty}^\infty K(t, \xi)x(\xi)d\xi = f(t),$$

we call such an integral equation a *singular integral equation.*†

This integral equation is classified into V type or F type, and further divided in each of them into the first kind, the second kind, and homogeneous, etc. In the V type there is a case where both upper and lower limits of the integral are variables; for example,

$$x(t) - \int_{-t}^{t} K(t, \xi)x(\xi)d\xi = f(t);$$

this is also called a singular integral equation (see Chapter 7).

When the kernel $K(t, \xi)$ of an integral equation is symmetric for t and ξ, that is,

$$K(t, \xi) = K(\xi, t),$$

we call the kernel a *symmetric kernel* (see Chapter 6). When a kernel is not symmetric, an integral equation in which t and ξ are substituted for each other is different from the original one. We call this form of equation an *associate equation.* For example, the associate equation of

$$\int_a^b K(t, \xi)x(\xi)d\xi = f(t) \qquad [F_1]$$

is

$$\int_a^b K(\xi, t)x(\xi)d\xi = f(t). \qquad [\tilde{F}_1]$$

Similarly, $[\tilde{F}_2]$, $[\tilde{F}_0]$, $[\tilde{V}_1]$, $[\tilde{V}_2]$, $[\tilde{V}_0]$ can be derived. In the case of a symmetric kernel, the associate equation is the same as the original equation.

† An integral equation with a singular kernel is sometimes called a singular integral equation.

When a functional is not linear, its regular part contains a high power of $x(t)$, and the functional generally takes the form

$$x(t) - \lambda\int_a^b H[t, \xi, x(\xi)]d\xi = f(t).$$

We call this kind of equation a *nonlinear integral equation*. For example,

$$x(t) - \lambda\int_a^b K(t, \xi)x^n(\xi)d\xi = f(t) \qquad n \neq 1$$

is nonlinear. If the exceptional part is nonlinear, the equation is to be treated as a nonlinear integral equation, even if its regular part is linear. Such an example is

$$x^n(t) - \lambda\int_a^b K(t, \xi)x(\xi)d\xi = f(t)$$

(see Chapter 9).

When there are two unknown functions, we get two simultaneous equations such as

$$F_1[x_1(\xi),\ x_2(\xi),\ t] = f_1(t) \quad \text{and} \quad F_2[x_1(\xi),\ x_2(\xi),\ t] = f_2(t).$$

Linear simultaneous equations are expressed in the following form:

$$x_1(t) - \lambda_1\int_a^b K_{11}(t, \xi)x_1(\xi)d\xi - \mu_1\int_a^b K_{12}(t, \xi)x_2(\xi)d\xi = f_1(t),$$

$$x_2(t) - \lambda_2\int_a^b K_{21}(t, \xi)x_1(\xi)d\xi - \mu_2\int_a^b K_{22}(t, \xi)x_2(\xi)d\xi = f_2(t)$$

(see Chapter 7).

When an unknown function is a function of more than two independent variables, the functional equation to be considered takes the form

$$F[u(\xi, \eta), x, y] = f(x, y).$$

An integral equation which contains a multiple integral

$$u(x, y) - \lambda\int_a^b\int_c^d K(x, y;\ \xi, \eta)u(\xi, \eta)d\xi d\eta = f(x, y)$$

is an example of this sort of equation.

Besides these, there are integral equations such as an integral equation containing a line integral in a complex plane,

$$x(z) - \lambda\int_c^d K(z, s)x(s)ds = f(z),$$

and a *difference-integral equation* such as

$$x(t-1) - \lambda\int_a^b K(t, \xi)x(\xi)d\xi = f(t)$$

(see Chapter 7).

TABLE 1.1. Linear functionals and integral equations with elementary continuity

$F[x(\xi), t]$	$\int_a^b K(t, \xi)x(\xi)d\xi$ (regular part) $+$ $a_0(t)x(t)$ (exceptional part); $\int_a^b x(\xi)K(\xi, t)d\xi + a_0(t)x(t)$ associate functional.
	$\int_a^b K(t, \xi)x(\xi)d\xi$ (elementary continuity of the zeroth order)
	$\int_a^b K(t, \xi)x(\xi)d\xi + a_0(t)x(t)$ (elementary continuity of the zeroth order)
	$\int_a^b K(t, \xi)x(\xi)d\xi + a_0(t)x(t) + a_1(t)x^{(1)}(t) + \cdots + a_n(t)x^{(n)}(t)$ (elementary continuity of the *n*th order)
Fredholm type	the first kind: $\int_a^b K(t, \xi)x(\xi)d\xi = f(t), \quad K \circ x = f$ [F₁]
	the second kind: $x(t) - \lambda\int_a^b K(t, \xi)x(\xi)d\xi = f(t), \quad (E - \lambda K) \circ x = f$ [F₂]
	the third kind: $a_0(t)x(t) + \int_a^b K(t, \xi)x(\xi)d\xi = f(t), \quad a_0x + K \circ x = f$ [F₃]
	homogeneous: $x(t) - \lambda\int_a^b K(t, \xi)x(\xi)d\xi = 0, \quad (E - \lambda K) \circ x = 0$ [F₀]
	integro-differential: $a_0(t)x(t) + a_1(t)x^{(1)}(t) + \cdots + a_n(t)x^{(n)}(t) + \int_a^b K(t, \xi)x(\xi)d\xi = f(t)$
Volterra type	the first kind: $\int_a^t K(t, \xi)x(\xi)d\xi = f(t), \quad K * x = f$ [V₁]
	the second kind: $x(t) - \int_a^t K(t, \xi)x(\xi)d\xi = f(t), \quad (E - K) * x = f$ [V₂]

TABLE 1.1 *Continued*

Volterra type	the third kind $$a_0(t)x(t) + \int_a^t K(t, \xi)x(\xi)d\xi = f(t), \qquad a_0x + K * x = f$$ [V_3]
	homogeneous $$x(t) - \int_a^t K(t, \xi)x(\xi)d\xi = 0, \qquad (E - K) * x = 0$$ [V_0]
	integro-differential $$a_0(t)x(t) + a_1(t)x^{(1)}(t) + \cdots + a_n(t)x^{(n)}(t) + \int_a^t K(t, \xi)x(\xi)d\xi = f(t)$$

Note that each of these has an associate integral equation, [F_1], ..., [V_1], ...

1.3.1. Integral equations and simultaneous equations

We have introduced an integral equation as the limit of an n-dimensional simultaneous equation when the dimensions are infinite, but, in addition, it is also useful in the study of an integral equation to treat it as the extension of simultaneous equations. In this method we divide an integration into a sum of finite functions. For example, in the finite integral $\int_a^b f(t)dt$ of a continuous function $f(t)$ in $t(a, b)$ we know that if we divide the interval into n equal subintervals of length h, and denote the end points by

$$t_0 = a, \quad t_1 = a + h, \quad \ldots, \quad t_{n-1} = a + (n + 1)h, \quad t_n = b,$$

then we can express the integral as the limit of the sum. Hence we have

$$\int_a^b f(t)dt = \lim_{n\to\infty} \sum_{i=1}^n f(t_i) \cdot h,$$

where

$$h = \frac{1}{n}(b - a).$$

Here, instead of integration on a continuous value of $f(t)$, summation is taken over distinct values of $f(t_i) \equiv f_i$. If we admit a finite sum instead of the limit $n \to \infty$, then we have substituted the finite sum of discontinuous values $h \sum f_i$ for the integral of the continuous function, $\int_a^b f(t)dt$. This is the inverse method of the 'principle of passage from discontinuity to continuity' which was discussed in 1.1.1.

For example, in Fredholm's integral equation of the second kind:

$$x(t) - \lambda \int_a^b K(t, \xi)x(\xi)d\xi = f(t), \qquad [F_2]$$

if we subdivide $t(a, b)$ into n parts and take the finite sum of distinct values of the integrand, we get

$$x(t) - \lambda \sum_{j=1}^{n} K(t, t_j)x(t_j)h = f(t).$$

Then, substituting the values $t_1, \ldots, t_{n-1}, t_n$ into t in turn, we obtain

$$x(t_i) - \lambda \sum_{j=1}^{n} K(t_i, t_j)x(t_j)h = f(t_i) \qquad (i = 1, 2, \ldots, n).$$

Here, if we make the substitutions

$$x(t_i) \to x_i, \quad K(t_i, t_j)h \to k_{ij}, \quad f(t_i) \to y_i,$$

we obtain the followng simultaneous equations:

$$\left.\begin{aligned} (1 - \lambda k_{11})x_1 - \lambda k_{12}x_2 - \cdots - \lambda k_{1n}x_n &= y_1, \\ -\lambda k_{21}x_1 + (1 - \lambda k_{22})x_2 - \cdots - \lambda k_{2n}x_n &= y_2, \\ \cdots \\ -\lambda k_{n1}x_1 - \lambda k_{n2}x_2 - \cdots + (1 - \lambda k_{nn})x_n &= y_n. \end{aligned}\right\} \qquad [F_2]$$

For Volterra's integral equation of the second kind,

$$x(t) - \int_a^t K(t, \xi)x(\xi)d\xi = f(t), \qquad [V_2]$$

if we subdivide $t(a, b)$ into n parts, $[V_2]$ yields

$$x(t) - \sum K(t, t_j)x(t_j)h = f(t).$$

Here if we put t_i into t, the summation over j is from $j = 1$ to $j = i$, and we get

$$x(t_i) - \sum_{j=1}^{j=i} K(t_i, t_j)x(t_j)h = f(t_i) \qquad (i = 1, 2, \ldots, n).$$

Here again the substitutions

$$x(t_i) \to x_i, \quad K(t_i, t_j)h \to k_{ij}, \quad f(t_i) \to y_i$$

lead to the simultaneous equations

$$\left.\begin{aligned} &x_0 = y_0,\\ &x_1 - k_{01}x_0 = y_1,\\ &x_2 - k_{02}x_0 - k_{12}x_1 = y_2.\\ &\quad\cdots\\ &x_n - k_{0n}x_0 - k_{1n}x_1 - \ldots - k_{n-1,n}x_{n-1} = y_n \end{aligned}\right\}. \qquad [\mathrm{V}_2]$$

It was Volterra who first converted integral equations to simultaneous equations and studied them. He made the work public in 1896.† In 1900, Fredholm used this method in the study of integral equations of the second kind, and obtained the solutions of integral equations in a general method by shifting the limit from the solution of simultaneous equations (here he used the principle of passage from discontinuity to continuity). With regard to this method, we shall have a minute explanation in the next chapter, in section 2.2. This sort of notion is important in the numerical solution of integral equations also (see Chapter 4). From 1904, D. Hilbert studied this sort of method and finally achieved his famous theory (Göttingen Nachrichten, 1904–1908). This theory was included in (Hilbert 1912).

1.4. Definitions and fundamental theorems

In the fundamental theories we treat only linear integral equations of the first kind, of the second kind, and of the homogeneous type. Unless otherwise stated all the variables are supposed to be real. This condition also applies to all the quantities we are going to treat; although not essential, this policy is adopted to simplify the treatment of the equations.

We denote an unknown function by $x(t)$, a kernel by $K(t, \xi)$ or $K(\xi, t)$, and a known function by $f(t)$. The functions $x(t)$ and $f(t)$ are defined in a closed interval $I = \{a \leqq t \leqq b\}$. In the case of Fredholm type, a kernel is defined in a closed square domain $S = \{a \leqq t \leqq b,\ a \leqq \xi \leqq b\}$, and in the case of Volterra type, it is defined in a closed triangular domain $T = \{a \leqq \xi \leqq t \leqq b\}$.

† Volterra (1895–96). The idea of searching for continuous values of an unknown function at discontinuous points first appeared in the study by J.C.F. Sturm (1803–1855) who derived the mathematical properties of differential equations of the second order converted into difference equations in 1836, and R.D. Carmichael (1922) showed the usefulness of this method in the study of solutions of various functional equations.

Therefore, in the case of Volterra type, if the kernel is zero in $a \leqq t \leqq \xi \leqq b$, we can consider that the kernel is defined in S. The kernel is assumed to be bounded in these domains.

Definition. Regularly distributed discontinuous points

Discontinuous points of a function $K(t, \xi)$, or $K(\xi, t)$ are said to be *regularly distributed* in S or T if the discontinuous points lie only on a finite number of smooth curves and that the curves cross any straight line parallel to the coordinate axis only a finite number of times. The kernel is assumed to be bounded in the closed domain and takes a finite determinate value at each of the discontinuous points. Therefore it takes only finite jumps at every discontinuous point.

Definition. The functional space $\mathscr{R}$

The set of functions $f(t, \xi)$ which are defined in S or T and

(i) bounded in the closed domain, and

(ii) continuous except for discontinuities which are regularly distributed,

form an entity which is called a functional space $\mathscr{R}$. The set in which condition (ii) is weakened to a condition of continuity in the closed domain is $\mathscr{R}_0$. Of course $\mathscr{R}_0$ is included in $\mathscr{R}$.

Next, we shall proceed to the convolution of functions introduced by Volterra.

Definition. Convolution of functions

Let $\varphi(t, \xi)$ and $\psi(t, \xi)$ be two functions which belong to $\mathscr{R}$. The integrals

$$\int_\xi^t \varphi(t, \tau)\psi(\tau, \xi)d\tau$$

and

$$\int_a^b \varphi(t, \tau)\psi(\tau, \xi)d\tau$$

are both functions of t, ξ, and we call these functions $H(t, \xi)$ and $F(t, \xi)$ respectively. The operation is called a *convolution of functions*; the former is called convolution of the first kind or convolution of finite variables, and the latter is called convolution of the second kind or convolution in a finite interval. We express these by the following symbols:

$$\varphi * \psi = H, \qquad \varphi \circ \psi = F.$$

We have the following lemma.

LEMMA 1.1 *Continuity*
Suppose that two functions $\varphi(t, \xi)$ and $\psi(t, \xi)$ are defined in S or T and bounded, and that the discontinuous points, if any, are regularly distributed; then each of

$$F(t, \xi) \equiv \int_a^b \varphi(t, \tau)\psi(\tau, \xi)d\tau \quad (\equiv \varphi \circ \psi)$$

and

$$H(t, \xi) \equiv \int_\xi^t \varphi(t, \tau)\psi(\tau, \xi)d\tau \quad (\equiv \varphi * \psi)$$

is continuous at all points either in S or T. (The convolution of elements which belong to $\mathscr{R}$ belongs to $\mathscr{R}_0$.)

Proof. Suppose that (t, ξ, τ) are rectangular coordinates of three-dimensional space;† then the integrand is bounded everywhere in the cube

$$a \leqq t \leqq b, \quad a \leqq \xi \leqq b, \quad a \leqq \tau \leqq b.$$

Therefore F and H are also bounded. Since discontinuous points in φ, ψ distribute regularly, discontinuous points of an integrand exist only on cylindrical surfaces which have such generation lines that are parallel to both the t-axis and the ξ-axis. Besides, lines that are parallel to that τ-axis, such as $t = t_0$, $\xi = \xi_0$, and these cylindrical surfaces can only have finite intersecting points. Therefore, when we substitute sufficiently small values for δ, δ',

$$\varphi(t + \delta, \tau)\psi(\tau, \xi + \delta') - \varphi(t, \tau)\psi(\tau, \xi)$$

always becomes smaller than any arbitrarily small quantity, except finite points which belong to $a \leqq \tau \leqq b$ on the lines parallel to the τ-axis. Since a finite number of discontinuous values do not affect the value of the finite integral,

$$\int_a^b \{\varphi(t + \delta, \tau)\psi(\tau, \xi + \delta') - \varphi(t, \tau)\psi(\tau, \xi)\}d\tau < \varepsilon(b - a).$$

Therefore F and H are continuous on t, ξ.

† In the case of a triangular domain we can put φ, $\psi = 0$ outside T in S and proceed similarly.

LEMMA 1.2 *Lemma of alternation of the order of integration*
When a function $\varphi(t, \xi)$ belongs to $\mathscr{R}$, in the domain S

$$\int_a^b dt \int_a^b \varphi(t, \xi)d\xi = \int_a^b d\xi \int_a^b \varphi(t, \xi)dt$$

holds, and in the domain T

$$\int_a^b dt \int_a^t \varphi(t, \xi)d\xi = \int_a^b d\xi \int_\xi^b \varphi(t, \xi)dt$$

holds.

Proof. We prove this lemma using the definition of integration and show that the left- and right-hand sides are each equal to a double integral.

(i) Existence of a double integral. Since the integrand $\varphi(t, \xi)$ belongs to $\mathscr{R}$ it is bounded in S, and even if there are discontinuous points they lie only on a finite number of smooth curves. The area of the set of discontinuous points is therefore null and the double integral over S,

$$\iint_S \varphi(t, \xi)dt\ d\xi \equiv L,$$

exists. Hence we divide S into n^2 small rectangles and set the maximum value of φ in the ijth section bounded by Δt_i, $\Delta\xi_j$ as M_{ij}, and the minimum value as m_{ij}. The area of the section is $\Delta t_i \Delta\xi_j$. Then the limits of $\sum_i \sum_j M_{ij}\Delta t_i \Delta\xi_j$ and $\sum_i \sum_j m_{ij}\Delta t_i \Delta\xi_j$ as $n \to \infty$ both exist and are equal to each other.

(ii) The use of $\Phi(t) = \int_a^b \varphi(t, \xi)d\xi$. Since φ has at most a finite number of discontinuous points on lines that are parallel to the ξ -axis, there exists an integral $\int_a^b \varphi(t, \xi)d\xi$ with the parameter t. We shall express this as $\Phi(t)$. Letting an arbitrary point that belongs to the ith section Δt_i in $t(a, b)$ be t_i, we have

$$\textstyle\sum_j m_{ij}\Delta\xi_j \leqq \Phi(t_i) \leqq \sum_j M_{ij}\Delta\xi_j.$$

Therefore, multiplying this by Δt_i and summing over i, we obtain

$$\textstyle\sum_i \sum_j m_{ij}\Delta t_i \Delta\xi_j \leqq \sum_i \Phi(t_i)\Delta t_i \leqq \sum_i \sum_j M_{ij}\Delta t_i \Delta\xi_j.$$

When $n \to \infty$, both sides of the inequality converge to L, and the middle term becomes $\int_a^b \Phi(t)dt$; therefore we have proved that

$$L = \int_a^b \Phi(t)dt = \int_a^b dt \int_a^b \varphi(t, \xi)d\xi.$$

Similarly, we have

$$L = \int_a^b d\xi \int_a^b \varphi(t, \xi) dt$$

Therefore, the lemma has been proved with regard to S. As for the theorem with regard to T, it can be dealt with sufficiently by amending the integral domain.

The theorem of alternation of the order of integration in a triangular domain is often called *Dirichlet's formula*. This lemma can be extended to the case where an integrand includes points that tend to infinity.

The following is an example.

LEMMA 1. 2a *Dirichlet's extended formula*

When $\varphi(t, \xi)$ belongs to a functional space, if all of λ, μ, ν are constants

$$0 \leqq \lambda < 1, \qquad 0 \leqq \mu < 1, \qquad 0 \leqq \nu < 1$$

in a triangular domain S, then

$$\int_a^b dt \int_a^t \frac{\varphi(t, \xi)}{(t-\xi)^\lambda (b-t)^\mu (\xi - a)^\nu} d\xi$$
$$= \int_a^b d\xi \int_a^b \frac{\varphi(t, \xi)}{(t-\xi)^\lambda (b-t)^\mu (\xi - a)^\nu} dt$$

holds.

The proof is omitted, but cf. (Whittaker and Watson [30]).

Suppose that a function E satisfies

$$f \circ E = E \circ f = f$$

or

$$f * E = E * f = f$$

for an arbitrary function $f(t, \xi)$ which belongs to $\mathscr{R}$. Since the functions E are not bounded, as will be explained later, they do not belong to the functional space $\mathscr{R}$; however, if we include E in $\mathscr{R}$ as an exception, it serves as a *unit element* in $\mathscr{R}$. There is always a unit element in $\mathscr{R}$ (see 1.4.1 below).

With respect to convolution, the law of distribution,

$$f \circ (g+h) = f \circ g + f \circ h, \qquad f * (g+h) = f * g + f * h,$$
$$(g+h) \circ f = g \circ f + h \circ f, \qquad (g+h) * f = g * f + h * f$$

and the law of combination,

$$f \circ (g \circ h) = (f \circ g) \circ h, \qquad f * (g * h) = (f * g) * h$$

hold.

Thus, if we take the convolution of functions as a multiplication of elements in the functional space $\mathscr{R}$, we can perform algebraic operations freely to some extent, but generally since

$$f \circ g \neq g \circ f, \qquad f * g \neq g * f,$$

the law of commutation does not apply to the product; consequently we have to take care these in symbolic operation. In cases where the law of commutation holds, we say the multiplication is commutative.

We shall consider $K \circ K$ or $K * K$ of the kernel $K(t, \xi)$. Since order does not matter here, there is no possibility of mistake if we use the notation $\mathring{K}^{2}$ and $\overset{*}{K}{}^{2}$. And since $\mathring{K}^{2} \circ K = K \circ \mathring{K}^{2}$, we can express this as $\mathring{K}^{3}$. We define $\mathring{K}^{n}$ and $\overset{*}{K}{}^{n}$ in like manner. Now K and $\mathring{K}^{n}$ or $\overset{*}{K}{}^{n}$ are commutative. That is,

$$K \circ \mathring{K}^{n} = \mathring{K}^{n} \circ K, \qquad K * \overset{*}{K}{}^{n} = \overset{*}{K}{}^{n} * K.$$

Furthermore,

$$a_1 K + a_2 \mathring{K}^{2} + a_3 \mathring{K}^{3} + \cdots + a_n \mathring{K}^{n} + \cdots,$$
$$a_1 K + a_2 \overset{*}{K}{}^{2} + a_3 \overset{*}{K}{}^{3} + \cdots + a_n \overset{*}{K}{}^{n} + \cdots$$

are commutative with K when they are uniformly convergent.

Since $|K| < M$,

$$|\mathring{K}^{2}| < (b - a)M^{2}, \quad |\mathring{K}^{3}| < (b - a)^{2}M^{3}, \quad \ldots,$$
$$|\mathring{K}^{n}| < (b - a)^{n-1}M^{n}, \quad \ldots;$$

$$|\overset{*}{K}{}^{2} < (t - \xi)M^{2}, \quad |\overset{*}{K}{}^{3}| < \frac{(t - \xi)^{2}}{2!}M^{3}, \quad \ldots,$$

$$|\overset{*}{K}{}^{n}| < \frac{(t - \xi)^{n-1}}{(n - 1)!}M^{n}, \quad \ldots$$

can be easily proved (see question 15 below).

In particular, if we put $K(t, \xi) = 1$, we have

$$\mathring{1}^{2} = b - a, \quad \mathring{1}^{3} = (b - a)^{2}, \quad \ldots, \quad \mathring{1}^{n} = (b - a)^{n-1}, \quad \ldots;$$

$$\overset{*}{1}{}^{2} = t - \xi, \quad \overset{*}{1}{}^{3} = \frac{(t - \xi)^{2}}{2!}, \quad \ldots, \quad \overset{*}{1}{}^{n} = \frac{(t - \xi)^{n-1}}{(n - 1)!}, \quad \ldots.$$

Example 1.1 $\mathring{K}^n$, $\overset{*}{K}^n$
Let us calculate $\mathring{K}^n$, $\overset{*}{K}^n$ for $K(t, \xi) = t\xi$ in $0 \leqq t, \xi \leqq 1$.
We have

$$\mathring{K}^2 = \int_0^1 t\tau \cdot \tau\xi \, d\tau = t\xi \int_0^1 \tau^2 d\tau = \frac{t\xi}{3},$$

$$\mathring{K}^3 = \int_0^1 t\tau \frac{\tau\xi}{3} d\tau = \frac{1}{3} t\xi \int_0^1 \tau^2 d\tau = \frac{t\xi}{3^2},$$

$$\ldots,$$

$$\mathring{K}^n = \frac{t\xi}{3^{n-1}}.$$

Further,

$$\overset{*}{K}^2 = \int_\xi^t t\tau \cdot \tau\xi d\tau = t\xi \left[\frac{\tau^3}{3}\right]_\xi^t = \frac{t\xi}{3}(t^3 - \xi^3),$$

$$\overset{*}{K}^3 = \int_\xi^t t\tau \cdot \frac{\tau\xi}{3}(\tau^3 - \xi^3) d\tau = \frac{t\xi}{3}\left\{\left[\frac{\tau^6}{6}\right]_\xi^t - \xi^3\left[\frac{\tau^3}{3}\right]_\xi^t\right\}$$

$$= \frac{t\xi}{3 \cdot 6}(t^3 - \xi^3)^2,$$

$$\overset{*}{K}^4 = \int_\xi^t t\tau \cdot \frac{\tau\xi}{3 \cdot 6}(\tau^3 - \xi^3)^2 d\tau = \frac{t\xi}{3 \cdot 6}\left\{\left[\frac{\tau^9}{9}\right]_\xi^t - \xi^3\left[\frac{\tau^6}{3}\right]_\xi^t\right.$$

$$\left. + \xi^6\left[\frac{\tau^3}{3}\right]_\xi^t\right\} = \frac{\tau\xi}{3 \cdot 6 \cdot 9}(t^3 - \xi^3)^3,$$

$$\ldots$$

$$\overset{*}{K}^n = \frac{t\xi}{3 \cdot 6 \cdot 9 \ldots (3n-3)}(t^3 - \xi^3)^{n-1} = \frac{t\xi}{3^{n-1}(n-1)!}(t^3 - \xi^3)^{n-1}.$$

In this case, it is evident that

$$|\mathring{K}^n| < 1, \quad |\overset{*}{K}^n| < \frac{(t - \xi)^{n-1}}{(n-1)!},$$

since $|K| < 1$.

1.4.1. ***Unit element*** $E(t, \xi)$

The *Delta function* is a function such that

$$\delta(t - \xi) = \begin{cases} 0 & (t \neq \xi), \\ \infty & (t = \xi), \end{cases}$$

and

$$\int_a^b \delta(t-\xi)d\xi = \int_a^b \delta(t-\xi)dt = 1.$$

Since this function is not bounded in S or T, it does not belong to the functional space $\mathscr{R}$. However, for an arbitrary function $f(t, \xi)$ in $\mathscr{R}$, we have

$$\int_a^b f(t, \tau)\delta(\tau-\xi)d\tau = f(t, \xi),$$

$$\int_a^b \delta(t-\tau)f(\tau, \xi)d\tau = f(t, \xi).$$

Thus if we include δ in $\mathscr{R}$, and substitute $\delta(t-\xi)$ as $E(t, \xi)$, we have

$$f \circ E = E \circ f = f.$$

Hence, a delta function serves as a unit element with regard to convolution. From now on, we assume that $\mathscr{R}$ always contains a unit element.

For example, for the following continuous functions which are symmetrical to $t = \xi$ with a parameter α:

$$\delta_\alpha(t-\xi) = \frac{1}{\alpha\sqrt{\pi}}\exp\left\{-\left(\frac{t-\xi}{\alpha}\right)^2\right\},$$

$$\delta_\alpha(t-\xi) = \frac{\alpha}{\pi}\sin^2\left(\frac{t-\xi}{\alpha}\right)\Big/(t-\xi)^2, \text{ etc.}$$

we get

$$\int_{-\infty}^{\infty} \delta_\alpha(t-\xi)dt = 1,$$

and

$$\lim_{\alpha\to 0} \delta_\alpha(t-\xi) = 0 \qquad (t \neq \xi).$$

Hence, $\lim_{\alpha\to 0} \delta_\alpha$ is a delta function as defined above.

1.4.2. Symbolic expression of integral equations

A linear integral equation is expressed symbolically by means of convolution of functions. For example, for

$$x(t) - \lambda\int_a^b K(t, \xi)x(\xi)d\xi = f(t), \qquad [F_2]$$

we can express the exceptional part as $E \circ x$, and the regular part as $-\lambda K \circ x$. Thus we can express $[F_2]$ as

$$(E - \lambda K) \circ x = f. \qquad [F_2]$$

Denoting $K(\xi, t)$ by $\tilde{K}$, we have the associate equation in the form

$$x \circ (E - \lambda \tilde{K}) = f. \qquad [\tilde{F}_2]$$

If we put $\xi = a$ and $x(t, a) = x(t)$ in the convolution of the first kind, $\int_\xi^t K(t, \tau)x(\tau, \xi)d\tau$, then, in

$$x(t) - \int_a^t K(t, \xi)x(\xi)d\xi = f(t),$$

the regular part is expressed by $- K * x$, *d*nd the equation can be expressed symbolically as

$$(E - K) * x = f, \qquad [V_2]$$

and its associate equation $[\tilde{V}_2]$ becomes

$$x * (E - \tilde{K}) = f. \qquad [\tilde{V}_2]$$

1.4.3. *Convolution of operational calculus*

Let $F_1(s)$ be the Laplace transform of $f_1(t)$ and $F_2(s)$ that of $f_2(t)$, then it is known that the Lapace transform of the convolution of f_1 and f_2, that is $\int_0^t f_1(t - \xi)f_2(\xi)d\xi$, can be expressed as $F_1(s)F_2(s)$. This is the convolution, or Faltung theorem. Here we note that the convolution $\int_0^t f_1(t - \xi)f_2(\xi)d\xi$ is a convolution of the first kind. That is, in the case of the set of functions $f(t - \xi)$, we have

$$f_1 * f_2 = \int_\xi^t f_1(t - \tau)f_2(\tau - \xi)d\tau.$$

Here, if we put 0 for ξ, and rewrite the variable of integration τ as ξ, we have

$$f_1 * f_2 = \int_0^t f_1(t - \xi)f_2(\xi)d\xi.$$

Hence the convolution theorem shows that the Laplace transform of the convolution of the first kind is equal to the product of the Laplace transform of each function. (See section 3.5 below.)

The same theorem holds for Fourier transforms. Suppose that the Fourier transform of $f_1(t)$ is $F_1(u)$, and that of $f_2(t)$ is $F_2(u)$, then the

Fourier transform of $\int_{-\infty}^{\infty} f_1(t-\xi)f_2(\xi)d\xi$ is $F_1(u)F_2(u)$. This is the convolution theorem of Fourier transforms.

Here the convolution $\int_{-\infty}^{\infty} f_1(t-\xi)f_2(\xi)d\xi$ is a convolution of the second kind. Therefore,

$$f_1 \circ f_2 = \int_{-\infty}^{\infty} f_1(t-\xi)f_2(\xi)d\xi.$$

Denote the Fourier sine transform of $f(t)$ by $F_s(u)$, and the Fourier cosine transform by $F_c(u)$. Then the convolution theorem becomes

$$\frac{1}{2}\int_0^{\infty} [f_1(|t-\xi|)+f_1(t+\xi)]f_2(\xi)d\xi = \int_0^{\infty} F_{1c}(u)F_{2c}(u)\cos(ut)du,$$

$$\frac{1}{2}\int_0^{\infty} [f_1(|t-\xi|)-f_1(t+\xi)]f_2(\xi)d\xi = \int_0^{\infty} F_{1c}(u)F_{2s}(u)\sin(ut)du,$$

$$\frac{1}{2}\int_0^{\infty} [f_2(|t-\xi|)-f_2(t+\xi)]f_1(\xi)d\xi = \int_0^{\infty} F_{1s}(u)F_{2c}(u)\sin(ut)du.$$

We also have a convolution theorem with regard to finite Fourier transforms. When $f(t)$ is defined in $0 \leqq t \leqq \pi$, we extend it to the interval $-\pi < t < \pi$ in the following two ways:

$$f_o(t) = \begin{cases} f(t), & 0 \leqq t \leqq \pi; \\ -f(-t), & -\pi < t < 0; \end{cases}$$

$$f_e(t) = \begin{cases} f(t), & 0 \leqq t \leqq \pi; \\ f(-t), & -\pi \leqq t < 0; \end{cases}$$

and call them the odd and even extensions, respectively. If we express the finite Fourier sine transform as $F_s(n)$ and the finite Fourier cosine transform as $F_c(n)$, the cosine transform of the convolution of the second kind, $f_{1o} \circ f_{2o} = \int_{-\pi}^{\pi} f_{1o} \circ (t-\xi)f_{2o}(\xi)d\xi$ is $-2F_{1s}(n)F_{2s}(n)$, which we express by

$$C\{f_{1o} \circ f_{2o}\} = -2F_{1s}(n)F_{2s}(n).$$

Besides this there are

$$C\{f_{1e} \circ f_{2e}\} = 2F_{1c}(n)F_{2c}(n), \qquad S\{f_{1o} \circ f_{2e}\} = 2F_{1s}(n)F_{2c}(n).$$

The last expression shows that the sine transform of $f_{1o} \circ f_{2e} = \int_{-\pi}^{\pi} f_{1o}(t-\xi)f_{2e}(\xi)d\xi$ is equal to $2F_{1s}(n)F_{2c}(n)$.

We can find the proofs of these in (Churchill 1944: pp. 274–276, Sneddon 1951: pp. 23–24 and 76–79), among others.

Exercises

1. Let two points (t_0, x_0), (t_1, x_1) in a vertical plane t–x be linked by a smooth curved line $x = x(t)$, and let a particle slide down along the curve under the influence of gravity without friction. Show that the required sliding time T is a functional of $x(t)$. (If we take the x-axis as perpendicularly downward, we have

$$T[x(t)] = \int_{t_0}^{t_1} \frac{(1 + x^{(1)2})^{\frac{1}{2}}}{(2g(x - x_0))^{\frac{1}{2}}} dt$$

(cf. section 10.8.2 below).)

2. Give an example of a functional in an applied case. [For example, the lift L to an aerofoil laid in two-dimensional potential flow is a functional of the planform of the profile, $x(\xi)$, having as its parameter the attack angle t, the angle to the direction of the main flow. There are also many other applications in this volume.]

3. Prove that the following functionals are linear:

$$\int_a^b K(t, \xi)x(\xi)d\xi + \sum_1^n \frac{1}{i^2} x^{(i)}(t), \qquad a < t < b \quad (i = 1, 2, \ldots, n),$$

$$\int_a^b K(t, \xi)x(\xi)d\xi + \sum_1^n \frac{1}{i^2} x^{(1)}(t_i), \qquad a < t_i < b \quad (i = 1, 2, \ldots, n),$$

$$\int_0^\infty e^{t\xi}x(\xi)d\xi + x^{(1)}(t), \qquad 0 < t,$$

$$\int_0^\pi \cot\frac{t - \xi}{2} x(\xi)d\xi + x(t), \qquad 0 < t < \pi.$$

What about the first two functionals when $n \to \infty$?

4. Point out regular parts and exceptional parts of the above functionals.

5. What are the degrees of elementary continuity in exercise 3? [$(n, 1, 1, 0)$-th degree]

6. Derive simultaneous equation for $x(t) - \lambda\int_a^b K(t - \xi)x(\xi)d\xi = f(t)$.

7. Derive simultaneous equations for $[F_1]$, $[V_1]$, $[F_3]$, $[V_3]$, $[F_0]$, $[V_0]$.

8. If the matrix of coefficients of simultaneous equations (f_2), (v_2) is

symmetrical to the diagonal element, what mathematical properties are found with regard to the kernel of an integral equation?
[$K(t, \xi) = K(\xi, t)$, Chapter 4]
9. Explain why the types of linear integral equations are restricted to only a few.
10. State the kinds of integral equations in the problems of Chapter 2, and explain the degrees of elementary continuity of functionals on the left-hand sides. Then, give the associated integral equations.
11. Prove that

$$\left(\int (x_1 - x_2)^2 d\xi\right)^{\frac{1}{2}}$$

can be taken as the distance between two elements $x_1(\xi)$ and $x_2(\xi)$ in the functional space, and show, using the principle of passage from discontinuity to continuity, that this definition is the extension of the definition of distance between two points in n-dimensional Euclidian space. (The above definition shows that it satisfies the postulate of distance in section. We call this distance the mean distance in abstract space.)

12. Prove that with this definition, the functional $\int K(t, \xi)x(\xi)\, d\xi$ is mean continuous. By Schwarz's inequality we have

$$|F_1 - F_2| \leqq \left(\int_a^b K^2(t, \xi)d\xi \int_a^b [x_1 - x_2]^2 d\xi\right)^{\frac{1}{2}}.$$

Mean continuous means that to any arbitrarily given number $\varepsilon > 0$ we can find η such that $|F_1 - F_2| < \varepsilon$ for all x_1, x_2 with $\int (x_1 - x_2)^2\, d\xi < \eta$.
13. Calculate iterated kernels $\overset{*}{K}{}^n$, $\overset{\circ}{K}{}^n$. The intervel is (0, 1) in every case.

$$K(t, \xi) = t + \xi + 1,$$
$$K(t, \xi) = t^2 + t\xi + \xi^2,$$
$$K(t, \xi) = \frac{1}{(t + \xi)}.$$

14. Prove that by Lemma 1.2

$$f \circ (g \circ h) = (f \circ g) \circ h, \qquad f * (g * h) = (f * g) * h$$

hold for three elements f, g, h which belong to $\mathscr{R}$.
15. When $|K| < M$, prove that

$$|\mathring{K}^n| < (b-a)^{n-1}M^n, \quad |\overset{*}{K}{}^n| < \frac{(t-\xi)^{n-1}}{(n-1)!}M^n.$$

16. When $a_1, a_2, \ldots, a_n$ are arbitrary constants, prove that elements of the form

$$a_1\mathring{K} + a_2\mathring{K}^2 + \cdots + a_n\mathring{K}^n, \qquad a_1\overset{*}{K} + a_2\overset{*}{K}{}^2 + \cdots + a_n\overset{*}{K}{}^n$$

are commutative for either the convolution of the first kind or that of the second kind.

17. When ψ, Ψ are zero, outside of T in S prove that $\Psi * \psi$ is zero outside T in S.

18. With regard to the functional of elementary continuity of the first order, state which of the following statements are correct. Provided $m < n$

(a) has elementary continuity of the nth order $(m < n)$;
(b) has elementary continuity of the lth order $(l < m)$;
(c) has elementary continuity of the sth order $(m \neq s)$.

19. Show that a variational problem

$$F[x(t)] = \int_a^b \left\{ \left[\frac{1}{2}x(t) - \int_a^b K(t, \xi)x(\xi)d\xi \right] x(t) + f(t)x^{(1)}(t) \right\} dt: \text{ min.}$$

can be reduced to an integral equation of Fredholm type:

$$x(t) - \int_a^b K(t, \xi)x(\xi)d\xi = f^{(1)}(t).$$

[Make use Euler's equation $\left. \frac{\partial}{\partial \alpha} F[x + \alpha\delta x] \right|_{\alpha=0} = 0$]

20. Let

$$f_1(t) = t^2, f_2(t) = 1 - e^{-t}.$$

Show

$$f_1 \times f_2 = 1/3t^3 - t^2 + 2t - 2 + 2e^{-t}$$

and

$$f_1 \times f_2 = f_2 \times f_1$$

calculate

$$f_1 \circ f_2 \text{ in } I(0, 1)$$

21. Show that the integral equation

$$x(t) - \int_{-\infty}^{t} \exp[-2(t-\xi)](t-\xi)x(\xi) = te^t$$

has a solution

$$x(t) = ate^t + be^t,$$

and determine a and b.

$$[a = 9/8, \ b = -3/64]$$

References

Carmichael, R.D.: (1922), Algebraic guides to transcendental problems. *Bull. Amer. Math. Soc.*, vol. 28.

Churchill, R.V.: (1944), *Modern Operational Mathematics in Engineering* (McGraw-Hill, New York).

Hadamard, J.: (1903), Sur les opérations fonctionnelles. *Comptes Rendus*, vol. 136.

Hilbert, D.: (1904–1908), Grundzüge einer allgemeine Theorie der linearen Integralgleichungen. *Gött. Nachr. math. phys.*, Klasse.

Hurwitz, W.A.: (1908), *Annals of Mathematics*, vol. 9.

Riesz, F.: (1909), Sur les opérations fonctionnelles linéaires. *Comptes Rendus*, vol. 149.

Sneddon, I.N.: (1951), *Fourier Transforsm* (McGraw-Hill, New York).

Volterra, V.: (1895–1896), Sulla inversione degli integrali definiti. *Atti Ac. Torino*, vol. 31.

Whittaker, E.T. and Watson, G.N.: (1953), *A Course of Modern Analysis*, Cambridge.

2 Methods of solution of integral equations

In this chapter we discuss the theorems of existence and uniqueness of continuous solutions with regard to integral equations. These are the fundamental problems in the theoretical study of integral equations, and, therefore, this is the most important chapter in Part I 'Fundamental Theories'. Many theories and solutions in Part II 'Special Theories' can be regarded as extensions of the basic cases treated in this chapter.

Integral equations of the second kind appear to be the easiest in the discussion where we consider Volterra's method by iterated kernels and Fredholm's method by determinants. The former is the method by which we can find a solution by algebraic operation using symbols of convolution; the latter method reduces integral equations to simultaneous equations of finite dimensions, and it is also the method used to extend the solution of simultaneous equations by Cramer's rule to infinite dimensions. We can obtain the same result by both methods with regard to [V_2]; however, for [F_2] the latter method gives a more general result. The latter method is important especially for the study of eigenfunctions and eigenvalues. Now [F_0] can have a continuous solution —eigenfunction—not 0 only when the parameter λ takes a special value, an eigenvalue. The only continuous solution for [V_0] is that which is identically 0.

In this chapter we study the solution of integral equations always comparing them with the method of solution of simultaneous equations. Some simultaneous equations are well known; with these, essential characteristics of the method of solution of integral equations will be made clear. The homogeneous integral equation is a special case of the integral equation of the second kind when its inhomogeneous term (disturbance function) is 0, and it is treated as a special case of the second kind. The integral equation of the first kind can be reduced to that of the second kind in special cases. However the general theory with regard to [F_1] has not been completed yet. As many notations as possible have been used in the expressions, in order to shorten expressions to a considerable extent, and at the same time to obtain a better theoretical perspective.

In the solution of integral equations, there is no essential difference between the case of Fredholm type and that of Volterra type, except that specialties such as the eigenfunction do not appear in the case of Volterra type.

2.1. Solution of integral equations of the second kind by iterated kernels

For Fredholm's integral equation of the second kind:

$$(E - \lambda K) \circ x = f, \qquad [F_2]$$

suppose that there exists an inverse element of $(E - \lambda K)$,† which is expressed as $(E - \lambda \Gamma)$. That is,

$$(E - \lambda K) \circ (E - \lambda \Gamma) = (E - \lambda \Gamma) \circ (E - \lambda K) = E. \qquad (2.1)$$

Now if we multiply $[F_2]$ by $(E - \lambda \Gamma)$ from the left, $[F_2]$ yields

$$(E - \lambda \Gamma) \circ (E - \lambda K) \circ x = (E - \lambda \Gamma) \circ f,$$

and the left-hand side is $E \circ x$ (that is, x), then we have

$$x = (E - \lambda \Gamma) \circ f. \qquad (2.2)$$

Therefore, we have only to find such an inverse element.

Now we put

$$(1 + \lambda z + \lambda^2 z^2 + \ldots)(1 - \lambda z) = 1$$

for (2.1) making use of algebraic analogy, and substituting E for 1, K for z and the convolution of K for the power of z, we get

$$(E + \lambda K + \lambda^2 \mathring{K}^2 + \ldots) \circ (E - \lambda K) = E, \qquad (2.1a)$$

then by comparing with (2.1) we obtain

$$\Gamma = -\{K + \lambda \mathring{K}^2 + \cdots + \lambda^{n-1} \mathring{K}^n + \cdots\}. \qquad (2.3)$$

In this process, we have simply set the form of solution making a formal operation. For the establishment of the solution, it is necessary to show that (2.3) is uniformly convergent, to make sure that (2.1a) or (2.1) really holds, and to prove that (2.2) is the unique solution that satisfies $[F_2]$.

† When K belongs to $\mathscr{R}$, the set of $E - \lambda K$ forms a field. Therefore the inverse element exists. (See question (6): cf. Chapter II, exercise (6).)

Hence, first of all, let us prove the following theorem.

THEOREM 2.1[F_2]

In Fredholm's integral equation of the second kind:

$$x(t) - \lambda\int_a^b K(t, \xi)x(\xi)d\xi = f(t), \qquad [F_2]$$

when the kernel $K(t, \xi)$ belongs to the functional space, $\mathscr{R}$ in the domain S,† and $f(t)$ is continuous in the interval I, a sufficient condition for the existence of a unique continuous solution is that

$$|\lambda|M(b-a) < 1. \qquad (2.4)$$

Here the solution is

$$x(t) = f(t) - \lambda\int_a^b \Gamma(t, \xi;\ \lambda)f(\xi)d\xi, \qquad (2.2a)$$

and the kernel of the solution, $\Gamma(t, \xi;\ \lambda)$, is

$$\Gamma(t, \xi;\ \lambda) = -\Big\{K(t, \xi) + \lambda\int_a^b K(t, \tau)K(\tau, \xi)d\tau$$
$$+ \lambda^2\int_a^b\int_a^b K(t, \tau_1)K(\tau_1, \tau_2)K(\tau_2, \xi)d\tau_1 d\tau_2 + \cdots\Big\}. \qquad (2.3a)$$

Proof. The proof is given in the order of convergence, reciprocity, and inversion.

(i) *Convergence.*

$$|\lambda^{n-1}\mathring{K}^n| \leqq M\{\lambda(b-a)M\}^{n-1},$$

(see § 1.4); therefore, when $|\lambda|(b-a)M < 1$, the right-hand side of Γ is absolutely convergent. Since all iterated kernels that are higher than $\mathring{K}^2$ are continuous, Γ belongs to $\mathscr{R}$ as well. Hence, if $f(t)$ is continuous, $x(t)$ is continuous by Lemma 1.2.

(ii) *Reciprocal theorem.*

If we multiply both sides of (2.3) by λK from the left we get

$$\lambda K \circ \Gamma = -\lambda K \circ \{K + \lambda\mathring{K}^2 + \cdots + \lambda^{n-1}\mathring{K}^n + \cdots\}$$
$$= -\{\lambda\mathring{K}^2 + \lambda^2\mathring{K}^3 + \cdots + \lambda^n\mathring{K}^{n+1} + \cdots\};$$

here the right-hand side is clearly $\Gamma + K$ from (2.3). Since the result is the same if we multiply from the right, then we have

† The condition $\mathscr{R}$ signifies boundedness and the regular distribution of discontinuous points. (See chapter I, § 1.4)

$$\Gamma + K = \lambda K \circ \Gamma = \lambda \Gamma \circ K \tag{2.5}$$

This formula (2.5) is called the *reciprocal theorem* (see the definition of a reciprocal function).

(iii) *Proof of a solution.* Substitute the solution (2.2) for the left-hand side of the given integral equation and we obtain

$$(E - \lambda K) \circ (E - \lambda \Gamma) \circ f = \{E - \lambda(K + \Gamma) + \lambda^2 K \circ \Gamma\} \circ f.$$

Applying the reciprocal theorem (2.5), the right-hand side becomes

$$E \circ f = f.$$

Thus it has been proved that (2.2) is a solution of $[F_2]$.

(iv) *Uniqueness.* We express $[F_2]$ in the form

$$x = f + \lambda K \circ x.$$

Substituting the formula for x on the right-hand side, we have

$$\begin{aligned} x &= f + \lambda K \circ (f + \lambda K \circ x), \\ &= f + \lambda K \circ f + \lambda^2 \mathring{K}^2 \circ x. \end{aligned}$$

Then substitute the first formula again for x on the right-hand side. Repetition of the process $(n - 1)$ times brings us to

$$x = f + \sum_1^{n-1} \lambda^i \mathring{K}^i \circ f + \lambda^n \mathring{K}^n \circ x.$$

Therefore, if $x(t)$ is continuous with $|x| < N$ in I, we get

$$|\lambda^n \mathring{K}^n \circ x| < N\lambda^n (b - a)^n \circ M^n.$$

We repeat the process infinitely often, and, since it has been supposed that $|\lambda|(b - a)M < 1$, we obtain $\lim_{n \to \infty} |\lambda^n \mathring{K}^n \circ x| = 0$; this indicates that all the solutions can be expressed in the form (2.4). Hence, uniqueness has been proved.

Note. In the proof of uniqueness we cannot use the following argument, which is usually adopted; that is, to set two solutions as x_1, x_2, to derive $(E - \lambda K) \circ (x_1 - x_2) = 0$ from

$$(E - \lambda K) \circ x_1 = f, \qquad (E - \lambda K) \circ x_2 = f$$

and then with this to prove that $x_1 - x_2 = 0$. This argument cannot be applied since we have not proved beforehand that there is no other solution than $x \equiv 0$ when $f = 0$. This will be established as a corollary.

Corollary [F_0]

In Fredholm's homogeneous integral equation:

$$x(t) = \lambda\int_a^b K(t, \xi)x(\xi)d\xi, \qquad [F_0]$$

if K belongs to $\mathscr{R}$ *and* $|\lambda|M(b-a) < 1$, *its only continuous solution is* $x \equiv 0$.

Example 2.1 [F_2]

$$x(t) - \lambda\int_0^1 t\xi x(\xi)d\xi = at,$$

where a is constant.

Since $K(t, \xi) = t\xi$ it follows that

$$\overset{\circ}{K}{}^n = \frac{t\xi}{3^{n-1}}, \text{ (see § 1.4, Example 1)}$$

If $\lambda \neq 3$, the kernel of the solution is

$$\Gamma = -\sum_1^\infty \lambda^{n-1}\frac{t\xi}{3^{n-1}} = -t\xi\left(\frac{1}{1-\frac{1}{3}\lambda}\right) = -\frac{3}{3-\lambda}t\xi.$$

The solution is

$$\begin{aligned} x(t) &= (E - \lambda\Gamma)\circ f = \left(1 + \frac{3\lambda}{3-\lambda}t\xi\right)\circ(at), \\ &= at + \frac{3\lambda a}{3-\lambda}\int_0^1 t\tau\,\tau\,d\tau, \\ &= at + \frac{\lambda}{3-\lambda}at = \left(1+\frac{\lambda}{3-\lambda}\right)at, \\ &= \frac{3}{3-\lambda}at. \end{aligned}$$

Also in the case of Volterra type we can carry out exactly the same calculations, and we can prove the following theorem for Theorem 2.1. As the principle and method of proof are the same as those above, however, we omit them here.

THEOREM 2.2 [V_2]

For Volterra's integral equation of the second kind:

$$x(t) - \int_a^t K(t, \xi)x(\xi)d\xi = f(t), \qquad [V_2]$$

if the kernel $K(t, \xi)$ belongs to the functional space $\mathscr{R}$ in the domain T, and if $f(t)$ is continuous in I, then there exists a unique continuous solution

$$x(t) = f(t) - \int_a^t G(t, \xi)x(\xi)d\xi,$$

and the kernel of the solution is

$$G(t, \xi) = -\Big\{K(t, \xi) + \int_\xi^t K(t, \tau)K(\tau, \xi)d\tau$$
$$+ \int_\xi^t \int_\xi^{\tau_1} K(t, \tau_1)K(\tau_1, \tau_2)K(\tau_2, \xi)d\tau_1 d\tau_2 + \cdots\Big\}.$$

Example 2.2 [V_2]

$$x(t) - \int_0^t t\xi x(\xi)d\xi = at$$

where a is constant.

Since $K(t, \xi) = t\xi$, it follows that $\overset{*}{K}{}^n = \dfrac{t\xi}{3^{n-1}(n-1)!}(t^3 - \xi^3)^{n-1}$ (see § 1.4, Example 1.1), the kernel of the solution is

$$G(t, \xi) = -\ t\xi \exp\left(\frac{1}{3}(t^3 - \xi^3)\right),$$

and the solution is

$$x(t) = at + \int_0^t t\xi \exp\left(\frac{1}{3}(t^3 - \xi^3)\right)a\xi\ d\xi = at \exp(t^3/3).$$

Definition. Iterated kernels

When the kernel $K(t, \xi)$ belongs to the functional space S, the convolution of the second kind in the domain S:

$$\overset{\circ}{K}{}^n\ (n = 1, 2, \ldots;\ \overset{\circ}{K}{}^1 = K(t, \xi))$$

or the convolution of the first kind in the domain T:

$$\overset{*}{K}{}^n\ (n = 1, 2, \ldots;\ \overset{*}{K}{}^1 = K(t, \xi))$$

is called an *iterated kernel* or an *iterated function.*†

With respect to the iterated kernels, by the definition of convolution the following naturally hold:

† The solution of Theorem 2.1 and Theorem 2.2 is sometimes called the solution by iterated kernel or Volterra's solution.

$$\mathring{K}^n = \int_a^b \int_a^b \cdots \int_a^b K(t, \tau_1)K(\tau_1, \tau_2) \cdots K(\tau_{n-1}, \xi)d\tau_1 \cdots d\tau_{n-1},$$

$$\overset{*}{K}^n = \int_a^t \int_a^{\tau_1} \cdots \int_a^{\tau_{n-2}} K(t, \tau_1)K(\tau_1, \tau_2) \cdots K(\tau_{n-1}, \xi)d\tau_1 \cdots d\tau_{n-1},$$

and

$$\begin{aligned} \mathring{K}^n &= K \circ \mathring{K}^{n-1} = \mathring{K}^{n-1} \circ K \\ &= \mathring{K}^i \circ \mathring{K}^{n-i} = \mathring{K}^{n-i} \circ \mathring{K}^i \qquad (i = 1, 2, \ldots, n-1), \\ \overset{*}{K}^n &= K * \overset{*}{K}^{n-1} = \overset{*}{K}^{n-1} * K \\ &= \overset{*}{K}^i * \overset{*}{K}^{n-i} = \overset{*}{K}^{n-1} * \overset{*}{K}^i \qquad (i = 1, 2, \ldots, n-1) \end{aligned}$$

(see Question 15 below).

For example, in Example 1.1 in § 1.4, if

$$K(t, \xi) = t\xi, \qquad I(1, 0),$$

then

$$\mathring{K}^i = \frac{t\xi}{3^{i-1}};$$

therefore,

$$\mathring{K}^i \circ \mathring{K}^{n-i} = \int_0^1 \frac{t\tau}{3^{i-1}} \frac{\tau\xi}{3^{n-i-1}} d\tau = \frac{t\xi}{3^{n-1}} = \mathring{K}^n.$$

And since in T by the convolution of the first kind

$$\overset{*}{K}^i = \frac{t\xi}{3^{i-1}(i-1)!}(t^3 - \xi^3)^{i-1},$$

it follows that

$$\begin{aligned} &\overset{*}{K}^i * \overset{*}{K}^{n-i} \\ &= \frac{1}{3^{n-2}} \frac{1}{(i-1)!\,(n-i-1)!} \int_\xi^t t\tau \cdot \tau\xi(t^3 - \tau^3)^{i-1}(\tau^3 - \xi^3)^{n-i-1} d\tau. \end{aligned}$$

Then we take the partial integral

$$\begin{aligned} &\overset{*}{K}^i * \overset{*}{K}^{n-i} \\ &= \frac{1}{3^{n-2}} \frac{1}{(i-2)!\,(n-i)!} \int_\xi^t t\tau \cdot \tau\xi(t^3 - \tau^3)^{i-2}(\tau^3 - \xi^3)^{n-i} d\tau. \end{aligned}$$

On repeating the process $(i-1)$ times we obtain

$$\frac{1}{3^{n-2}}\,\frac{1}{(n-2)!}\int_{\xi}^{t} t\tau\cdot\tau\xi(\tau^3-\xi^3)^{n-2}d\tau.$$

Hence,

$$\overset{*}{K}{}^{i} * \overset{*}{K}{}^{n-i} = \frac{t\xi}{3^{n-1}(n-1)!}(t^3-\xi^3)^{n-1} = \overset{*}{K}{}^{n}.$$

Definition. Reciprocal Functions

Suppose that there are two functions $K(t, \xi)$, $\Gamma(t, \xi)$ which belong to the functional space $\mathscr{R}$; when the reciprocal theorem (2.5), that is

$$K + \Gamma = \lambda K \circ \Gamma = \lambda \Gamma \circ K$$

holds in S, we state that the two functions are *reciprocal*, and that they are *reciprocal functions*. In addition, when there are two functions $K(t, \xi)$, $G(t, \xi)$, and the reciprocal theorem:

$$K + G = K * G = G * K \tag{2.6}$$

holds in T, we say that they are also reciprocal and that they are reciprocal functions.

The reciprocal theorems (2.5), (2.6) are symmetrical in K, Γ or K, G. The right-hand side is continuous from LEMMA 1.1, therefore even if there are discontinuous points in K, the continuous part on the right-hand side and those of Γ, G complement each other, and, therefore, the sum of reciprocal functions becomes continuous.

For example, with regard to $K(t, \xi) = t\xi$, $I(0, 1)$, if $\lambda \neq 3$, the reciprocal function is

$$\Gamma(t, \xi; \lambda) = -\,t\xi\left\{1 + \frac{\lambda}{3} + \frac{\lambda^2}{3^2} + \cdots\right\} = -\,\frac{3t\xi}{3-\lambda},$$

then, the reciprocal theorem becomes

$$K + \Gamma = -\,\lambda K \circ \Gamma = -\,\lambda\frac{t\xi}{3-\lambda}.$$

Besides, since

$$G(t, \xi) = -\,t\xi\left\{1 + \frac{1}{1!}\,\frac{t^3-\xi^3}{3} + \frac{1}{2!}\left(\frac{t^3-\xi^3}{3}\right)^2 + \cdots\right\}$$

$$= -t\xi\,\exp\left(\frac{1}{3}(t^3-\xi^3)\right),$$

the reciprocal theorem becomes

$$K + G = K * G = t\xi\left[1 - \exp\left\{\frac{1}{3}(t^3 - \xi^3)\right\}\right].$$

The existence and uniqueness of a reciprocal function itself can be derived directly from Theorems 2.1 and 2.2, as well.

(1.1) *Existence of a reciprocal function (Fredholm type)*

When $K(t, \xi)$ belongs to the functional space $\mathscr{R}$, if

$$|\lambda| M(b - a) < 1$$

in the domain S, there exists a function $\Gamma(t, \xi)$ which is reciprocal to K; it is given by (2.3).

(2.1) *Existence of a reciprocal function (Volterra type)*

When $K(t, \xi)$ belongs to the functional space $\mathscr{R}$, in the domain T there exists a function $G(t, \xi)$ which is always reciprocal to K; it is given by (2.4).

(1.2) *Uniqueness of a reciprocal function (Fredholm type)*

When $K(t, \xi)$ belongs to the functional space $\mathscr{R}$, a function in the domain S which is reciprocal to K is unique.

Proof. If there exist two reciprocal functions k_1, k_2, the remainder $k_1 - k_2 (\equiv \sigma)$ is continuous, since $K + k_1$ and $K + k_2$ are continuous functions. Here

$$\sigma = \lambda K \circ \sigma.$$

Take this as an integral equation in σ. It satisfies the condition of Theorem 2.1; therefore, there is no other solution than $\sigma = 0$. (See Theorem 2.1, corollary.) Thus, uniqueness of the reciprocal function has been proved.

(2.2) *Uniqueness of a reciprocal function (Volterra type)*

When $K(t, \xi)$ belongs to the functional space $\mathscr{R}$, a function which is reciprocal to K in the domain T is unique.

[The proof is omitted since it is obvious.]

In Theorems 2.1 and 2.2 we proved that there exists a reciprocal function, and we used it to prove existence and uniqueness of a solution. On the other hand, we can introduce the theorem of existence and uniqueness of a solution by assuming the existence of the reciprocal theorem.

(1.3) *For Fredholm's integral equation of the second kind*:

$$(E - \lambda K) \circ x = f, \qquad [F_2]$$

when the kernel $K(t, \xi)$ *belongs to the functional space* $\mathscr{R}$, *and* $f(t)$ *is continuous in* I, *the necessary and sufficient condition for the existence of a continuous solution is that the reciprocal function of the kernel* $\Gamma(t, \xi; \lambda)$ *exists. Here the solution is*

$$x = (E - \lambda \Gamma) \circ f,$$

and it is unique.

(Again we omit the proof.)

(2.3) *For Volterra's integral equation of the second kind*:

$$(E - K) * x = f, \qquad [V_2]$$

when the kernel $K(t, \xi)$ *belongs to the functional space* $\mathscr{R}$, *and* $f(t)$ *is continuous, the reciprocal function of the kernel,* $G(t, \xi)$, *always exists, and the solution is*

$$x = (E - G) * f.$$

This solution is unique.

(The proof is omitted.)

2.1.1. Discontinuous solution

In this section we have shown that in an integral equation of the second kind, when the domain S or I is bounded and

(i) the kernel K (t, ξ) belongs to $\mathscr{R}$, and

(ii) f(t) is continuous,

then there necessarily exists a solution, and it is continuous and unique.

But we use the word unique here in the sense of uniqueness of a continuous solution; we do not mean to deny the existence of any other, discontinuous solution. For example,

$$x(t) = \int_0^t \xi^{\tau-\xi} x(\xi) d\xi$$

has but one solution

$$x(t) \equiv 0$$

as a continuous solution, but additionally there exist an infinite number of discontinuous solutions:

$$x(t) = Ct^{t-1}.$$

(Here C is a non-zero constant.)

Of course, it is possible that if (i) and (ii) are not satisfied, there does not exist a continuous solution; but, since these conditions are not necessary, this does not mean that no continuous solution exists. Take, for example, the equation

$$x(t) - \int_0^1 x(\xi) \log \left(\frac{1}{t\xi}\right)^3 d\xi = t^2 + \log t - \frac{8}{27}.$$

By the integral formula

$$\int_0^1 \xi^m \log\left(\frac{1}{\xi}\right)^n d\xi = \frac{\Gamma(n+1)}{(m+1)^{n+1}}$$

this clearly has a continuous solution (Bocher 1914, p. 17)

$$x(t) = t^2.$$

In any case, with regard to the kernel belonging to $\mathscr{R}$, various theorems in this section treat the existence and uniqueness of solutions, restricting the sphere in continuous solution.

2.1.2. *Method of successive substitution*

As we have already given in (iv) of the the proof of Theorem 2.1, first we express [F_2] in the form

$$x = f + \lambda K \circ x,$$

then substituting the equation for x on the right-hand side,

$$\begin{aligned} x &= f + \lambda K \circ (f + \lambda K \circ x) \\ &= f + \lambda K \circ f + \lambda^2 \mathring{K}^2 \circ x. \end{aligned}$$

Substituting successively for x in a similar manner, we obtain

$$x = f + \sum_1^{n-1} \lambda^i \mathring{K}^i \circ f + \lambda^n \mathring{K}^n \circ x.$$

If $|\lambda|(b-a)M < 1$, $\lim_{n\to\infty} |\lambda^n \mathring{K}^n \circ x| = 0$, then we have

$$x = f + \sum_1^{\infty} \lambda^i \mathring{K}^i \circ f;$$

that is,

$$x(t)=f(t)+\lambda\int_a^b K(t,\xi)f(\xi)d\xi+\lambda^2\int_a^b\int_a^b K(t,\xi)K(\xi,\tau)f(\tau)d\tau d\xi$$
$$+\lambda^3\int_a^b\int_a^b\int_a^b K(t,\xi)K(\xi,\tau_1)K(\tau_1,\tau_2)f(\tau_2)d\tau_2 d\tau_1 d\xi$$
$$+\cdots+\lambda^n\int_a^b\int_a^b\cdots\int_a^b K(t,\xi)K(\xi,\tau_1)\cdots K(\tau_{n-2},\tau_{n-1})$$
$$\times f(\tau_{n-1})d\tau_{n-1}\cdots d\tau_1 d\xi.$$

In the case of Volterra type, similarly we obtain

$$x(t)=f(t)+\int_a^t K(t,\xi)f(\xi)d\xi+\int_a^t\int_a^\xi K(t,\xi)K(\xi,\tau)f(\tau)d\tau d\xi$$
$$+\int_a^t\int_a^\xi\int_a^{\tau_1} K(t,\xi)K(\xi,\tau_1)K(\tau_1,\tau_2)f(\tau_2)d\tau_2 d\tau_1 d\xi$$
$$+\cdots+\int_a^t\int_a^\xi\int_a^{\tau_1}\cdots\int_a^{\tau_{n-1}} K(t,\xi)K(\xi,\tau_1)\cdots K(\tau_{n-2},\tau_{n-1})$$
$$\times f(\tau_{n-1})d\tau_{n-1}\cdots d\tau_1 d\xi.$$

We call this method of solution the *method of successive substitution.* For example, with regard to Example 1.1,

$$\begin{aligned}x(t)&=at+\lambda\int_0^1 t\xi\, x(\xi)d\xi\\&=at+\lambda\int_0^1 t\xi\, a\xi\, d\xi+\lambda^2\int_0^1\int_0^1 t\xi\,\xi\tau\, at\, d\tau\, d\xi+\cdots\\&=at\left(1+\frac{\lambda}{3}+\frac{\lambda^2}{3^2}+\cdots\right)\\&=\frac{3}{3-\lambda}at.\end{aligned}$$

And for Example 2.2 we have

$$\begin{aligned}x(t)&=at+\int_0^t t\xi\, a\xi\, d\xi+\int_0^t t\xi\, d\xi\int_0^\xi \xi\tau\, a\tau\, d\tau+\cdots\\&=at\left(1+\frac{t^3}{3}+\frac{t^6}{3\cdot 6}+\cdots\right)\\&=at\exp(t^3/3).\end{aligned}$$

2.1.3. *Method of successive approximation*

The *method of successive approximation* as a solution of the integral equation of the second kind is intrinsically the same and in 2.1.2. For example in $[F_2]$ we express

$$x = f + \lambda K \circ x$$

and take an arbitrary continuous function x_0 as the zeroth approximation. The first approximation x_1 is

$$x_1 = f + \lambda K \circ x_0,$$

then using this x_1, we represent the second approximation as

$$x_2 = f + \lambda K \circ x_1.$$

The same rule applies throughout. Generally we can express the nth approximation, using the $(n-1)$th approximation x_{n-1}, in

$$x_n = f + \lambda K \circ x_{n-1}.$$

This is what is called the perturbation method.

Putting $f + \lambda K \circ x_{n-2}$ into x_{n-1} on the right-hand side, and substituting successively in the inverse order, we get

$$x_n = f + (\lambda K + \lambda^2 \mathring{K}^2 + \ldots + \lambda^{n-1} \mathring{K}^{n-1}) \circ f + \lambda^n \mathring{K}^n \circ x_0.$$

This is the same as in the case of Theorem 2.1; that when $n \to \infty$, we have $|\lambda^n \mathring{K}^n \circ x_0| \to 0$. Hence this process ultimately shows that if we repeat infinitely often the above method of successive approximations, the perturbation method always converges as the approximation of the zeroth order with the adoption of an arbitrary continuous function, and that this convergence results in the solution of the integral equation.

We have used the method on $[F_2]$, but the same can be applied to $[V_2]$.

2.2. Solution of integral equations of the second kind by determinants

If we pass an integral equation from continuity to discontinuity, it becomes a set of simultaneous equations of an infinite order and, by

Cramer's rule, the solution can be expressed with a determinant for the case of finite order. If we pass this solution for finite order from discontinuity to continuity, moving the order to an infinite sphere, we can expect to obtain the solution of the integral equation. This notion stands as the base of Fredholm's method of solution. To prove that the solution thus obtained is the solution of the integral equation, we have to treat it in a more precise way. This is the proof of Theorem 2.1, in which we are going to use Hadamard's theorem.

Now, according to Cramer's rule on the solution of linear simultaneous equations, in linear simultaneous equations with unknown quantities $x_1, x_2, \ldots, x_n$,

$$\left.\begin{aligned} a_{11}x_1 + a_{12}x_2 + \cdots + a_{1n}x_n &= b_1, \\ a_{21}x_1 + a_{22}x_2 + \cdots + a_{2n}x_n &= b_2, \\ \cdots \quad \cdots \quad \cdots \quad \cdots & \\ a_{n1}x_1 + a_{n2}x_2 + \cdots + a_{nn}x_n &= b_n \end{aligned}\right\} \tag{a}$$

use the determinant of coefficients

$$D = \begin{vmatrix} a_{11} & a_{12} & \cdots & a_{1n} \\ a_{21} & a_{22} & \cdots & a_{2n} \\ & \cdots & \cdots & \\ a_{n1} & a_{n2} & & a_{nn} \end{vmatrix},$$

and a determinant D_i, a version of D in which the ith column is replaced by $b_1, b_2, \ldots, b_n$. Then, when $D \neq 0$, we come to have only one pair of solutions:

$$x_i = D_i/D \qquad (i = 1, 2, \ldots, n).$$

When we adopt vector symbols, we can represent the above simultaneous equations using a row vector $\boldsymbol{a}_i(a_{i1}, a_{i2}, \ldots, a_{in})$ and a constant-term vector $\boldsymbol{b}(b_1, b_2, \ldots, b_n)$, in

$$\boldsymbol{a}_1x_1 + \boldsymbol{a}_2x_2 + \cdots + \boldsymbol{a}_nx_n = \boldsymbol{b}. \tag{$\boldsymbol{a}$}$$

Here, the condition $D \neq 0$ corresponds the row vectors $a_1, a_2, \ldots, a_n$ being linearly independent. If the solution of the simultaneous equations $x_1, x_2, \ldots, x_n$ exists, ($\boldsymbol{a}$) shows that the constant-term vector $\boldsymbol{b}$ is the linear combination of these row vectors.

Now let us apply this method to the case of Fredholm's integral equation of the second kind. We take the simultaneous equations

$$\left.\begin{aligned} (1-\lambda k_{11})x_1 - \lambda k_{12}x_2 - \cdots - \lambda k_{1n}x_n &= y_1, \\ -\lambda k_{21}x_1 + (1-\lambda k_{22})x_2 - \cdots - \lambda k_{2n}x_n &= y_2, \\ \cdots \\ -\lambda k_{n1}x_1 - \lambda k_{n2}x_2 - \cdots + (1-\lambda k_{nn})x_n &= y_n \end{aligned}\right\} \qquad (\mathrm{f}_2)$$

corresponding to

$$x(t) - \lambda\int_a^b K(t, \xi)x(\xi)d\xi = f(t) \qquad [\mathrm{F}_2]$$

(see 1.3.1). By Cramer's rule, when $D \neq 0$, we get the solution

$$x_i = D_i/D.$$

Here D is the determinant of coefficients, and D_i is a determinant, a version of D, in which the ith column is replaced by $y_1, y_2, \ldots, y_n$. They are,

$$D = \begin{vmatrix} 1-\lambda k_{11} & -\lambda k_{12} \ldots & -\lambda k_{1n} \\ -\lambda k_{21} & 1-\lambda k_{22} \ldots & -\lambda k_{2n} \\ \cdots & \cdots & \\ -\lambda k_{n1} & -\lambda k_{n2} \ldots & 1-\lambda k_{nn} \end{vmatrix},$$

$$D_i = \begin{vmatrix} 1-\lambda k_{11} & -\lambda k_{12} \ldots & y_1 \ldots & -\lambda k_{1n} \\ -\lambda k_{21} & 1-\lambda k_{22} \ldots & y_2 \ldots & -\lambda k_{2n} \\ \cdots & \cdots & \cdots & \cdots \\ -\lambda k_{i1} & -\lambda k_{i2} \ldots & y_i \ldots & -\lambda k_{in} \\ \cdots & \cdots & \cdots & \cdots \\ -\lambda k_{n1} & -\lambda k_{n2} \ldots & y_n \ldots & 1-\lambda k_{nn} \end{vmatrix}.$$

When $D = 0$, there usually exists no solution and the result is either indeterminate or inconsistent. The detailed explanation of this case is given in § 2.4; here we consider the case of $D \neq 0$.

Now D is a polynomial of λ of nth degree. If we expand this in the order of ascending powers of λ, by Maclaurin's expansion the expression becomes

$$D = D_0 + \frac{\lambda}{1!}D_0^{(1)} + \frac{\lambda^2}{2!}D_0^{(2)} + \cdots + \frac{\lambda^n}{n!}D_0^{(n)},$$

$$D_0^{(m)} \equiv \left[\frac{d^m D}{d\lambda^m}\right]_{\lambda=0},$$

where $D_0^{(m)}$ is the mth derivative of D in λ when $\lambda = 0$. It is evident that we can get

$$D_0 = 1; \qquad D_0^{(1)} = -\sum_1^n{}_i\, k_{ii} \equiv -\sum k\binom{i}{i},$$

$$D_0^{(2)} = \textstyle\sum_i\sum_j \begin{vmatrix} k_{ii} & k_{ij} \\ k_{ji} & k_{jj} \end{vmatrix} \equiv \sum_i\sum_j k\binom{i\ j}{i\ j},$$

$$D_0^{(3)} = -\textstyle\sum_i\sum_j\sum_r \begin{vmatrix} k_{ii} & k_{ij} & k_{ir} \\ k_{ji} & k_{jj} & k_{jr} \\ k_{ri} & k_{rj} & k_{rr} \end{vmatrix} \equiv -\sum_i\sum_j\sum_r k\binom{i\ j\ r}{i\ j\ r}.$$

The notation $k(\vdots)$ is an abbreviation expressed only by the suffixes of the elements of the determinant. Consequently, the determinant of coefficients is

$$D = 1 - \lambda\textstyle\sum_i k\binom{i}{i} + \frac{\lambda^2}{2!}\sum_i\sum_j k\binom{i\ j}{i\ j} - \frac{\lambda^2}{3!}\sum_i\sum_j\sum_r k\binom{i\ j\ r}{i\ j\ r} + \cdots$$
$$+ (-1)^m\frac{\lambda^m}{m!}\underbrace{\textstyle\sum_i\sum_j \cdots \sum_u}_{(m \text{ times})} k\binom{i\ j \ldots u}{i\ j \ldots u} +$$
$$\cdots + (-1)^n\frac{\lambda^n}{n!}\underbrace{\textstyle\sum_i\sum_j \cdots \sum_v}_{(n \text{ times})} k\binom{i\ j \ldots v}{i\ j \ldots v},$$

where the last term is expressed formally in a general form. But this can clearly be expressed as

$$(-1)^n\frac{\lambda^n}{n!}\begin{vmatrix} k_{11} & k_{12} & \ldots & k_{1n} \\ k_{21} & k_{22} & \ldots & k_{2n} \\ \ldots & \ldots & & \\ k_{n1} & k_{n2} & \ldots & k_{nn} \end{vmatrix},$$

and we can express this as

$$D = 1 + \sum_1^n{}_\nu (-1)^\nu \frac{\lambda^\nu}{\nu!}\Bigl\{\underbrace{\textstyle\sum_i\sum_j \cdots \sum_s}_{\nu \text{ times}} k\binom{i\ j \ldots s}{i\ j \ldots s}\Bigr\}. \qquad (2.7)$$

Next, expanding D_i in the ith row, we take the cofactor of D at $-\lambda k_{1i}$ as the coefficient of y_1, the cofactor of D at $-\lambda k_{2i}$ as the coefficient of y_2, ..., the cofactor of the diagonal element $1 - \lambda k_{ii}$ of D as the coefficient of y_i, ..., and the cofactor of D at $-\lambda k_{ni}$ as the coefficient of y_n. These cofactors become

$$-\frac{1}{\lambda}\frac{\partial D}{\partial k_{1i}},\quad -\frac{1}{\lambda}\frac{\partial D}{\partial k_{2i}},\quad \ldots,\quad -\frac{1}{\lambda}\frac{\partial D}{\partial k_{ni}},$$

where the coefficient of y_i is the cofactor of D at $1-\lambda k_{ii}$, which is a determinant quite similar to D except that it is one column and one row less than D.

Therefore, if we substitute D' for it, D_i is expressed as

$$D_i = y_i D' - \sum_{i\neq j} y_j \frac{\partial D}{\partial k_{ji}}\frac{1}{\lambda}.$$

Then dividing both sides by D, we get

$$x_i = y_i\frac{D'}{D} - \frac{1}{\lambda D}\sum_j y_j \frac{\partial D}{\partial k_{ji}}, \tag{2.8}$$

where since D' is similar to D, if $n\to\infty$, $D'\to D$, then we have $D'/D\to 1$.

Further, there follows the calculation of $\partial D/\partial k_{ji}$. This is the λ times of the cofactor of D at λk_{ji}. For example, when $j=2$, $i=1$, with $\partial D/\partial k_{21} = \lambda D_{12}$, we have

$$D_{12} = \begin{vmatrix} -\lambda k_{12} & -\lambda k_{13} \ldots & -\lambda k_{1n} \\ -\lambda k_{32} & 1-\lambda k_{33} \ldots & -\lambda k_{3n} \\ \ldots & \ldots & \ldots \\ -\lambda k_{n2} & -\lambda k_{n1} \ldots & 1-\lambda k_{nn} \end{vmatrix}.$$

Therefore, as in the preceding process we can express D_{12} in a linear equation of $(n-1)$th degree:

$$D_{12} = D_0 + \frac{\lambda}{1!}D_0^{(1)} + \ldots + \frac{\lambda^{n-1}}{(n-1)!}D_0^{(n-1)},$$

$$D_0^{(m)} = \left[\frac{d^m D}{d\lambda^m}\right]_{\lambda=0}.$$

Arranging this equation with the original notation,

$$D_0 = 0,\qquad D_0^{(1)} = -k_{12} = -k\binom{1}{2},\qquad D_0^{(2)} = \sum_{i=1}^{n}\begin{vmatrix} k_{12} & k_{1i} \\ k_{i2} & k_{ii}\end{vmatrix}$$

$$= \sum_i k\begin{pmatrix}1 & i\\ 2 & i\end{pmatrix},\qquad D_0^{(3)} = -\sum_{i=1}^{n}\sum_{j=1}^{n}\begin{vmatrix} k_{12} & k_{1i} & k_{1j} \\ k_{i2} & k_{ii} & k_{ij} \\ k_{j2} & k_{ji} & k_{jj}\end{vmatrix}$$

$$= -\Sigma_i\Sigma_j k\begin{pmatrix}1 & i & j\\ 2 & i & j\end{pmatrix}, \dots$$

Hence, we have

$$D_{12} = -\lambda k\begin{pmatrix}1\\2\end{pmatrix} + \frac{\lambda^2}{2!}\Sigma k\begin{pmatrix}1 & i\\ 2 & i\end{pmatrix} - \frac{\lambda^3}{3!}\Sigma\Sigma k\begin{pmatrix}1 & i & j\\ 2 & i & j\end{pmatrix}$$

$$+ \cdots + (-1)^m \frac{\lambda^m}{m!}\underbrace{\Sigma\Sigma\dots\Sigma}_{(m-1)\text{ times}} k\begin{pmatrix}1 & i & j & \dots & l\\ 2 & i & j & \dots & l\end{pmatrix}$$

$$+ \cdots + (-1)^{n-1}\frac{\lambda^{n-1}}{(n-2)!}\Sigma\Sigma\cdots\Sigma k\begin{pmatrix}1 & i & j & \dots & u\\ 2 & i & j & \dots & u\end{pmatrix},$$

where, for the same reason as before, the last term is

$$(-1)^{n-1}\frac{\lambda^{n-1}}{(n-1)!}\begin{vmatrix} k_{12} & k_{13} & \dots & k_{1n}\\ k_{32} & k_{33} & \dots & k_{3n}\\ \dots & & \dots & \\ k_{n2} & k_{n3} & \dots & k_{nn}\end{vmatrix} = (-1)^{n-1}\frac{\lambda^{n-1}}{(n-1)!}k\begin{pmatrix}1 & 3 & \dots & n\\ 2 & 3 & \dots & n\end{pmatrix}$$

Therefore, if in general we put

$$\frac{\partial D}{\partial k_{ji}} = -\lambda^2 D'_{ij},$$

then (2.8) becomes

$$x_i = y_i\frac{D'}{D} + \frac{\lambda}{D}\Sigma_j D'_{ij}y_j \tag{2.8a}$$

and D'_{ij} is

$$D'_{ij} = k_{ij} + \sum_{\nu=1}^{n-2}(-1)^\nu\frac{\lambda^\nu}{\nu!}\Big\{\underbrace{\Sigma_k\Sigma_l\cdots\Sigma_r}_{(\nu-1)\text{ times}} k\begin{pmatrix}j & k & l & \dots & r\\ i & k & l & \dots & r\end{pmatrix}. \tag{2.9}$$

Let us consider the case when the number of simultaneous equations grows to infinity, that is, when $n \to \infty$. For this purpose by the 'principle of passage from discontinuity to continuity' it will be enough to make the substitutions

$$i \to t, \quad j \to \xi, \quad \Sigma \to \int_a^b, \quad k \to K.$$

Here,

$$\underbrace{\Sigma_i\Sigma_j \cdots \Sigma_s}_{\nu \text{ times}} k\begin{pmatrix} i\, j \ldots s \\ i\, j \ldots s \end{pmatrix}$$

signifies the sum of the determinant

$$\begin{vmatrix} k_{ii}\; k_{ij} \ldots k_{is} \\ k_{ji}\; k_{jj} \ldots k_{js} \\ k_{si}\; k_{sj} \ldots k_{ss} \end{vmatrix}$$

for all the values of $i, j, \ldots, s$. For the calculation in infinite dimensions, since we substitute the integration over continuous variables for the sum of discontinuous parameters, we can make the substitutions

$$k_{ii} \to K(\xi_1, \xi_1), \quad k_{ij} \to K(\xi_1, \xi_2), \quad \ldots, \quad k_{ss} \to K(\xi_\nu, \xi_\nu),$$

and when we use the notation

$$\begin{vmatrix} K(\xi_1, \xi_1)\; K(\xi_1, \xi_2) \ldots K(\xi_1, \xi_\nu) \\ K(\xi_2, \xi_1)\; K(\xi_2, \xi_2) \ldots K(\xi_2, \xi_\nu) \\ \cdots \qquad \cdots \\ K(\xi_\nu, \xi_1)\; K(\xi_\nu, \xi_2) \ldots K(\xi_\nu, \xi_\nu) \end{vmatrix} \equiv K\begin{pmatrix} \xi_1\; \xi_2 \ldots \xi_\nu \\ \xi_1\; \xi_2 \ldots \xi_\nu \end{pmatrix}$$

as above, then the ν-ple sum can be expressed in the form of a ν-ple integral:

$$\underbrace{\int_a^b \int_a^b \cdots \int_a^b}_{(\nu \text{ times})} K\begin{pmatrix} \xi_1\; \xi_2 \ldots \xi_\nu \\ \xi_1\; \xi_2 \ldots \xi_\nu \end{pmatrix} d\xi_1 d\xi_2 \ldots d\xi_\nu.$$

Therefore in (2.7) when $n \to \infty$, write $D \to \Delta$; then Δ becomes the power series of λ:

$$\Delta = 1 + \sum_1^\infty{}_\nu (-1)^\nu \frac{\lambda^\nu}{\nu!} \int_a^b \int_a^b \cdots \int_a^b K\begin{pmatrix} \xi_1\; \xi_2 \ldots \xi_\nu \\ \xi_1\; \xi_2 \ldots \xi_\nu \end{pmatrix} d\xi_1 d\xi_2 \ldots d\xi_\nu \tag{2.10}$$

Then, paying attention to $D'/D \to 1$, $y_i \to f(t)$, $\partial D/\partial k_{ji} \to \Delta'(\xi, t)$, we know that (2.8) yields

$$x(t) = f(t) - \frac{1}{\lambda \Delta} \int_a^b f(\xi)\Delta'(\xi, t) d\xi. \tag{2.11}$$

Write $D'_{ji} \to \Delta\begin{pmatrix} t \\ \xi \end{pmatrix}; \lambda\Big)$, then $\Delta'(\xi, t) = -\lambda^2 \Delta\begin{pmatrix} t \\ \xi \end{pmatrix}; \lambda\Big)$, and (2.8a) becomes

$$x(t) = f(t) + \frac{\lambda}{\Delta(\lambda)} \int_a^b \Delta\binom{t}{\xi}; \lambda\Big) f(\xi) d\xi, \tag{2.12}$$

where if we substitute

$$\Delta\binom{t}{\xi}; \lambda\Big) \Big/ \Delta(\lambda) = -\Gamma(t, \xi; \lambda), \tag{2.13}$$

(2.12) can be expressed as

$$x(t) = f(t) - \lambda \int_a^b \Gamma(t, \xi; \lambda) f(\xi) d\xi. \tag{2.12a}$$

Now $\Gamma(t, \xi; \lambda)$ is the kernel of the solution. When, in (2.9), $n \to \infty$,

$$\Delta\binom{t}{\xi}; \lambda\Big) = K(t, \xi) + \sum_{1}^{\infty}{}_{\nu}(-1)^{\nu} \frac{\lambda^{\nu}}{\nu!} \int_a^b \int_a^b \cdots \int_a^b K\begin{pmatrix} t & \xi_1 & \xi_2 & \cdots & \xi_\nu \\ \xi & \xi_1 & \xi_2 & \cdots & \xi_\nu \end{pmatrix} d\xi_1\, d\xi_2 \ldots d\xi_\nu. \tag{2.14}$$

This is called *Fredholm's minor, minor of the integral equation,* or *minor of the kernel* $K(t, \xi)$. (In some books (Bocher 1914), etc., this is called the adjoint.)

Table 2.1 is a table indicating all these results arranged systematically.

Thus, it is a very important step in the study of integral equations to consider an integral equation as a set of simultaneous equations in ∞ dimensions, to formulate it in n dimensions, and with the result to solve the integral equation as $n \to \infty$. It is also worthy of much attention that an integral equation and the simultaneous equations correspond to quite a simple 'principle of passage from discontinuity to continuity' or vice versa. Now with regard to the fact that we can arrive at a solution by this method, let us go through an often used example.

Example 2.3 [F_2]

In

$$x(t) - \lambda \int_0^1 t\xi\, x(\xi) d\xi = at,$$

since from the calculation of the determinant of the kernel in (2.10)

$$K\begin{pmatrix} \xi_1 & \xi_2 & \cdots & \xi_\nu \\ \xi_1 & \xi_2 & \cdots & \xi_\nu \end{pmatrix} = 0$$

TABLE 2.1. Correspondence of integral equations and simultaneous equations [F_2]

Integral equation in the ∞ th dimension	Simultaneous equation in the nth dimension
$x(t) - \lambda\int_a^b K(t, \xi)x(\xi)d\xi = f(t)$ [F_2]	$\begin{bmatrix} 1-\lambda k_{11} & -\lambda k_{12} \dots & -\lambda k_{1n} \\ -\lambda k_{21} & 1-\lambda k_{22} \dots & -\lambda k_{2n} \\ \dots & \dots & \dots \\ -\lambda k_{n1} & -\lambda k_{n2} \dots & 1-\lambda k_{nn} \end{bmatrix} \begin{bmatrix} x_1 \\ x_2 \\ \bullet \\ x_n \end{bmatrix} = \begin{bmatrix} y_1 \\ y_2 \\ \bullet \\ y_n \end{bmatrix}$ (f_2)
determinant: $\Delta(\lambda) = 1 + \sum_1^\infty{}_\nu(-1)^\nu \frac{\lambda^\nu}{\nu!}\int_a^b\int_a^b \cdots \int_a^b K\begin{pmatrix}\xi_1 & \xi_2 & \cdots & \xi_\nu \\ \xi_1 & \xi_2 & \cdots & \xi_\nu\end{pmatrix} d\xi_1\, d\xi_2 \dots d\xi_\nu$ (2.10)	$x_i = D_i/D$ [$D \neq 0$]
$x(t) = f(t) - \frac{1}{\lambda\Delta}\int_a^b f(\xi)\Delta'(\xi, t)d\xi$ (2.11)	$D = 1 + \sum_1^n{}_\nu(-1)^\nu\frac{\lambda^\nu}{\nu!}\left\{\Sigma\Sigma \cdots \Sigma k\begin{pmatrix} i\, j \dots s \\ i\, j \dots s\end{pmatrix}\right\}$ (2.7)
$\left(\Delta'(\xi, t) = -\lambda^2\Delta\begin{pmatrix} t \\ \xi \end{pmatrix}; \lambda\right)$	$x_i = y_i\frac{D'}{D} - \frac{1}{\lambda D}\Sigma_j y_j \frac{\partial D}{\partial k_{ji}}$ (2.8)
$= f(t) + \frac{\lambda}{\Delta}\int_a^b \Delta\left(\begin{matrix} t \\ \xi \end{matrix}; \lambda\right) f(\xi)d\xi$ (2.12)	$\left(\frac{\partial D}{\partial k_{ji}} = -\lambda^2 D'_{ji}\right)$
$= f(t) - \lambda\int_a^b \Gamma(t, \xi; \lambda) f(\xi)d\xi$ [$\Delta \neq 0$] (2.12a)	$= y_i\frac{D'}{D} + \frac{\lambda}{D}\Sigma_j y_j D'_{ji}$ (2.8a)
minor: $\Delta\left(\begin{matrix} t \\ \xi \end{matrix}; \lambda\right) = K(t, \xi) + \sum_1^\infty{}_\nu(-1)^\nu\frac{\lambda^\nu}{\nu!}\int_a^b\int_a^b \cdots \int_a^b K\begin{pmatrix} t & \xi_1 & \xi_2 & \cdots & \xi_\nu \\ \xi & \xi_1 & \xi_2 & \cdots & \xi_\nu\end{pmatrix} d\xi_1\, d\xi_2 \dots d\xi_\nu$ (2.14)	cofactor: $D'_{ij} = k_{ij} + \sum_{\nu=1}^{n-2}(-1)^\nu\frac{\lambda^\nu}{\nu!}\left\{\Sigma\Sigma \cdots \Sigma k\begin{pmatrix} j\, k\, l \dots r \\ i\, k\, l \dots r\end{pmatrix}\right\}$ (2.9)

for $\nu \geqq 2$, we have

$$\Delta(\lambda) = 1 - \lambda\int_0^1 \xi_1{}^2 d\xi = 1 - \frac{\lambda}{3}.$$

And since in the minor, all the terms for $\nu \geqq 1$ vanish, in (2.14) we get

$$\Delta\left(\begin{matrix} t \\ \xi \end{matrix}; \lambda\right) = t\xi.$$

Therefore, when $\lambda \neq 3$, from (2.12a) the solution is

$$x(t) = at + \lambda\Big/\left(1 - \frac{\lambda}{3}\right)\int_0^1 t\xi\, a\xi\, d\xi = \frac{3at}{3 - \lambda}.$$

From (2.13) the kernel of the solution is

$$\Gamma(t, \xi; \lambda) = \frac{3t\xi}{3-\lambda}.$$

To prove the theory of Fredholm's solution exactly, we use Hadamard's lemma on the maximum of the determinant which we shall give in the following.

LEMMA 2.1 *Hadamard's lemma*

If all the elements of a determinant of the nth order:

$$A = \begin{vmatrix} a_{11} & a_{12} & \ldots & a_{1n} \\ a_{21} & a_{22} & \ldots & a_{2n} \\ \ldots & \ldots & \ldots & \\ a_{n1} & a_{n2} & \ldots & a_{nn} \end{vmatrix}$$

are positive, and satisfy the condition that

$$a_{i1}^2 + a_{i2}^2 + \ldots + a_{in}^2 = 1 \qquad (i = 1, 2, \ldots, n), \tag{2.15}$$

then

$$|A| \leqq 1.$$

The geometrical significance of this theorem can be introduced from the case when $n = 2, 3$. The analogies are as follows.

The case $n = 2$. Given a rhombus OP_1QP_2 as illustrated in Fig. 2.1, where O is the origin, two points near the origin are $P_1(a_{11}, a_{12})$, $P_2(a_{21}, a_{22})$, and the length of each side is 1, then since $OP_1 = OP_2 = 1$, we get

$$a_{11}^2 + a_{12}^2 = 1 \quad \text{and} \quad a_{21}^2 + a_{22}^2 = 1.$$

The absolute value of a determinant

$$A = \begin{vmatrix} a_{11} & a_{12} \\ a_{21} & a_{22} \end{vmatrix}$$

stands for the area of the rhombus. When the area is a maximum, it is the case of a regular square; and since in this case the area is 1, generally $|A| \leqq 1$.

The case $n = 3$. Given an equilateral parallel hexahedron $OP_1P_2P_3 \ldots Q$, where O is the vertex, and three points near the vertex are P_1 (a_{11}, a_{12}, a_{12}), P_2, P_3, and the length of an edge is 1, then since $OP_1 = OP_2 = OP_3 = 1$, we get

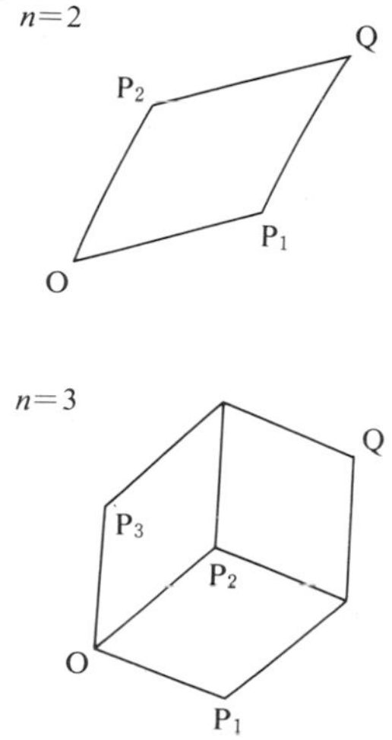

Fig. 2.1. Geometrical signification of Hadamard's lemma

$$a_{11}^2 + a_{12}^2 + a_{13}^2 = 1,$$

etc., where the absolute value of a determinant

$$A = \begin{vmatrix} a_{11} & a_{12} & a_{13} \\ a_{21} & a_{22} & a_{23} \\ a_{31} & a_{32} & a_{33} \end{vmatrix}$$

stands for the volume of this hexahedron. Since when the length of an edge is constant, the volume of a right hexahedron becomes a maximum and the volume is 1, we have $|A| \leqq 1$ in general.

The general case. In the case of a parallel polyhedron $OP_1P_2 \ldots P_n \ldots Q$, if O is the origin of the coordinates, a neighbouring point to O is $P_i(a_{i1}, a_{i2}, \ldots, a_{in})$, and the length of each edge is 1, then the conditions (2.15) hold and the absolute value of a determinant A stands for the volume of this polyhedron.

Now, to prove this theorem, we have only to show that both the maximum and the minimum of A take the value ± 1. Now A is continuous over a_{ij}, but under the conditions (2.15), the domain of a_{ij} is bounded and closed. Therefore A is expected to take a maximum and a minimum within this domain.† Hence, we have only to solve

† A continuous function is bounded in a bounded closed domain and it takes both a maximum and a minimum within the domain.

a conditional extremum problem under the conditions (2.15), regarding A as a function of a_{ij}. Take λ_i as Lagrange's multiplier and suppose that

$$F \equiv A + \frac{1}{2}\sum_{i=1}^{n} \lambda_i(a_{i1}^2 + a_{i2}^2 + \ldots + a_{in}^2 - 1),$$

where the necessary condition is that

$$\frac{\partial F}{\partial a_{ij}} = \frac{\partial A}{\partial a_{ij}} + \lambda_i a_{ij} = 0.$$

Here, keeping in mind that $\partial A/\partial a_{ij}$ is A_{ij} and the cofactor of a_{ij} in the determinant A, we get

$$A_{ij} + \lambda_i a_{ij} = 0 \qquad (i, j = 1, 2, \ldots, n) \tag{2.16}$$

as the necessary condition. Let us multiply both sides of this equation by a_{ij} and take the discrete sum over i. Then in

$$\sum_{j=1}^{n} a_{ij}A_{ij} + \lambda_i \sum_{j=1}^{n} a_{ij}^2 = 0,$$

$\sum a_{ij}A_{ij} = A$ and $\sum a_{ij2}^2 = 1$; hence, we get $A + \lambda_i = 0$, that is,

$$\lambda_i = -A \qquad (i = 1, 2, \ldots, n).$$

Substituting this in (2.16),

$$A_{ij} = Aa_{ij},$$

where the determinant made up of only cofactors A_{ij}, called the adjoint, is equal to A^{n-1}. Since the determinant created by substituting Aa_{ij} for each element of A_{ij} is equal to A^{n+1}, then

$$A^{n+1} = A^{n-1}.$$

The extremum of A must satisfy this equation. Therefore its maximum is $+1$ and minimum is -1. Hence we have proved that

$$|A| \leqq 1.$$

With regard to the element a_{ij}, if we remove conditions (2.15) and instead put the boundary as only

$$|a_{ij}| \leqq M,$$

still the determinant A is bounded. This result corresponds to the following theorem.

LEMMA 2.2 *Hadamard's theorem*
When all the elements of a determinant of the nth order

$$A = \begin{vmatrix} a_{11} & a_{12} & \dots & a_{1n} \\ a_{21} & a_{22} & \dots & a_{2n} \\ \dots & \dots & \dots & \\ a_{n1} & a_{n2} & \dots & a_{nn} \end{vmatrix}$$

are real and bounded,

$$|a_{ij}| \leqq M,$$

then

$$|A| \leqq M^n \sqrt{n^n}.$$

Proof. First we make a sum as we did in (2.15) and express it in

$$a_{i1}^2 + a_{i2}^2 + \dots + a_{in}^2 = \sigma_i \qquad (i = 1, 2, \dots, n).$$

In this sum, if $\sigma_i = 0$, all the elements on the ith row vanish and the theorem naturally holds at $A = 0$.

Therefore we set all of $\sigma_i \neq 0$, and $\sigma_i > 0$. Then in a determinant

$$\frac{A}{\sqrt{\sigma_1 \sigma_2 \dots \sigma_n}} = \begin{vmatrix} \dfrac{a_{11}}{\sqrt{\sigma_1}} & \dots & \dfrac{a_{1n}}{\sqrt{\sigma_1}} \\ & \dots & \\ \dfrac{a_{n1}}{\sqrt{\sigma_n}} & \dots & \dfrac{a_{nn}}{\sqrt{\sigma_n}} \end{vmatrix},$$

since all the elements are real and the previous condition

$$\frac{a_{i1}^2}{\sigma_i} + \frac{a_{i2}^2}{\sigma_i} + \dots + \frac{a_{in}^2}{\sigma_i} = 1$$

holds, then by the lemma we have

$$|A| \leqq \sqrt{\sigma_1 \sigma_2 \dots \sigma_n}.$$

Since $|a_{ij}| \leqq M$, we obtain

$$\sigma_i \leqq nM^2 \qquad (i = 1, 2, \dots, n).$$

Hence, we have

$$|A| \leqq M^n \sqrt{n^n}.$$

THEOREM 2.3 [F_2]
For Fredholm's integral equation of the second kind:

$$x(t) - \lambda \int_a^b K(t, \xi)x(\xi)d\xi = f(t). \qquad [F_2]$$

When the kernel $K(t, \xi)$ *belongs to the functional space* $\mathscr{R}$ *in S,* $K(t, t)$ *is integrable in the interval I, and* $f(t)$ *is continuous in I, then the necessary and sufficient condition for the existence of a unique continuous solution is that Fredholm's determinant* $\Delta(\lambda)$ *is not zero in I. In this case the solution is*

$$x(t) = f(t) - \lambda \int_a^b \Gamma(t, \xi; \lambda) f(\xi) d\xi, \tag{2.12}$$

and the kernel of the solution $\Gamma(t, \xi; \lambda)$ *is*

$$\Gamma(t, \xi; \lambda) = -\Delta\begin{pmatrix} t \\ \xi \end{pmatrix}; \lambda\Big/\Delta(\lambda); \tag{2.13}$$

provided

$$\Delta(\lambda) = 1 + \sum_1^\infty {}_\nu (-1)^\nu \frac{\lambda^\nu}{\nu!} \int_a^b \int_a^b \cdots \int_a^b K\begin{pmatrix} \xi_1 & \xi_2 & \cdots & \xi_\nu \\ \xi_1 & \xi_2 & \cdots & \xi_\nu \end{pmatrix} d\xi_1\, d\xi_2 \ldots d\xi_\nu \tag{2.10}$$

$$\Delta\left(\begin{matrix} t \\ \xi \end{matrix}; \lambda\right) = K(t, \xi) + \sum_1^\infty {}_\nu (-1)^\nu \frac{\lambda^\nu}{\nu!} \int_a^b \int_a^b \cdots$$
$$\int_a^b K\begin{pmatrix} t & \xi_1 & \xi_2 & \cdots & \xi_\nu \\ \xi & \xi_1 & \xi_2 & \cdots & \xi_\nu \end{pmatrix} d\xi_1\, d\xi_2 \ldots d\xi_\nu. \tag{2.14}$$

Proof. The proof will be given in the order of convergence, reciprocity, and inversion as in the proof of Theorem 2.1.

(i) We are going to prove that Fredholm's determinant $\Delta(\lambda)$, and minor $\Delta\left(\begin{matrix} t \\ \xi \end{matrix}; \lambda\right)$ converge absolutely and uniformly. With respect to (2.14) the integrability of $K(\xi_i, \xi_i)$, etc., is supposed. Since the integrand

$$K\begin{pmatrix} t & \xi_1 & \cdots & \xi_\nu \\ \xi & \xi_1 & \cdots & \xi_\nu \end{pmatrix}$$

is a functional determinant with $K(t, \xi)$, etc., as its elements and consists of $\nu + 1$ columns and $\nu + 1$ rows, where all the elements are smaller than M in their absolute values, then by Hadamard's theorem,

$$\left| K\begin{pmatrix} t & \xi_1 & \xi_2 & \cdots & \xi_\nu \\ \xi & \xi_1 & \xi_2 & \cdots & \xi_\nu \end{pmatrix} \right| < M^{\nu+1}(\nu + 1)^{\frac{1}{2}(\nu+1)}.$$

Therefore, the majorant series for the positive-term series made of the absolute value of each term of $\Delta\binom{t}{\xi}; \lambda)$ is

$$\sum_{1}^{\infty}{}_{\nu} \frac{|\lambda|^{\nu}}{\nu!}(b-a)^{\nu} M^{\nu+1}(\nu+1)\cdots(\nu+1)^{\frac{1}{2}(\nu+1)}.$$

Since the ratio of the νth term and the $(\nu-1)$th term of this series is

$$|\lambda|(b-a)M\sqrt{\left(1+\frac{1}{\nu}\right)^{\nu}}\,\frac{\nu+1}{\nu}\,\frac{1}{\sqrt{(\nu+1)}},$$

the ratio converges to zero when $\nu \to \infty$. Hence, by the ratio test of the positive-term series, we can prove that this majorant series is absolutely convergent. Therefore, $\Delta\left(\begin{matrix} t \\ \xi \end{matrix}; \lambda\right)$ is uniformly convergent.

We can prove the convergence of (2.10) in quite the same manner. This time the integrability of $K(t, t)$ is necessary (see Question 2 below).

Thus Fredholm's determinant $\Delta(\lambda)$, and minor $\Delta\left(\begin{matrix} t \\ \xi \end{matrix}; \lambda\right)$ are absolutely and uniformly convergent. In (2.14), by Lemma 1.1, the integral of each of the second and ensuing terms is the product of $K(t, \xi)$ and a continuous function of t, ξ, or a continuous function of t, ξ (see Question 3 below). Therefore, at the continuous points of $K(t, \xi)$, $\Delta\left(\begin{matrix} t \\ \xi \end{matrix}; \lambda\right)$ is a bounded continuous function with parameter λ, and the character of discontinuity coincides with that of $K(t, \xi)$ (provided that each term of a uniformly and absolutely convergent series is a continuous function, and the sum is also a continuous function). Hence when $\Delta(\lambda) \neq 0$, $\Gamma(t, \xi; \lambda)$ is a bounded function with parameter λ, and discontinuity is brought about only by the kernel $K(t, \xi)$; therefore, $\Gamma(t, \xi; \lambda)$ belongs to the functional space $\mathscr{R}$ as before.

(ii) Now we shall prove the reciprocal theorem

$$\Gamma + K = \lambda K \circ \Gamma = \lambda \Gamma \circ K. \tag{2.17}$$

With regard to the integrand of (2.14), if we expand the first row, we get

$$K\left(\begin{matrix} t & \xi_1 & \xi_2 & \dots & \xi_\nu \\ \xi & \xi_1 & \xi_2 & \dots & \xi_\nu \end{matrix}\right) = \begin{vmatrix} K(t, \xi) & K(t, \xi_1) & K(t, \xi_2) & \dots & K(t, \xi_\nu) \\ K(\xi_1, \xi) & K(\xi_1, \xi_1) & K(\xi_1, \xi_2) & \dots & K(\xi_1, \xi_\nu) \\ & \dots & \dots & & \dots \\ K(\xi_\nu, \xi) & K(\xi_\nu, \xi_1) & K(\xi_\nu, \xi_2) & \dots & K(\xi_\nu, \xi_\nu) \end{vmatrix}$$

$$= K(t, \xi)K\begin{pmatrix}\xi_1 & \xi_2 & \dots & \xi_\nu \\ \xi_1 & \xi_2 & \dots & \xi_\nu\end{pmatrix} - K(t, \xi_1)K\begin{pmatrix}\xi_1 & \xi_2 & \dots & \xi_\nu \\ \xi & \xi_2 & \dots & \xi_\nu\end{pmatrix}$$

$$+ \dots + (-1)^\nu K(t, \xi_\nu)K\begin{pmatrix}\xi_1 & \xi_2 & \dots & \xi_\nu \\ \xi & \xi_1 & \dots & \xi_{\nu-1}\end{pmatrix}.$$

Therefore, substituting the above expanded formula for the integrand on the right-hand side of (8), we can write

$$\Delta\begin{pmatrix}t \\ \xi\end{pmatrix}; \lambda\Big) = K(t, \xi) + \sum_1^\infty{}_\nu(-1)^\nu \frac{\lambda^\nu}{\nu!}\Big\{K(t, \xi)\int_a^b\int_a^b \dots \int_a^b K\begin{pmatrix}\xi_1 & \xi_2 & \dots & \xi_\nu \\ \xi_1 & \xi_2 & \dots & \xi_\nu\end{pmatrix} d\xi_1\, d\xi_2 \dots d\xi_\nu\Big\}$$

$$+ \sum_1^\infty{}_\nu(-1)^\nu \frac{\lambda^\nu}{\nu!}\Big\{-\int_a^b\int_a^b \dots \int_a^b K(t, \xi_1) \times K\begin{pmatrix}\xi_1 & \xi_2 & \dots & \xi_\nu \\ \xi & \xi_2 & \dots & \xi_\nu\end{pmatrix} d\xi_1\, d\xi_2 \dots d\xi_\nu$$

$$+ \dots$$

$$+ (-1)^\nu \int_a^b\int_a^b \dots \int_a^b K(t, \xi_\nu)K\begin{pmatrix}\xi_1 & \xi_2 & \dots & \xi_\nu \\ \xi & \xi_1 & \dots & \xi_{\nu-1}\end{pmatrix} \times d\xi_1\, d\xi_2 \dots d\xi_\nu\Big\}. \tag{2.18}$$

Then, by the rule for interchanging rows of a determinant, the last term in the parenthesis on the right-hand side, for example, becomes

$$K\begin{pmatrix}\xi_1 & \xi_2 & \dots & \xi_{\nu-1} & \xi_\nu \\ \xi & \xi_1 & \dots & \xi_{\nu-2} & \xi_{\nu-1}\end{pmatrix} = -K\begin{pmatrix}\xi_1 & \xi_2 & \dots & \xi_\nu & \xi_{\nu-1} \\ \xi & \xi_1 & \dots & \xi_{\nu-2} & \xi_{\nu-1}\end{pmatrix}$$

$$= \dots = (-1)^{\nu-1}K\begin{pmatrix}\xi_\nu & \xi_1 & \xi_2 & \dots & \xi_{\nu-1} \\ \xi & \xi_1 & \xi_2 & \dots & \xi_{\nu-1}\end{pmatrix}.$$

In arranging the second term and the following $\nu - 1$ terms in the parenthesis, if we change the variables and put τ for all that are in the first K, and $\xi_1, \xi_2, \dots, \xi_{\nu-1}$ for all in the determinant, then each term becomes

$$-\int_a^b\int_a^b \dots \int_a^b K(t, \tau)K\begin{pmatrix}\tau & \xi_1 & \xi_2 & \dots & \xi_{\nu-1} \\ \xi & \xi_1 & \xi_2 & \dots & \xi_{\nu-1}\end{pmatrix} d\tau\, d\xi_1\, d\xi_2 \dots d\xi_{\nu-1}.$$

If we express this as

$$-\int_a^b K(t, \tau)d\tau \int_a^b \dots \int_a^b K\begin{pmatrix}\tau & \xi_1 & \xi_2 & \dots & \xi_{\nu-1} \\ \xi & \xi_1 & \xi_3 & \dots & \xi_{\nu-1}\end{pmatrix} d\xi_1\, d\xi_2 \dots d\xi_{\nu-1},$$

there appear ν of the same terms as in the second sum in (2.18); then by arranging them we get

$$+\sum_{1}^{\infty}{}_{\nu}(-1)^{\nu-1}\frac{\lambda^{\nu-1}}{(\nu-1)!}\lambda\int_a^b K(t,\tau)d\tau$$
$$\times\int_a^b\cdots\int_a^b K\begin{pmatrix}\tau & \xi_1 & \xi_2 & \dots & \xi_{\nu-1}\\ \xi & \xi_1 & \xi_2 & \dots & \xi_{\nu-1}\end{pmatrix}d\xi_1\,d\xi_2\dots d\xi_{\nu-1}.$$

As this is absolutely and uniformly convergent, we can change the order of total sum and integration, and we have

$$+\lambda\int_a^b K(t,\tau)d\tau\Big\{\sum_{1}^{\infty}{}_{\nu}(-1)^{\nu-1}\frac{\lambda^{\nu-1}}{(\nu-1)!}$$
$$\int_a^b\cdots\int_a^b K\begin{pmatrix}\tau & \xi_1 & \xi_2 & \dots & \xi_{\nu-1}\\ \xi & \xi_1 & \xi_2 & \dots & \xi_{\nu-1}\end{pmatrix}d\xi_1\,d\xi_2\dots d\xi_{\nu-1}\Big\}.$$

Further, dividing the contents of the parenthesis into the term of $\nu=1$and the sum of the rest, and putting $\nu-1=\mu$, the above formula becomes

$$\Big\{K(\tau,\xi)+\sum_{1}^{\infty}{}_{\mu}(-1)^{\mu}\frac{\lambda^{\mu}}{\mu!}\int_a^b\cdots\int_a^b K\begin{pmatrix}\tau & \xi_1 & \xi_2 & \dots & \xi_{\mu}\\ \xi & \xi_1 & \xi_2 & \dots & \xi_{\mu}\end{pmatrix}d\xi_1\,d\xi_2\dots d\xi_{\mu}\Big\}$$
$$=\Delta\begin{pmatrix}\tau\\ \xi\end{pmatrix};\lambda\Big).$$

And the first term and the first sum become

$$K(t,\xi)\Big\{1+\sum_{1}^{\infty}{}_{\nu}(-1)^{\nu}\frac{\lambda^{\nu}}{\nu!}\int_a^b\int_a^b\cdots\int_a^b K\begin{pmatrix}\xi_1 & \xi_2 & \dots & \xi_{\nu}\\ \xi_1 & \xi_2 & \dots & \xi_{\nu}\end{pmatrix}d\xi_1\,d\xi_2\dots d\xi_{\nu}\Big\}$$
$$=K(t,\xi)\Delta(\lambda).$$

Therefore (2.18) becomes

$$\Delta\Big(\begin{matrix}t\\ \xi\end{matrix};\lambda\Big)=K(t,\xi)\Delta(\lambda)+\lambda\int_a^b K(t,\tau)\Delta\Big(\begin{matrix}\tau\\ \xi\end{matrix};\lambda\Big)d\tau. \tag{2.19}$$

Similarly we can prove that

$$\Delta\Big(\begin{matrix}t\\ \xi\end{matrix};\lambda\Big)=K(t,\xi)\Delta(\lambda)+\lambda\int_a^b \Delta\Big(\begin{matrix}t\\ \tau\end{matrix};\lambda\Big)K(\tau,\xi)d\tau \tag{2.19a}$$

(see Question 4). If we set $\Delta(\lambda)\neq 0$, then

$$\Delta\Big(\begin{matrix}t\\ \xi\end{matrix};\lambda\Big)\Big/\Delta(\lambda)=-\Gamma(t,\xi;\lambda);$$

hence the equation

$$K(t, \xi) + \Gamma(t, \xi; \lambda) = \lambda \int_a^b K(t, \tau)\Gamma(\tau, \xi; \lambda)d\tau$$
$$= \lambda \int_a^b \Gamma(t, \tau; \lambda)K(\tau, \xi)d\tau \qquad (2.17a)$$

can be proved.

(iii) *Inversion.* Here we are going to prove that (2.12) is really the solution of $[F_2]$ using the reciprocal theorem. Put (2.12), that is, $x = (E - \lambda\Gamma) \circ f$, into x in $(E - \lambda K) \circ x$ on the left-hand side of $[F_2]$, then from (2.17), we have

$$(E - \lambda K) \circ (E - \lambda\Gamma) \circ f = (E - \lambda K - \lambda\Gamma + \lambda^2 K \circ \Gamma) \circ f$$
$$= E \circ f$$
$$= f.$$

Therefore, it turns out that (2.12) satisfies the integral equation $[F_2]$.

Under the assumption that f is continuous and Γ belongs to $\mathscr{R}$, then $x(t)$ is continuous in Γ.

And since if we multiply $[F_2]$ by $(E - \lambda\Gamma)$ from the left we get

$$x = (E - \lambda\Gamma) \circ f,$$

then we may assume that all solutions can be given in this form. Hence when $f = 0$ there is no other solution than $x \equiv 0$. Now suppose that there are two solutions $x = x_1(t)$, $x = x_2(t)$, if we put $\sigma(t) = x_1(t) - x_2(t)$, then $\sigma(t)$ satisfies $\sigma = \lambda K \circ \sigma$. Therefore we must have $\sigma \equiv 0$ (see Theorem 2.1, corollary). Thus uniqueness can be proved. If we use the already established Theorems 2.1 and 2.2 and 2.3, then uniqueness can be proved directly from the existence of the kernel of the solution Γ.

Theorem 2.3 contains the following important theorems.

(1) *Existence of* $\Delta(\lambda)$, $\Delta\begin{pmatrix} t \\ \xi \end{pmatrix}; \lambda\Big)$

If $K(t, \xi)$ belongs to the functional space $\mathscr{R}$ in the domain S and $K(t, t)$ is integrable, the series of (2.10) λ *is absolutely convergent and* (2.14) *is absolutely and uniformly convergent.*

(2) *Reciprocal theorem*

If $K(t, \xi)$ belongs to the functional space $\mathscr{R}$ in the domain S and $K(t, t)$ is integrable, the following reciprocal theorem holds:

$$\Delta\binom{t}{\xi};\lambda\Big) = K(t, \xi)\Delta(\lambda) + \lambda\int_a^b K(t, \tau)\Delta\binom{\tau}{\xi};\lambda\Big)d\tau.$$

This is sometimes called Fredholm's first fundamental relation;

$$\Delta\binom{t}{\xi};\lambda\Big) = K(t, \xi)\Delta(\lambda) + \lambda\int_a^b \Delta\binom{t}{\tau};\lambda\Big)K(\tau, \xi)d\tau$$

is sometimes called Fredholm's second fundamental relation.

(3) *Existence of the kernel of a solution*

If $K(t, \xi)$ belongs to the functional space $\mathscr{R}$ in the domain S, $K(t, \xi)$ is integrable in $t(a, b)$ and Fredholm's determinant $\Delta(\lambda)$ does not become zero, then the kernel of a solution $\Gamma(t, \xi; \lambda)$ exists and belongs to $\mathscr{R}$. *It is expressed as*

$$\Gamma(t, \xi; \lambda) = -\Delta\binom{t}{\xi};\lambda\Big)\Big/\Delta(\lambda).$$

Note that Theorem 2.3 proved above requires, as has been explained, the condition that $K(t, t)$ is integrable in the interval I. Without this condition, Fredholm's determinant $\Delta(\lambda)$ and minor $\Delta\binom{t}{\xi};\lambda\Big)$ become insignificant. But even if we change the value of $K(t, \xi)$ arbitrarily on a line $t = \xi$, the regular part of $[F_2]$ does not change, nor does the solution $x(t)$; therefore, we can change the value on $t = \xi$ of $K(t, \xi)$ to the form which is bounded and always integrable. Even if there occur some discontinuities on $t = \xi$, it does not violate the condition of $\mathscr{R}$, since no line of discontinuity is included in the part parallel to the coordinate axis. Hence we adopt the following definition.

Definition. Modified kernel, modified determinant, modified adjoint

When $K(t, \xi)$ belongs to the functional space $\mathscr{R}$ in the domain S, a function $K_0(t, \xi)$, which is defined by

$$K_0(t, \xi) = \begin{cases} K(t, \xi), & t \neq \xi; \\ 0 & , t = \xi \end{cases}$$

is called a *modified kernel*; K_0 belongs to $\mathscr{R}$. When we use $K_0(t, \xi)$ in place of $K(t, \xi)$, Fredholm's determinant is called a *modified determinant of K* and Fredholm's minor is called a *modified minor of K*. We express these by Δ_0, $\Delta_0\binom{t}{\xi};\lambda\Big)$.

We can omit the condition of integrability of $K(t, t)$ in Theorem 2.3 if we use the modified determinant, etc.

Corollary

In [F_2], *the kernel* $K(t, \xi)$ *belongs to the functional space* $\mathscr{R}$ *in the domain* S, *and* $f(t)$ *is continuous in the interval* I. *The necessary and suffiicient condition for the existence of a unique solution is that Fredholm's modified determinant* $\Delta_0(\lambda)$ *be not zero.*

The solution is then

$$x(t) = f(t) - \lambda \int_a^b \Gamma_0(t, \xi;\ \lambda) f(\xi) d\xi \tag{2.12a}$$

and the kernel of the solution is

$$\Gamma_0(t, \xi;\ \lambda) = -\,\Delta_0\!\left(\begin{matrix} t \\ \xi \end{matrix};\ \lambda\right)\Big/\Delta_0(\lambda). \tag{2.13a}$$

Example 2.4 [F_2]

Now let us calculate Fredholm's determinant and minor with regard to Fredholm's integral equation of the second kind:

$$x(t) - \lambda \int_0^1 t\xi\, x(\xi) d\xi = at.$$

First, from (2.10) we have

$$\Delta(\lambda) = 1 - \lambda \int_0^1 t^2 dt + \frac{\lambda^2}{2!} \int_0^1 \int_0^1 \begin{vmatrix} t_1^2 & t_1 t_2 \\ t_2 t_1 & t_2^2 \end{vmatrix} dt_1 dt_2$$

$$- \frac{\lambda^2}{3!} \int_0^1 \int_0^1 \int_0^1 \begin{vmatrix} t_1^2 & t_1 t_2 & t_1 t_3 \\ t_2 t_1 & t_2^2 & t_2 t_3 \\ t_3 t_1 & t_3 t_2 & t_3^2 \end{vmatrix} dt_1 dt_2 dt_3 + \ldots = 1 - \frac{\lambda}{3}.$$

Then from (2.14) we have

$$\Delta\!\left(\begin{matrix} t \\ \xi \end{matrix};\ \lambda\right) = t\xi - \lambda \int_0^1 \begin{vmatrix} t\xi & tt_1 \\ t_1\xi & t_1^2 \end{vmatrix} dt_1$$

$$+ \frac{\lambda^2}{2!} \int_0^1 \int_0^1 \begin{vmatrix} t\xi & tt_1 & tt_2 \\ t_1\xi & t_1^2 & t_1 t_2 \\ t_2\xi & t_2 t_1 & t_2^2 \end{vmatrix} dt_1 dt_2 - \ldots = t\xi.$$

Therefore, from (2.12) the solution is

$$x(t) = at + \frac{\lambda}{1 - \frac{\lambda}{3}} \int_0^1 a\xi\, t\xi\, d\xi = \frac{3}{3 - \lambda} at.$$

2.2.1. *The case of Volterra type*

The idea given in this section can also be applied to the case of Volterra type.

For Volterra's integral equation of the second kind:

$$x(t) - \int_a^t K(t, \xi)x(\xi)d\xi = f(t); \qquad [V_2]$$

we shall consider the simultaneous equations:

$$\left.\begin{aligned} &x_1 = y_1, \\ &x_2 - k_{21}x_1 = y_2, \\ &\dots \\ &x_n - k_{n1}x_1 - \dots - k_{n\,n-1}\,x_{n-1} = y_n \end{aligned}\right\} . \qquad [v_2]$$

The solution of the simultaneous equations may be expressed by

$$x_i = D_i/D.$$

Here D is a determinant of the coefficients, and D_i is derived from D by replacing the ith column by $y_1, y_2, \dots, y_n$:

$$D = \begin{vmatrix} 1 & 0 & \dots 0 \\ -k_{21} & 1 & \dots 0 \\ \dots & & \dots \\ -k_{n1} & -k_{n2} & \dots 1 \end{vmatrix} = 1, \qquad D_i = \begin{vmatrix} 1 & 0 & \dots y_1 \dots 0 \\ -k_{21} & 1 & \dots y_2 \dots 0 \\ \dots & & \dots \\ -k_{n1} & -k_{n2} & \dots y_n \dots 1 \end{vmatrix}.$$

Since clearly $D = 1$,

$$x_i = D_i.$$

Expanding this in the ith column we obtain

$$x_i = \sum_1^n {}_j g_{ij} y_j.$$

This g_{ij} is the cofactor of the element of the jth row and ith column of the determinant D. Where g_{ii} is a determinant similar to D and is made up of $n - 1$ rows and columns, then the value is equal to 1. Then for g_{ij} $(j > i)$, of all the diagonal elements only those of the ith column are zero and the rest are all 1, and one side of the diagonal line is all zero; therefore, we have

$$g_{ij} = \begin{cases} 1 & (j = i), \\ 0 & (j > i). \end{cases}$$

Hence

$$x_i = y_i + \sum_{1}^{i-1} {}_j g_{ij} y_j. \tag{2.20}$$

Now when $j < i$,

$$g_{ij} = (-1)^{i+j} \left| \begin{array}{c|c|c} (1) & (0) & (0) \\ \hline (k') & (k) & (0) \\ \hline (k'') & (k''') & (1) \end{array} \right|,$$

where the parts (1) are matrices in which all the diagonal parts are **1** and the upper right-hand sides are composed of zeros, the (0) parts are matrices in which all elements are 0, and (k), (k'), (k''), (k''') are matrices of elements k_{rs}, etc. Expanding g_{ij} in the $(j-1)$-th column, only the central part remains, and we have

$$g_{ij} = (-1)^{i+j}(k)$$

$$= (-1)^{i+j} \begin{vmatrix} -k_{j+1\,j} & 1 & 0 & 0 \ldots 0 \\ -k_{j+2\,j} & -k_{j+2\,j+1} & 1 & 0 \ldots 0 \\ -k_{j+3\,j} & -k_{j+3\,j+1} & -k_{j+3\,j+2} & 1 \ldots 0 \\ \ldots & \ldots & \ldots & \ldots \\ -k_{i\,j} & -k_{i\,j+1} & -k_{i\,j+2} & -k_{i\,i-1} \end{vmatrix}. \tag{2.21}$$

Expand with respect to the lowest row and

$$g_{ij} = k_{ij} + \sum_{j+1}^{i-1} k_{ir}\, g_{rj}. \tag{2.22}$$

Then expand with respect to the first column, and we get

$$g_{ij} = \sum_{j+1}^{i-1} g_{ir} k_{rj} + k_{ij}. \tag{2.23}$$

While, since g_{ij} is a polynomial of k_{rs}, etc., if we rewrite the equation dividing it into terms of the first degree, second degree, ..., $(i-j)$-th degree, as

$$g_{ij} = g_{ij}^{(1)} + g_{ij}^{(2)} + \ldots + g_{ij}^{(i-j)}, \tag{2.24}$$

then comparing it with (2.21), we get

$$\left.\begin{aligned} g_{ij}^{(1)} &= k_{ij}, \\ g_{ij}^{(2)} &= \sum_{j+1}^{i-1} g_{ir}^{(1)} g_{rj}^{(1)}, \\ &\dots, \\ g_{ij}^{(n)} &= \sum_{j+1}^{i-1} g_{ir}^{(n-1)} g_{rj}^{(1)}. \end{aligned}\right\} \qquad (2.25)$$

Here, if we increase the degree to infinity as $n \to \infty$, and make the following substitutions:

$$i \to t, \quad j \to \xi, \quad \sum_1^i{}_j \to \int_a^t d\xi, \quad k \to K, \quad y \to f$$

and

$$g_{ij} \to -G(t, \xi)$$

then (2.20) gives the solution

$$x(t) = f(t) - \int_a^t G(t, \xi) f(\xi) d\xi,$$

TABLE 2.2. Correspondence of integral equations and simultaneous equations [V_2]

Integral equations of the ∞th dimension	Simultaneous equations of the nth dimension
$x(t) - \int_a^t K(t, \xi) x(\xi) d\xi = f(t)$ [V_2]	$\begin{bmatrix} 1 & 0 & \dots 0 \\ -k_{21} & 1 & \dots 0 \\ \dots & & \dots \\ -k_{n1} & -k_{n2} & \dots 1 \end{bmatrix} \begin{bmatrix} x_1 \\ x_2 \\ \cdot \\ x_n \end{bmatrix} = \begin{bmatrix} y_1 \\ y_2 \\ \cdot \\ y_n \end{bmatrix}$ [V_2]
$x(t) = f(t) - \int_a^t G(t, \xi) f(\xi) d\xi$	$x_i = D_i/D$, $[D = 1]$ $x_i = y_i + \sum_1^{i-1}{}_j g_{ij} y_j$ (2.20)
reciprocal function: $G(t, \xi) = -\sum_{i=1}^{\infty} \overset{*}{K}{}^i$	$g_{ij} = g_{ij}^{(1)} + g_{ij}^{(2)} + \dots + g_{ij}^{(i-j)}$ (2.24)
iterated kernel: $\overset{*}{K}{}^n(t, \xi) = \int_\xi^t \overset{*}{K}{}^{n-1}(t, \tau) \overset{*}{K}(\tau, \xi) d\tau$	$g_{ij}^{(1)} = k_{ij}, \dots, g_{ij}^{(n)} = \sum_{j+1}^{i-1} g_{ir}^{(n-1)} g_{rj}^{(1)}$ (2.25)
reciprocal theorem: $K(t, \xi) + G(t, \xi)$ $= \int_\xi^t K(t, \tau) G(\tau, \xi) d\tau$ $= \int_\xi^t G(t, \tau) K(\tau, \xi) d\tau$	$k_{ij} - g_{ij} = -\sum_{j+1}^{i-1} k_{ir} g_{rj}$ (2.22) $= -\sum_{j+1}^{i-1} g_{ir} k_{rj}$ (2.23)

As can be understood from the table, in the case of [V_2], the solution of a determinant agrees with that of an iterated kernel studied in the previous section.

and (2.22), (2.23) yield the reciprocal theorem:

$$K(t, \xi) + G(t, \xi) = \int_{\xi}^{t} K(t, \tau)G(\tau, \xi)d\tau$$
$$= \int_{\xi}^{t} G(t, \xi)K(\tau, \xi)d\tau,$$

(2.25) gives iterated kernels:

$$\overset{*}{K} = - K(t, \xi), \overset{*}{K}{}^2 = \int_{\xi}^{t} K(t, \tau)K(\tau, \xi)d\tau, \ldots$$

and (2.24) gives a reciprocal function:

$$G(t, \xi) = - \sum_{1}^{\infty} \overset{*}{K}{}^i$$

All of these are arranged in Table 2.2.

2.2.2. *Two existence theorems*

Theorem 2.1 in the last section and Theorem 2.3 in the present section furnish uniqueness–existence theorems of the continuous solution of $[F_2]$, but they differ from each other in their conditions. Though integrability of $K(t, t)$ is not essential, the condition that

$$|\lambda| M(b - a) < 1$$

in the former theorem is replaced by

$$\Delta(\lambda) \neq 0$$

in the latter theorem. In this point they are different from the uniqueness–existence theorem of the continuous solution of $[V_2]$ in Theorem 2.2. In the case of $[V_2]$ we cannot obtain new theorems if our treatment is by way of simultaneous equations, as in 2.2.1.

Now, for example, in $[F_2]$:

$$x(t) - \lambda \int_{0}^{1} x(\xi)d\xi = t,$$

since $M = 1$, $(b - a) = 1$, from the previous theorem

$$|\lambda| < 1.$$

But as with Fredholm's determinant

$$\Delta(\lambda) = 1 - \lambda,$$

as long as $\lambda \neq 1$ there exists a solution

$$x(t) = t + \frac{1}{2}\,\frac{\lambda}{1-\lambda}.$$

Consequently, for example, when $\lambda = 4$ there exists a continuous solution

$$x(t) = t - \frac{2}{3}.$$

Therefore, it seems that the conditions of Theorem 2.3 are more extensive than those of Theorem 2.1. Now, to prove this we are going to show that when $|\lambda| M(b-a) < 1$, we never have $\Delta(\lambda) = 0$.

In the previous proof we have shown that $\Delta(\lambda)$ is absolutely convergent over λ. Therefore, if we consider λ to be a variable, $\Delta(\lambda)$ is uniformly convergent. If we extend λ to a complex number and take

$$f(\lambda) \equiv 1/\Delta(\lambda),$$

then $\Delta(\lambda)$ is an analytic function of λ; hence $f(\lambda)$ is analytic as well.

Since $\Delta(0) = 1$, we have $f(0) = 1$. Therefore in the neighbourhood of $\lambda = 0$ we may represent

$$f(\lambda) = 1 + a_1\lambda + a_2\lambda^2 + \ldots.$$

While we express

$$\Delta(\lambda) = 1 + b_1\lambda + b_2\lambda^2 + \ldots,$$

and

$$a_1 = -b_1,\ a_2 = -(b_2 + b_1{}^2),\ a_3 = -(b_3 + b_1 b_2 + b_2 b_1 + b_1{}^3),\ \ldots.$$

Generally, we can write this as

$$a_n = -\left(\sum b_h{}^r + \sum (b_i)^l (b_j)^m\right),$$

provided

$$hr = n, \quad il + jm = n; \qquad (h, i, j;\ r, l, m = 1, 2, \ldots, n;\ i \geqq j).$$

As we had in the proof of the theorem

$$|b_\nu| < \frac{1}{\nu!}\sqrt{\nu^\nu}\ M^\nu (b-a)^\nu \leqq M^\nu (b-a)^\nu,$$

then, $|a_n| \leqq \sum |b_h|^r + \sum |b_i|^l |b_j|^m$, where each term on the right-hand side is smaller than $M^n(b-a)^n$ and the number of terms is not more than $2n^2$, and ultimately

$$|a_n| < 2n^2M^n(b-a)^n.$$

Here, taking the ratio of the positive-term series made up of the absolute value of each term of $f(\lambda)$,

$$\frac{|a_{n+1}\lambda^{n+1}|}{|a_n\lambda^n|} = |\lambda|M(b-a)\left(\frac{n+1}{n}\right)^2;$$

then if

$$|\lambda|M(b-a) < 1,$$

$f(\lambda)$ converges absolutely and uniformly over λ. Hence $f(\lambda)$ is analytic in the interior of a circle around the origin in $|\lambda| < 1/M(b-a)$. This means that the zero point of $\Delta(\lambda)$ is not included in this domain. Since it is natural that there exists no zero point even on real axes in this circle, if $|\lambda|M(b-a) < 1$, as was the case in Theorem 2.1, we have $\Delta(\lambda) \neq 0$ in this range of values of λ as a matter of course. Therefore, Theorem 2.3 is more general as the condition for the existence of a solution.

2.2.3. Reciprocal theorem

In proof (ii) of Theorem 2.3 we introduced the reciprocal theorems:

$$\Delta\left(\begin{matrix}t\\ \xi\end{matrix}; \lambda\right) = K(t, \xi)\Delta(\lambda) + \lambda\int_a^b K(t, \tau)\Delta\left(\begin{matrix}\tau\\ \xi\end{matrix}; \lambda\right)d\tau \qquad (2.19)$$

and

$$\Delta\left(\begin{matrix}t\\ \xi\end{matrix}; \lambda\right) = K(t, \xi)\Delta(\lambda) + \lambda\int_a^b \Delta\left(\begin{matrix}t\\ \tau\end{matrix}; \lambda\right)K(\tau, \xi)d\tau. \qquad (2.19a)$$

Here let us consider what significance they have in the case of simultaneous equations of nth dimension.

From the fundamental characteristics of a determinant—the sum of the results of multiplying each element of an arbitrary row by the cofactor of the element of some other row is 0—we can sum the results of multiplying each element of the jth row by the cofactor of each element of the ith row in D in § 2.2. Where, if we write the cofactor of $-\lambda k_{ri}$ as $\lambda D_{ri}'$ and write $\lambda D_{ii}'$ as D', since it is similar to D, then

$$(1-\lambda k_{jj})\lambda D_{ji}' - \lambda k_{ij}D' - \sum_{r=1}^{n}{}_{\Delta}\lambda^2 K_{rj}D_{ri}' = 0,$$

where $\sum_{\Delta}$ denotes the sum over r with the exception of $r = j$. Hence

$$\lambda D_{ji}' = \lambda\, k_{ij} D' + \lambda^2 \sum_{1r}^{n} K_{rj} D_{ri}'.$$

Here we divide both sides by λ and move the equation into the infinite dimension, making the substitutions

$$D_{ji}' \to \Delta\begin{pmatrix} t \\ \xi \end{pmatrix}; \lambda\Big), \quad k_{ij} \to K(t, \xi), \quad D' \to \Delta(\lambda), \quad \sum_{1r} \to \int_a^b d\tau$$

as we did before; then

$$\Delta\Big(\begin{matrix} t \\ \xi \end{matrix}; \lambda\Big) = K(t, \xi)\Delta(\lambda) + \lambda \int_a^b \Delta\Big(\begin{matrix} t \\ \tau \end{matrix}; \lambda\Big) K(\tau, \xi) d\tau. \qquad (2.19\text{a})$$

The same method applies to (2.19).

For example, when $K(t, \xi) = t\xi$ in $I(0, 1)$, then, as we had in Example 2.4,

$$\Delta(\lambda) = 1 - \frac{\lambda}{3}, \qquad \Delta\Big(\begin{matrix} t \\ \xi \end{matrix}; \lambda\Big) = t\xi;$$

then (2.19) becomes

$$t\xi\Big(1 - \frac{\lambda}{\xi}\Big) + \lambda \int_0^1 t\tau \cdot \tau\xi \, d\tau = t\xi,$$

and thus the reciprocal theorem holds. We can prove (2.19a) and (2.17) similarly.

2.2.4. Extension of reciprocal theorem

With regard to the determinant of the coefficients of (f_2), if we take a minor and cofactor instead of the elements of a row and cofactors, and move the dimension to the finite one, we get an extended reciprocal theorem. This theorem will be necessary in the next section.

THEOREM 2.4 *Extension of the reciprocal theorem*

When $K(t, \xi)$ belongs to the functional space $\mathscr{R}$ in the domain S and $K(t, t)$ is integrable in I, we obtain

$$\Delta\Big(\begin{matrix} t_1 & t_2 & \dots & t_n \\ \xi_1 & \xi_2 & \dots & \xi_n \end{matrix}; \lambda\Big) = K(t_1, \xi_1)\Delta\Big(\begin{matrix} t_2 & t_3 & \dots & t_n \\ \xi_2 & \xi_3 & \dots & \xi_n \end{matrix}; \lambda\Big)$$

$$- K(t_1, \xi_2)\Delta\Big(\begin{matrix} t_2 & t_3 & \dots & t_n \\ \xi_1 & \xi_3 & \dots & \xi_n \end{matrix}; \lambda\Big) + \dots$$

$$+(-1)^{n-1}K(t_1, \xi_n)\Delta\begin{pmatrix} t_2 & t_3 & \dots & t_n \\ \xi_1 & \xi_2 & \dots & \xi_{n-1} \end{pmatrix}; \lambda\Big)$$

$$+\lambda\int_a^b K(t_1, \tau)\Delta\begin{pmatrix} \tau_1 & t_2 & \dots & t_n \\ \xi_1 & \xi_2 & \dots & \xi_n \end{pmatrix}; \lambda\Big)d\tau \tag{2.26}$$

and

$$\Delta\begin{pmatrix} t_1 & t_2 & \dots & t_n \\ \xi_1 & \xi_2 & \dots & \xi_n \end{pmatrix}; \lambda\Big) = K(t_1, \xi_1)\Delta\begin{pmatrix} t_2 & t_3 & \dots & t_n \\ \xi_2 & \xi_3 & \dots & \xi_n \end{pmatrix}; \lambda\Big)$$

$$-K(t_2, \xi_1)\Delta\begin{pmatrix} t_1 & t_3 & \dots & t_n \\ \xi_2 & \xi_3 & \dots & \xi_n \end{pmatrix}; \lambda\Big) + \dots$$

$$+(-1)^{n-1}K(t_n, \xi_1)\Delta\begin{pmatrix} t_1 & t_2 & \dots & t_{n-1} \\ \xi_2 & \xi_3 & \dots & \xi_n \end{pmatrix}; \lambda\Big)$$

$$+\lambda\int_a^b \Delta\begin{pmatrix} t_1 & t_2 & \dots & t_n \\ \tau & \xi_2 & \dots & \xi_n \end{pmatrix}; \lambda\Big)K(\tau, \xi_1)d\tau, \tag{2.26a}$$

provided

$$\Delta\begin{pmatrix} t_1 & t_2 & \dots & t_n \\ \xi_1 & \xi_2 & \dots & \xi_n \end{pmatrix}; \lambda\Big) = K\begin{pmatrix} t_1 & t_2 & \dots & t_n \\ \xi_1 & \xi_2 & \dots & \xi_n \end{pmatrix} + \sum_{1}^{\infty}{}_{\nu}(-1)^{\nu}\frac{\lambda^{\nu}}{\nu!}\int_a^b\int_a^b$$

$$\dots\int_a^b K\begin{pmatrix} t_1 & \dots & t_n & \tau_1 & \dots & \tau_\nu \\ \xi_1 & \dots & \xi_n & \tau_1 & \dots & \tau_\nu \end{pmatrix}d\tau_1 d\tau_2 \dots d\tau_\nu. \tag{2.27}$$

Proof. The plan of the proof is similar to that of (ii) for Theorem 2.3; that is, to expand the determinant of (2.21) and to lead it to (2.26), summing over $K(t_1, \xi_1)$, $K(t_1, \xi_2)$, ...

First, for the first term of (2.27),

$$K\begin{pmatrix} t_1 & t_2 & \dots & t_n \\ \xi_1 & \xi_2 & \dots & \xi_n \end{pmatrix} = K(t_1, \xi_1)K\begin{pmatrix} t_2 & \dots & t_n \\ \xi_2 & \dots & \xi_n \end{pmatrix} - K(t_1, \xi_2)K\begin{pmatrix} t_2 & \dots & t_n \\ \xi_1 & \dots & \xi_n \end{pmatrix} + \dots$$

$$+(-1)^{n-1}K(t_1, \xi_n)K\begin{pmatrix} t_2 & \dots & t_n \\ \xi_1 & \dots & \xi_{n-1} \end{pmatrix}.$$

Next, the general term on the right-hand side yields

$$(-1)^{\nu}\frac{\lambda^{\nu}}{\nu!}\int_a^b\int_a^b\dots\int_a^b K\begin{pmatrix} t_1 & t_2 & \dots & t_n & \tau_1 & \dots & \tau_\nu \\ \xi_1 & \xi_2 & \dots & \xi_n & \tau_1 & \dots & \tau_\nu \end{pmatrix}d\tau_1 \dots d\tau_\nu$$

$$= (-1)^{\nu}\frac{\lambda^{\nu}}{\nu!}\Big\{K(t_1, \xi_1)\int_a^b\int_a^b\dots\int_a^b K\begin{pmatrix} t_2 & \dots & t_n & \tau_1 & \dots & \tau_\nu \\ \xi_2 & \dots & \xi_n & \tau_1 & \dots & \tau_\nu \end{pmatrix}d\tau_1 \dots d\tau_\nu$$

$$-K(t_1, \xi_2)\int_a^b\int_a^b\dots\int_a^b K\begin{pmatrix} t_2 & \dots & t_n & \tau_1 & \dots & \tau_\nu \\ \xi_1 & \dots & \xi_n & \tau_1 & \dots & \tau_\nu \end{pmatrix}d\tau_1 \dots d\tau_\nu + \dots$$

$$+(-1)^{n-1}K(t_1,\xi_n)\int_a^b\int_a^b\cdots\int_a^b K\begin{pmatrix}t_2 \ldots t_n & \tau_1 \ldots \tau_\nu\\ \xi_1 \ldots \xi_{n-1} & \tau_1 \ldots \tau_\nu\end{pmatrix}d\tau_1 \ldots d\tau_\nu$$

$$+(-1)^n\int_a^b\int_a^b\cdots\int_a^b K(t_1,\tau_1)K\begin{pmatrix}t_2 \ldots t_n & \tau_1\ \tau_2 \ldots \tau_\nu\\ \xi_1 \ldots \xi_{n-1} & \xi_n\ \tau_2 \ldots \tau_\nu\end{pmatrix}d\tau_1 \ldots d\tau_\nu$$

$$+\ldots+(-1)^{n+\nu-1}\int_a^b\int_a^b\cdots\int_a^b K(t_1,\tau_\nu)$$

$$\times K\begin{pmatrix}t_2 \ldots t_n & \tau_1 \ldots \tau_\nu\\ \xi_1 \ldots \xi_{n-1} & \xi_n \ldots \tau_{\nu-1}\end{pmatrix}d\tau_1 \ldots d\tau_\nu\Bigg\}.$$

In the last νth term, if we replace the variables common to the first factor with τ and permute the rows in the second factor so that the rows that contain τ always come first, then all the terms take the same form. For example, the first article of the νth term becomes

$$+(-1)\int_a^b K(t_1,\tau)d\tau\int_a^b\cdots\int_a^b K\begin{pmatrix}\tau & t_2 \ldots t_n & \tau_2 \ldots \tau_\nu\\ \xi_1 & \xi_2 \ldots \xi_n & \tau_2 \ldots \tau_\nu\end{pmatrix}d\tau_2 \ldots d\tau_\nu.$$

The other terms take the same form if we represent integration variables as $\tau_2 \ldots \tau_\nu$.

Then the right-hand side of (2.27), being arranged over $K(t_1, \xi_1)$, $K(t_1, \xi_2)$, ..., runs as follows,

$$K(t_1,\xi_1)\Bigg\{K\begin{pmatrix}t_2 \ldots t_n\\ \xi_2 \ldots \xi_n\end{pmatrix}+\Sigma(-1)^\nu\frac{\lambda^\nu}{\nu!}\int_a^b\int_a^b\cdots\int_a^b$$

$$K\begin{pmatrix}t_2 \ldots t_n\ \tau_1 \ldots \tau_\nu\\ \xi_2 \ldots \xi_n\ \tau_1 \ldots \tau_\nu\end{pmatrix}d\tau_1 \ldots d\tau_\nu\Bigg\}-K(t_1,\xi_2)\Bigg\{K\begin{pmatrix}t_2 \ldots t_n\\ \xi_1 \ldots \xi_n\end{pmatrix}$$

$$+\Sigma(-1)^\nu\frac{\lambda^\nu}{\nu!}\int_a^b\int_a^b\cdots\int_a^b K\begin{pmatrix}t_2 \ldots t_n\ \tau_1 \ldots \tau_\nu\\ \xi_1 \ldots \xi_n\ \tau_1 \ldots \tau_\nu\end{pmatrix}d\tau_1 \ldots d\tau_\nu\Bigg\}$$

$$+\ldots$$

$$+(-1)^{n-1}K(t_1,\xi_n)\Bigg\{K\begin{pmatrix}t_2 \ldots t_n\\ \xi_1 \ldots \xi_{n-1}\end{pmatrix}$$

$$+\Sigma(-1)^\nu\frac{\lambda^\nu}{\nu!}\int_a^b\int_a^b\cdots\int_a^b K\begin{pmatrix}t_2 \ldots t_n & \tau_1 \ldots \tau_\nu\\ \xi_1 \ldots \xi_{n-1} & \tau_1 \ldots \tau_\nu\end{pmatrix}d\tau_1 \ldots d\tau_\nu\Bigg\}$$

$$+\lambda\int_a^b K(t,\tau)d\tau\Bigg\{\sum_1^\infty{}_\nu(-1)^{\nu-1}\frac{\lambda^{\nu-1}}{(\nu-1)!}\underbrace{\int_a^b\cdots\int_a^b}_{(\nu-1)\text{ terms}}$$

$$K\begin{pmatrix}\tau & t_2 \ldots t_n & \tau_2 \ldots \tau_\nu\\ \xi_1 & \xi_2 \ldots \xi_n & \tau_2 \ldots \tau_\nu\end{pmatrix}d\tau_2 \ldots d\tau_\nu\Bigg\}.$$

where the change of order of the integral of τ in the last term and the total sum over ν are guaranteed by the absolute convergence of this series. Compare (2.21) again with this and (2.26) can be proved.

Expression (2.27) is called *Fredholm's minor of the nth degree.* For example, if $K(t, \xi) = t + \xi$ in $I(0, 1)$, then for (2.27),

$$\Delta\begin{pmatrix} t_1 & t_2 \\ \xi_1 & \xi_2 \end{pmatrix}; \lambda\Big) = K\begin{pmatrix} t_1 & t_2 \\ \xi_1 & \xi_2 \end{pmatrix} + \sum(-1)^\nu \frac{\lambda^\nu}{\nu!}\int_0^1 \cdots$$

$$\int_0^1 K\begin{pmatrix} t_1 & t_2 & \tau_1 & \ldots & \tau_\nu \\ \xi_1 & \xi_2 & \tau_1 & \ldots & \tau_\nu \end{pmatrix} d\tau_1 \ldots d\tau_\nu.$$

Since we have

$$K\begin{pmatrix} t_1 & t_2 \\ \xi_1 & \xi_2 \end{pmatrix} = (t_2 - t_1)(\xi_1 - \xi_2),\ K\begin{pmatrix} t_1 & t_2 & \tau_1 \\ \xi_1 & \xi_2 & \tau_1 \end{pmatrix} = 0,$$

it follows that

$$\Delta\left(\begin{matrix} t_1 & t_2 \\ \xi_1 & \xi_2 \end{matrix}; \lambda\right) = (t_2 - t_1)(\xi_1 - \xi_2).$$

Next, the right-hand side of (2.26) becomes

$$\begin{aligned} K(t_1, \xi_1)\Delta\left(\begin{matrix} t_2 \\ \xi_2 \end{matrix}; \lambda\right) &- K(t_1, \xi_2)\Delta\left(\begin{matrix} t_2 \\ \xi_1 \end{matrix}; \lambda\right) + \lambda\int_0^1 K(t_1, \tau)\Delta\left(\begin{matrix} \tau & t_2 \\ \xi_1 & \xi_2 \end{matrix}; \lambda\right)d\tau \\ &= (t_1 + \xi_1)\left\{(t_2 + \xi_2) - \lambda\left(\frac{t_2 + \xi_2}{2}\right) + \lambda t_2\xi_2 + \frac{\lambda}{3}\right\} \\ &- (t_1 + \xi_2)\left\{(t_2 + \xi_1) - \lambda\left(\frac{t_2 + \xi_1}{2}\right) + \lambda t_2\xi_1 + \frac{\lambda}{3}\right\} \\ &+ \lambda\int_0^1 (t_1 + \tau)(t_2 - \tau)(\xi_1 - \xi_2)d\tau. \end{aligned}$$

The result of this calculation is $(t_2 - t_1)(\xi_1 - \xi_2)$. Hence (2.26) can hold in this case.

Then, as to (2.26), this time,

$$\begin{aligned} \Delta\left(\begin{matrix} t_1 & t_2 \\ \xi_1 & \xi_2 \end{matrix}; \lambda\right) &= K(t_1, \xi_1)\Delta\left(\begin{matrix} t_2 \\ \xi_2 \end{matrix}; \lambda\right) - K(t_2, \xi_1)\Delta\left(\begin{matrix} t_1 \\ \xi_2 \end{matrix}; \lambda\right) \\ &+ \lambda\int_a^b \Delta\left(\begin{matrix} t_1 & t_2 \\ \tau & \xi_2 \end{matrix}; \lambda\right)K(\tau, \xi)d\tau, \end{aligned}$$

and, as above, the result of calculation is $(t_2 - t_1)(\xi_1 - \xi_2)$ and this equation holds.

2.3. Methods of solution of homogeneous integral equations

Now for a homogeneous integral equation let us consider the homogeneous simultaneous equations

$$\left.\begin{array}{l} a_{11}x_1 + a_{12}x_2 + \cdots + a_{1n}x_n = 0, \\ a_{21}x_1 + a_{22}x_2 + \cdots + a_{2n}x_n = 0, \\ \cdots \quad \cdots \quad \cdots \quad \cdots \\ a_{n1}x_1 + a_{n2}x_2 + \cdots + a_{nn}x_n = 0. \end{array}\right\} \qquad (\mathrm{a}_0)$$

Here, we take the determinant of the coefficients D, and we have the following two results.

(i) When $D \neq 0$ there exists only one group of solutions $x_1 = x_2 = \ldots = x_n = 0$

(ii) When the rank of D is r, there exist $\nu = n - r$ linearly independent elementary solutions:

$$\boldsymbol{x}_1, \boldsymbol{x}_2, \ldots, \boldsymbol{x}_\nu,$$

and an arbitrary solution of (a_0) is expressed by their linear combination

$$\boldsymbol{x} = c_1\boldsymbol{x}_1 + c_2\boldsymbol{x}_2 + \ldots + c_\nu\boldsymbol{x}_\nu.$$

That the rank of D is r means that within the minor of D, though there are non-zero elements in those of the rth dimension, all those of more than the $(r+1)$-th dimension become zero.

When we write (a_0) using a column vector in

$$\boldsymbol{a}_1x_1 + \boldsymbol{a}_2x_2 + \ldots + \boldsymbol{a}_nx_n = 0, \qquad (\mathbf{a}_0)$$

a vector, whose components are a group of solutions of the above vector:

$$x_{1i}, x_{2i}, \ldots, x_{ni},$$

is called a solution vector and we represent it by $\boldsymbol{x}_i$.

In the case of infinite dimension, a homogeneous integral equation:

$$x(t) - \lambda \int_a^b K(t, \xi)x(\xi)d\xi = 0 \qquad [\mathrm{F}_0]$$

corresponds to $(\mathbf{a}_0)$, and

$$\Delta(\lambda) = 1 + \sum(-1)^{\nu}\frac{\lambda^{\nu}}{\nu!}\int_a^b\int_a^b\cdots\int_a^b K\begin{pmatrix}\xi_1 & \xi_2 & \cdots & \xi_\nu\\ \xi_1 & \xi_2 & \cdots & \xi_\nu\end{pmatrix}d\xi_1\ldots d\xi_\nu$$

corresponds to the determinant of the coefficients D, $\Delta\begin{pmatrix}t\\ \xi\end{pmatrix}; \lambda\Big)$ corresponds to the cofactor $\lambda\,\partial D/\partial k_{ji}$ of the $(n-1)$th dimension of D, and

$$\Delta\begin{pmatrix}t_1 & t_2 & \ldots & t_\nu\\ \xi_1 & \xi_2 & \ldots & \xi_\nu\end{pmatrix}; \lambda\Big)$$

corresponds to the cofactor of the $(\nu = n - r)$-th dimension. Therefore, when the order—in the case of a determinant of infinite dimension we call it index—of $\Delta(\lambda)$ is ν, we can make

$$\Delta\begin{pmatrix}t_1 & t_2 & \ldots & t_p\\ \xi_1 & \xi_2 & \ldots & \xi_p\end{pmatrix}; \lambda\Big),$$

so that it becomes identically zero when $p < \nu$; and for the case of $p \geqq \nu$, we can make it so that there is one which does not become zero. Further, since from the formula of $\Delta(\lambda)$

$$\Delta^{(m)}(\lambda) = 0, \quad m \geqq \nu,$$

when the index of $\Delta(\lambda)$ is ν, $\Delta(\lambda) = 0$ has multiple roots of at least more than $(\nu - 1)$-th dimension; hence, the following theorem corresponds to a homogeneous integral equation.

THEOREM 2.5 [F_0]

In Fredholm's homogeneous integral equation

$$x(t) = \lambda\int_a^b K(t, \xi)x(\xi)d\xi, \qquad [F_0]$$

if the kernel $K(t, \xi)$ belongs to the functional space $\mathscr{R}$ in the domain S, and $K(t, t)$ is integrable in the interval I, and

(i) *if $\Delta(\lambda) \neq 0$, the continuous solution is only $x(t) \equiv 0$ and, furthermore this is unique: further, if $K(t, \xi)$ is continuous in I on t,*

(ii) *if $\Delta(\lambda) = 0$, $\Delta^{(1)}(\lambda) \neq 0$, there exist continuous solutions that are not identically zero, and if ξ' is properly selected,*

$$\varphi(t) = \Delta\begin{pmatrix}t\\ \xi'\end{pmatrix}; \lambda\Big).$$

(iii) *If $\Delta(\lambda) = 0$, $\Delta^{(i)}(\lambda) = 0$ $(i < m)$, $\Delta^{(m)}(\lambda) \neq 0$, there exist ν continuous solutions $(0 < \nu \leqq m)$ not identically zero, and when t_1', t_2', ..., t_ν'; ξ_1', ξ_2', ..., ξ_ν' are properly chosen, they run as*

$$\left.\begin{aligned}
\varphi_1(t) &= \Delta\begin{pmatrix} t & t_2' & \dots & t_\nu' \\ \xi_1' & \xi_2' & \dots & \xi_\nu' \end{pmatrix}; \lambda\Big),\\
\varphi_2(t) &= \Delta\begin{pmatrix} t_1' & t & \dots & t_\nu' \\ \xi_1' & \xi_2' & \dots & \xi_\nu' \end{pmatrix}; \lambda\Big),\\
&\dots,\\
\varphi_\nu(t) &= \Delta\begin{pmatrix} t_1' & t_2' & \dots & t \\ \xi_1' & \xi_2' & \dots & \xi_\nu' \end{pmatrix}; \lambda\Big),
\end{aligned}\right\} \tag{2.28}$$

which are linearly independent, and all the solutions of an arbitrary [F_0] *can be expressed by their linear combination*:

$$x(t) = \sum_{i=1}^{\nu} c_i \, \varphi_i(t), \tag{2.29}$$

where c_i is a complex number not identically zero.

Proof. We can deduce (i) directly from Theorem 2.3, (ii) is proved directly by means of the reciprocal theorem, and (iii) by the extended reciprocal theorem.

(i) If we consider [F_0] to be a special case when $f(t) \equiv 0$ in [F_3], then it satisfies all the conditions of Theorem 2.3, and hence the solution

$$x(t) = f(t) - \lambda \int_a^b \Gamma(t, \xi; \lambda) f(\xi) d\xi,$$

where $f(t) \equiv 0$ therefore $x(t) \equiv 0$ is naturally a unique continuous solution.

(ii) In the reciprocal theorem, if $\Delta(\lambda) = 0$,

$$\Delta\left(\begin{matrix} t \\ \xi \end{matrix}; \lambda\right) = \lambda \int_a^b K(t, \tau) \Delta\left(\begin{matrix} \tau \\ \xi \end{matrix}; \lambda\right) d\tau,$$

where ξ is is an arbitrary value in I. Choose the value of ξ so that $\Delta\left(\begin{matrix} t \\ \xi \end{matrix}; \lambda\right)$ does not become identically zero, and let it be ξ'; then the above equation shows that $\Delta\left(\begin{matrix} t \\ \xi' \end{matrix}; \lambda\right)$ is one of the solutions of [F_0]. Since the right-hand side of (2.14) is a continuous function of t when $K(t, \xi)$ is continuous on t in $t(a, b)$, the above $\Delta\left(\begin{matrix} t \\ \xi' \end{matrix}; \lambda\right)$ is one of the solutions of [F_0].

Besides, it is obvious from the equation (2.10), that if $\Delta^{(1)}(\lambda) \neq 0$, $\Delta\left(\begin{matrix} t \\ \xi \end{matrix}; \lambda\right)$ cannot be identically zero.

(iii) From (2.10) and (2.27), in

$$\Delta^{(p)}(\lambda) = (-1)^p \int_a^b \cdots \int_a^b \Delta\begin{pmatrix} \tau_1 & \tau_2 & \dots & \tau_p \\ \tau_1 & \tau_2 & \dots & \tau_p \end{pmatrix}; \lambda\Big) d\tau_1 d\tau_2 \dots d\tau_p, \tag{2.30}$$

we have $\Delta^{(m)}(\lambda) \neq 0$, hence we can choose a positive integer ν, $0 < \nu \leqq m$, such that

$$\Delta\left(\begin{matrix} t_1 & t_2 & \dots & t_p \\ \xi_1 & \xi_2 & \dots & \xi_p \end{matrix}; \lambda\right)$$

is identically zero for $p < \nu$ and not identically zero for $p \geqq \nu$. By selecting suitable values of t_i', ξ_i' in the interval (a, b), we may have

$$\Delta\left(\begin{matrix} t & t_2' & \dots & t_\nu' \\ \xi_1' & \xi_2' & \dots & \xi_\nu' \end{matrix}; \lambda\right) \neq 0.$$

By the extended reciprocal theorem (Theorem 2.4), this becomes

$$\Delta\left(\begin{matrix} t & t_2' & \dots & t_\nu' \\ \xi_1' & \xi_2' & \dots & \xi_\nu' \end{matrix}; \lambda\right) = \lambda \int_a^b K(t, \tau) \Delta\left(\begin{matrix} \tau & t_2' & \dots & t_\nu' \\ \xi_1' & \xi_2' & \dots & \xi_\nu' \end{matrix}; \lambda\right) d\tau.$$

This equation shows that $\varphi_1(t)$ in (2.28) is a solution not identically zero of $[F_0]$. When $K(t, \xi)$ is continuous in I, $\varphi_i(t)$ is also continuous in this interval, as has been shown. We can prove in a like manner that another solution $\varphi_1(t)$ is a solution not identically zero of $[F_0]$.

(iv) Now we are going to prove linear independence. Here we have only to show

$$c_1\varphi_1(t) + c_2\varphi_2(t) + \cdots + c_\nu\varphi_\nu(t) \neq 0$$

as long as all c_i are not zero. Write the conjugate complex number of c_i as $\bar{c}_i$ and consider

$$\left\{\sum_1^\nu{}_j \bar{c}_j\, K(t_j', t)\right\} \circ \{c_i\, \varphi_i(t)\}.$$

From (2.28) we have

$$K(t_j', \tau) \circ \varphi_1(\tau) = \frac{1}{\lambda} \Delta\left(\begin{matrix} t_j' & t_2' & \dots & t_\nu' \\ \xi_1' & \xi_2' & \dots & \xi_\nu' \end{matrix}; \lambda\right),$$

then when $j = 1$, the right-hand side is not zero by assumption. We express this as $(1/\lambda)\Delta_0$. When $j \neq 1$, in a determinant contained in the formula of Δ, the jth column becomes equal to the first column and then all columns become zero; hence the right-hand side of this formula becomes zero.

Generally

$$K(t_j', \tau) \circ \varphi_i(\tau) = \begin{cases} \frac{1}{\lambda}\Delta_0 & (j = i), \\ 0 & (j \neq i). \end{cases}$$

Hence

$$\left\{\sum_{1}^{\nu}{}_j \bar{c}_j K(t_j', t)\right\} \circ \{\sum c_i \varphi_i(t)\} = \frac{\Delta_0}{\lambda}\sum_{1}^{\nu} c_i \bar{c}_i.$$

Therefore, this equation does not become zero so long as all the c_i are non-zero. Hence

$$\sum c_i \varphi_i(t) \neq 0.$$

(v) Lastly, we are going to prove that any arbitrary solution of $[F_0]$ can be expressed by a linear combination of $\varphi_i(t)$. Now we express the integral equation of the first kind as

$$(E - \lambda K) \circ x = 0; \qquad [F_0]$$

then take an arbitrary kernel H which belongs to the functional space $\mathscr{R}$, and multiply $[F_2]$ by $(E - \lambda H)$ from the left:

$$(E - \lambda H) \circ (E - \lambda K) \circ x = 0,$$

hence we have

$$x = \lambda(H + K - \lambda H \circ K) \circ x. \tag{2.31}$$

Here if we express H especially as

$$H = -\frac{1}{\Delta_0}\Delta\begin{pmatrix} t & t_1' & \dots & t_\nu' \\ \xi & \xi_1' & \dots & \xi_\nu' \end{pmatrix}; \lambda\Big),$$

then from (2.26a),

$$\begin{aligned} \Delta\begin{pmatrix} t & t_1' & \dots & t_\nu' \\ \xi & \xi_1' & \dots & \xi_\nu' \end{pmatrix}; \lambda\Big) &= K(t, \xi)\Delta\begin{pmatrix} t_1' & \dots & t_\nu' \\ \xi_1' & \dots & \xi_\nu' \end{pmatrix}; \lambda\Big) \\ &\quad - K(t_1', \xi)\Delta\begin{pmatrix} t & t_2' & \dots & t_\nu' \\ \xi_1' & \xi_2' & \dots & \xi_\nu' \end{pmatrix}; \lambda\Big) + \dots \\ &\quad + (-1)^\nu K(t_\nu', \xi)\Delta\begin{pmatrix} t & t_1' & \dots & t_{\nu-1}' \\ \xi_1' & \xi_2' & \dots & \xi_\nu' \end{pmatrix}: \lambda\Big) \\ &\quad + \lambda\int_a^b \Delta\begin{pmatrix} t & t_1' & \dots & t_\nu' \\ \tau & \xi_1' & \dots & \xi_\nu' \end{pmatrix}; \lambda\Big)K(\tau, \xi)d\tau. \end{aligned}$$

Exchanging the rows in the right-hand side of the determinant makes all the signs attached to the second and following terms become minus; then we have

$$\Delta = K\Delta_0 - K(t_1, \xi)\varphi_1(t)\Delta_0 - K(t_2, \xi)\varphi_2(t)\Delta_0 - \dots$$
$$- K(t_\nu, \xi)\varphi_\nu(t)\Delta_0 - \lambda\Delta_0 H \circ K.$$

Since Δ is equal to $-\Delta_0 H$,

$$H + K - \lambda H \circ K = \sum_1^\nu {}_i K(t_i, \xi)\varphi_i(t). \tag{2.32}$$

Hence from (2.31) and (2.32),

$$x = \lambda \sum_1^\nu {}_i \varphi_i(t) \int_a^b K(t_i, \xi)x(\xi)d\xi.$$

Therefore if $\int_a^b K(t_i, \xi)x(\xi)d\xi = c_i$, then

$$x(t) = \sum_i c_i \varphi_i(t);$$

and thus a bounded continuous solution can be expressed by a linear combination of $\varphi_i(t)$.

We can obtain the following theorem on the associated integral equation $[\tilde{\mathrm{F}}_0]$ in the same method using the second fundamental formula.

THEOREM 2.6 $[\tilde{\mathrm{F}}_0]$

With regard to the associate integral equation of $[\mathrm{F}_0]$:

$$x(t) = \lambda \int_a^b x(\xi)\, K(\xi, t)d\xi \tag*{$[\tilde{\mathrm{F}}_0]$}$$

if $\Delta^{(1)}(\lambda) \neq 0$,

$$\tilde{\varphi}(t) = \Delta\begin{pmatrix} \xi' \\ t \end{pmatrix}; \lambda\Big);$$

and if $\Delta^{(m)}(\lambda) \neq 0$, *for* $0 < \nu \leqq m$,

$$\tilde{\varphi}_1(t) = \Delta\begin{pmatrix} \xi_1' & \xi_2' & \dots & \xi_\nu' \\ t & t_2' & \dots & t_\nu' \end{pmatrix}; \lambda\Big), \qquad \tilde{\varphi}_2(t) = \Delta\begin{pmatrix} \xi_1' & \xi_2' & \dots & \xi_\nu' \\ t_1' & t & \dots & t_\nu' \end{pmatrix}; \lambda\Big), \dots,$$

$$\tilde{\varphi}_\nu(t) = \Delta\begin{pmatrix} \xi_1' & \xi_2' & \dots & \xi_\nu' \\ t_1' & t_2' & \dots & t \end{pmatrix}; \lambda\Big)$$

are the continuous solutions, where $\tilde{\varphi}_1, \tilde{\varphi}_2, \dots, \tilde{\varphi}_\nu$ *are linearly independent, and all the solutions of* $[\tilde{F}_0]$ *can be expressed by the linear combination of them.*

(We omit the proof.)

Example 2.5 $[\tilde{F}_0]$

For Fredholm's homogeneous integral equation:

$$x(t) - \lambda \int_0^1 t\xi \, x(\xi) d\xi = 0,$$

as has been given in Example 2.4,

$$\Delta(\lambda) = 1 - \frac{\lambda}{3};$$

therefore, when $\lambda \neq 3$, we have no other solution than

$$x(t) \equiv 0.$$

Since when $\lambda = 3$ we have $\Delta^{(1)}(\lambda) \neq 0$, there exists a continuous solution not identically zero, and that is

$$\Delta\begin{pmatrix} t \\ \xi \end{pmatrix}; \lambda\Big) = t\xi;$$

hence

$$x(t) = \xi_1 t,$$

where ξ_1 is a constant. It is evident that this is a solution. However, ξ_1 can take a value outside the interval (0, 1), and

$$x(t) = at, \qquad a > 1$$

can stand as another solution.

Example 2.6 $[F_0]$

With regard to $[F_0]$:

$$x(t) - \lambda \int_0^1 (t - 3\xi) x(\xi) d\xi = 0.$$

We have

$$\Delta(\lambda) = \left(1 + \frac{\lambda}{2}\right)^2.$$

When $\lambda = -2$,

$$\Delta(\lambda) = 0, \qquad \Delta^{(1)}(\lambda) = 0, \qquad \Delta^{(2)}(\lambda) \neq 0.$$

Then from (2.28),

$$\varphi_1(t) = \Delta\begin{pmatrix} t & t_0' \\ \xi_1' & \xi_2' \end{pmatrix}; \lambda\Big) = 3(t - t_2')(\xi_1' - \xi_2'),$$

$$\varphi_2(t) = \Delta\begin{pmatrix} t_1' & t \\ \xi_1' & \xi_2' \end{pmatrix}; \lambda\Big) = 3(t_1' - t)(\xi_1' - \xi_2').$$

Since ξ_1', ξ_2', t_1', t_2' are suitably chosen constants, both φ_1 and φ_2 give no solution but

$$\varphi(t) = At + B;$$

where A, B are constant. Put this into the integral equation,

$$B = -A.$$

Therefore the solution of (2.28) becomes

$$\varphi(t) = A(t - 1),$$

where A is a constant. To be sure, in this case, there is $\Delta\begin{pmatrix} t \\ \xi_1 \end{pmatrix}; \lambda\Big)$ which does not become identically zero, and

$$\Delta\begin{pmatrix} t \\ \xi_1 \end{pmatrix}; \lambda\Big) = 6\left(\xi_1 - \frac{1}{3}\right)(t - 1);$$

that is, $A(t - 1)$.

Theorem 2.5 contains the following theorems.

(1) *Existence of the solution of* $[F_0]$

If $K(t, \xi)$ belongs to the functional space $\mathscr{R}$ in the domain S and $K(t, t)$ is integrable, the necessary and sufficient condition for $[F_0]$ *to have a continuous solution not identically zero is that λ is the root of the kernel K.*

This theorem corresponds to the following theorem on (a_0).

The necessary and sufficient condition for the simultaneous equations (a_0) *to have solutions not identically zero is that the determinant of coefficients becomes zero.*

Definition. Eigenvalue and eigenfunction

The root of the analytic entire function $\Delta(\lambda)$ of λ is called a *root*, a *characteristic constant*, or an *eigenvalue*, of the kernel $K(t, \xi)$; the solution of the homogeneous integral equation $[F_0]$ is called a *characteristic function*, an *eigenfunction*, or a *fundamental function* of the kernel $K(t, \xi)$. When an arbitrary solution of the eigenfunction $[F_0]$ can be expressed by the linear composition of $\varphi_1(t)$, $\varphi_2(t)$, ..., we say that the set of this eigenfunctions makes a *complete system.*

We call the number of linearly independent eigenfunctions the *index* of λ for an eigenvalue λ, and the degree of the multiple root of an eigenvalue λ in Fredholm's determinant the *multiplicity* of λ.

We have omitted the proof of Theorem 2.6, but the following theorem can be easily proved by comparing the expressions of the eigenfunction and the eigenvalue of the conjugate function $K(\xi, t)$ with those which correspond to them in Theorem 2.5.

(2) *Coincidence of eigenvalues*

If $K(t, \xi)$ belongs to the functional space $\mathscr{R}$ in the domain S and $K(t, t)$ is integrable, the eigenvalue of $K(t, \xi)$ with index v and multiplicity m becomes the eigenvalue of a conjugate function $K(\xi, t)$ with the same index and multiplicity. The inverse of this is also true (see Question 7).

As for the orthogonality of an eigenfunction (see Theorem 6.2 below), we have the following theorem.

THEOREM 2.7. *Orthogonality of eigenfunctions*

If $K(t, \xi)$ belongs to the functional space $\mathscr{R}$ in the domain S and $K(t, t)$ is integrable in the interval I, and if λ_1, λ_2 are the different eigenvalues for K, $\tilde{\varphi}_1(t)$ is an eigenfunction of $K(t, \xi)$ for λ_1, and $\tilde{\varphi}_2(t)$ is an eigenfunction of $\tilde{K}(\xi, t)$ for λ_2, and we have

$$\int_a^b \varphi_1(t)\tilde{\varphi}_2(t)dt = 0.$$

Proof. We introduce $\varphi_1 \circ \tilde{\varphi}_2 = 0$ using symbolic notation, and write $K(\xi, t)$ as $\tilde{K}$.

By the assumption of the theorem $\lambda_1 \neq \lambda_2$, and we have

$$\varphi_1 = \lambda_1 K \circ \varphi_1, \qquad \tilde{\varphi}_2 = \lambda_2 \tilde{\varphi}_2 \circ \tilde{K}.$$

Then multiply the first equation by $\lambda_2\tilde{\varphi}_2$ from the right and the second equation by $\lambda_1\varphi_1$ from the left, to obtain

$$\lambda_2\varphi_1 \circ \tilde{\varphi}_2 = \lambda_1\lambda_2 K \circ \varphi_1 \circ \tilde{\varphi}_2 = \lambda_1\lambda_2 \int_a^b \int_a^b K(\tau, \xi)\varphi_1(\xi)\tilde{\varphi}_2(\tau)d\xi d\tau,$$

$$\lambda_1\varphi_1 \circ \tilde{\varphi}_2 = \lambda_1\lambda_2\varphi_1 \circ \tilde{\varphi}_2 \circ \tilde{K} = \lambda_1\lambda_2 \int_a^b \int_a^b \varphi_1(\tau)\tilde{\varphi}_2(\xi)K(\xi, \tau)dd\tau\xi.$$

Subtracting the second equation from the first results in

$$(\lambda_2 - \lambda_1)\, \varphi_1 \circ \tilde{\varphi}_2 = 0.$$

Since $\lambda_1 \neq \lambda_2$ it follow that

$$\varphi_1 \circ \tilde{\varphi}_2 = 0.$$

2.3.1. *Index and multiplicity of an eigenvalue*

That the multiplicity of an eigenvalue is m means that λ is the m-ple root of $\Delta(\lambda) = 0$; then $\Delta^{(p)}(\lambda) = 0$ $(p < m)$, $\Delta^{(m)}(\lambda) \neq 0$. Here there is no need, for the integrand, to assume that

$$\Delta\begin{pmatrix} \tau_1 \dots \tau_p \\ \tau_1 \dots \tau_p \end{pmatrix}; \lambda \Big) \equiv 0$$

even if $\Delta^{(p)} = 0$. But whenever this equation holds, $\Delta^{(p)}(\lambda) = 0$ also holds. Therefore a certain number ν exists (with $0 < \nu \leqq m$), which the integral becomes identically zero for $p < \nu$ and not identically zero for $p \geqq \nu$. When the number ν coincides with that of an eigenfunction the following theorem holds from Theorem 2.5.

(1) *Index and multiplicity*

If the kernel $K(t, \xi)$ belongs to the functional space $\mathscr{R}$ in the domain S and $K(t, t)$ is integrable in the interval I, the index of the root of the kernel never exceeds the multiplicity of the root. That is,

$$0 < \nu \leqq m.$$

Here the multiplicity m is finite. If it is infinite, $\Delta^{(i)}(\lambda) = 0$ $(i = 1, 2, \dots)$ and $\Delta(\lambda)$ becomes identically zero according to Taylor's theorem.

But there is no limit to the number of eigenvalues. For

$$K(t, \xi) = t\xi^2,\ I(0, 1);\ \Delta(\lambda) = 1 - \frac{\lambda}{4},$$

there is one eigenvalue, but for

$$K(t, \xi) = t\xi + \xi^2, \quad I(0, 1); \qquad \Delta(\lambda) = 1 - \tfrac{2}{3}\lambda - \tfrac{1}{36}\lambda^2,$$

there are two, and for

$$K(t, \xi) = \begin{cases} t, \ \xi > t \\ \xi, \ \xi < t \end{cases}; \ I(0, 1); \ \Delta(\lambda) = \cos\sqrt{\lambda},$$

there exist an infinite number of eigenvalues $\lambda_n = \frac{1}{4}(2n-1)^2\pi^2$. Here the eigenfunctions

$$\sqrt{2}\sin\frac{(2n-1)\pi}{2}t \qquad (n = 1, 2, \ldots)$$

are infinite in number, but the multiplicity of each eigenvalue is 1 and the index is 1. And there corresponds one eigenfunction to each eigenvalue (see Question 20).

2.3.2. *Order of solution of homogeneous integral equations*

We solve a homogeneous integral equation in the following order.

(i) Calculate Fredholm's determinant $\Delta(\lambda)$ of the kernel $K(t, \xi)$.

(ii) Solve $\Delta(\lambda) = 0$ to determine the eigenvalues.

(iii) Examine the multiplicity of each eigenvalue. Suppose it to be m.

(iv) Evaluate Fredholm's minors

$$\Delta\begin{pmatrix} t \\ \xi_1 \end{pmatrix}; \lambda\Big), \ \Delta\begin{pmatrix} t & t_2' \\ \xi_1' & \xi_2' \end{pmatrix}; \lambda\Big), \ \ldots \ \Delta\begin{pmatrix} t & t_2' & \ldots & \xi_p \\ \xi_1' & \xi_2' & \ldots & \xi_p \end{pmatrix}; \lambda\Big) \ldots$$

in the due order, and let

$$\Delta\begin{pmatrix} t & t_2' & \ldots & t_\nu' \\ \xi_1' & \xi_2' & \ldots & \xi_\nu' \end{pmatrix}; \lambda\Big)$$

be the first determinant in the above series which is not identically zero; then ν is the index of the eigenvalue with $\nu \leqq m$.

(v) The eigenfunctions are then given by (2.28)

$$\varphi_1(t), \ \varphi_2(t), \ \ldots, \ \varphi_\nu(t).$$

(vi) The general solution of $[F_0]$ is expressed by linear combinations of these eigenfunctions,

$$x(t) = c_1\varphi_1(t) + c_2\varphi_2(t) + \ldots + c_\nu\varphi_\nu(t).$$

2.4. Method of solution of Fredholm's integral equation of the second kind (in the case of the eigenvalue)

In § 2.2, we considered the methods of solution of $[F_2]$ in the case of $\Delta(\lambda) \neq 0$, while in this section we are going to treat the case where this condition does not hold. Here again we consider the simultaneous equation labelled (**a**), or its vector expression:

$$\boldsymbol{a}_1 x_1 + \boldsymbol{a}_2 x_2 + \cdots + \boldsymbol{a}_n x_n = \boldsymbol{b}. \tag{a}$$

If a vector $\tilde{\boldsymbol{r}}$ is orthogonal to all of $\boldsymbol{a}_1, \boldsymbol{a}_2, \ldots, \boldsymbol{a}_n$, it is necessarily orthogonal to $\boldsymbol{b}$. Since

$$\boldsymbol{a}_1 \tilde{\boldsymbol{r}} = 0,\ \boldsymbol{a}_2 \tilde{\boldsymbol{r}} = 0,\ \ldots,\ \boldsymbol{a}_n \tilde{\boldsymbol{r}} = 0,$$

$\tilde{\boldsymbol{r}}$ is the solution vector of the homogeneous simultaneous equations:

$$\left.\begin{aligned} a_{11}x_1 + a_{21}x_2 + \cdots + a_{n1}x_n &= 0, \\ a_{12}x_1 + a_{22}x_3 + \cdots + a_{n2}x_n &= 0, \\ \cdots \quad \cdots \quad \cdots & \\ a_{1n}x_1 + a_{2n}x_2 + \cdots + a_{nn}x_n &= 0. \end{aligned}\right\} \tag{a_0}$$

We use ($\mathbf{a}_0$) for a transposed equation corresponding to the transposed determinant (in which rows and columns are interchanged) of the coefficient determinant of (**a**). Therefore, for the existence of the solution of (**a**), it is necessary that the constant-term vector is orthogonal to an arbitrary solution vector of the transposed homogeneous equation. If the solution vector of the transposed homogeneous equation is $\tilde{\boldsymbol{r}}_i(r_{i1}, r_{i2}, \ldots, r_{in})$, then

$$\tilde{\boldsymbol{r}}_i \cdot \boldsymbol{b} = \boldsymbol{0},$$

viz.

$$r_{i1}b_1 + r_{i2}b_2 + \cdots + r_{in}b_n = 0 \quad (i = 1, 2, \ldots, \nu).$$

It is well known that this condition is also sufficient.

In the simultaneous equations (**a**) with $\Delta(\lambda) = 0$, the constant-term vector is required to be orthogonal to an arbitrary solution vector of the transposed homogeneous equation. If the fundamental solutions of the homogeneous integral equation ($\mathbf{a}_0$) are $\boldsymbol{x}_1, \boldsymbol{x}_2, \ldots, \boldsymbol{x}_\nu$, and a particular solution of (a) is $\boldsymbol{x}_a$, then an arbitrary solution of (a) is expressed by

$$x = \boldsymbol{x}_a + c_1\boldsymbol{x}_1 + c_2\boldsymbol{x}_2 + \cdots + c_\nu\boldsymbol{x}_\nu.$$

This is also well known.

We have the following theorem with regard to [F_2].

THEOREM 2.8 Existence of solution [F_2]

In Fredholm's integral equation of the second kind:

$$x(t) - \lambda \int_a^b K(t, \xi)x(\xi)d\xi = f(t), \qquad [F_2]$$

if λ is the eigenvalue of the kernel $K(t, \xi)$, and the index is ν, the kernel $K(t, \xi)$ belongs to the functional space $\mathscr{R}$ in the domain S, $K(t, t)$ is integrable in the interval I, and $f(t)$ is continuous in I, then the necessary and sufficient condition for the existence of a continuous solution is that $f(t)$ is orthogonal to the eigenfunction $\tilde{\varphi}_i(t)$ of the associated kernel $\tilde{K}(\xi, t)$:

$$\int_a^b f(t)\tilde{\varphi}_i(t)dt = 0 \quad (i = 1, 2, \ldots, \nu). \tag{2.33}$$

In this, the solution contains the linear combination of the eigenfunction $\varphi_i(t)$ of the kernel $K(t, \xi)$, and

$$x(t) = f(t) + \lambda \int_a^b H(t, \xi)f(\xi)d\xi + c_1\varphi_1(t) + c_2\varphi_2(t) + \ldots + c_\nu\varphi_\nu(t), \tag{2.34}$$

where

$$H(t, \xi) = \Delta\begin{pmatrix} t & t_1' & t_2' & \ldots & t_n' \\ \xi & \xi_1' & \xi_2' & \ldots & \xi_n' \end{pmatrix}; \lambda\Big) \Big/ \Delta\begin{pmatrix} t_1' & t_2' & \ldots & t_n' \\ \xi_1' & \xi_2' & \ldots & \xi_n' \end{pmatrix}. \tag{2.35}$$

Proof. We use symbolic notation here.

(i) $f \circ \tilde{\varphi}_i = 0$ *is necessary.* Since $\tilde{\varphi}_i$ is the solution of the associated integral equation $x = \lambda x \circ K$,

$$\tilde{\varphi}_i \circ (E - \lambda K) = 0 \qquad (i = 1, 2, \ldots, n).$$

Suppose that [F_2] has a solution, then multiply

$$(E - \lambda K) \circ x = f$$

by $\tilde{\varphi}_i$ from the left, and we get

$$\tilde{\varphi}_i \circ (E - \lambda K) \circ x = \tilde{\varphi}_i \circ f.$$

Since the left-hand side is zero from the first equation, we obtain

$$\tilde{\varphi}_i \circ f = 0 \qquad (i = 1, 2, \ldots n).$$

(ii) *Expression of the solution.*
Using H from the proof of Theorem 2.5, we put

$$H + K - \lambda H \circ K \equiv F,$$

then

$$K = F - H + \lambda H \circ K.$$

Rewriting $[F_2]$ and substituting K on the right-hand side from the above expression we have

$$\begin{aligned} x &= f + \lambda K \circ x \\ &= f + \lambda(F - H + \lambda H \circ K) \circ x \\ &= f - \lambda H \circ (E - \lambda K) \circ x + \lambda F \circ x. \end{aligned}$$

Here if we take (2.35) as H, $\mathrm{F} \circ x$ becomes a linear combination in φ_i.

Since $(E - \lambda K) \circ x = f$ from $[F_2]$, the solution of $[F_2]$, if it exists, takes the form

$$x = f - \lambda H \circ f + \sum c_i \varphi_i(t) = (E - \lambda H) \circ f + \sum c_i \varphi_i.$$

(iii) $f \circ \tilde{\varphi}_i = 0$ *is sufficient.*
If we put (2.34) in the left-hand side of $[F_2]$, then,

$$\begin{aligned} &(E - \lambda K) \circ (E - \lambda H) \circ f + (E - \lambda K) \circ (\textstyle\sum c_i \varphi_i) \\ &\quad = \{E - \lambda(K + H - \lambda K \circ H)\} \circ f + \textstyle\sum c_i (E - \lambda K) \circ \varphi_i. \end{aligned}$$

Since φ_i is an eigenfunction, the last term vanishes. As in (2.32), $(K + H - \lambda K \circ H)$ is $\sum_i K(\xi, t_i)\tilde{\varphi}_i(t)$ (see Question 8), therefore if

$$\tilde{\varphi}_i \circ f = f \circ \tilde{\varphi}_i = 0,$$

the above expression becomes f and (2.34) becomes the solution of $[F_2]$. Thus, $f \circ \tilde{\varphi}_i = 0$ is a sufficient condition for the existence of the solution.

This concludes the proof of the theorem. One can see that the symbolic notation can express the proof concisely.

2.4.1. *Correspondence of the solutions of integral equations and those of simultaneous equations*

We have already noted that a linear integral equation can be treated as the limit of simultaneous equations of the first degree of infinite dimension (in § 1.4, [2]). In § 2.2–2.4 we have studied the solution of integral equations according to this notion. Through this study it became clear that if we pay attention to the fact that Fredholm's determinant $\Delta(\lambda)$ of an integral equation corresponds to the determinant D of the coefficients of simultaneous equations, and if the index of $\Delta(\lambda)$ corresponds to the order of D, then the solutions of inhomogeneous and homogeneous simultaneous equations, and the solutions of homogeneous integral equations and those of the second kind have a similar correspondence between them.

In the case of Volterra type, since $\lambda = 1$ and $\Delta(\lambda) = 1$, the homogeneous integral equation [V_0] has no continuous solution other than the identically zero solution; therefore, no singular case such as an eigenfunction appears. And, as has been noted in the case of Volterra type, there is consequently no difference between the two methods of solution, one leading to simultaneous equations and the other to an iterated kernel in § 2.1.

2.5. Methods of solution of the integral equation of the first kind

The integral equation of the first kind:

$$\int_a^b K(t, \xi)x(\xi)d\xi = f(t), \text{ or } K \circ x = f \qquad [F_1]$$

is simple in its form but the solution method is difficult compared with that of the second kind. This is because there does not exist a so-called inverse element of K (see Question 6), such that $K^{-1} \circ K = E$, except in some special cases. For instance, when $K = (2/\pi)^{\frac{1}{2}} \sin t\xi$, $a = 0, b = \pi$, we have $K^{-1} = (2/\pi)^{\frac{1}{2}} \sin t\xi$. In general, we treat an integral equation of the first kind by changing it to that of the second kind. The results are summarized in Table 2.3.

Volterra type. Changing Volterra's integral equation of the first kind:

TABLE 2.3. Correspondence of integral equations and simultaneous equations of the first degree (summary)

	Linear integral equations		Simultaneous equations of the first degree	
	$\Delta(\lambda) \neq 0$	$\Delta(\lambda) = 0$, index ν	$D \neq 0$	$D = 0$, order ν
	$x(t) - \lambda \int_a^b K(t, \xi) x(\xi) d\xi = f(t)$ [F_2]		$\sum_{j=1}^{n} a_{ij} x_j = b_i$, $\boldsymbol{a}_1 x_1 + \boldsymbol{a}_2 x_2 + \cdots + \boldsymbol{a}_n x_n = \boldsymbol{b}$ (**a**)	
Inhomogeneous type	Unique solution exists. $x(t) = f(t) + \frac{\lambda}{\Delta(\lambda)} \int_a^b \Delta\left(\begin{matrix} t \\ \xi \end{matrix}; \lambda\right) f(\xi) d\xi$	General solution does not exist. Condition of the existence of solution: $f \circ \tilde{\varphi} = 0$ Provided $\tilde{\varphi}(t)$ is the eigen function of [$\tilde{F}_0$]: ∞^ν of solutions exist. $x(t) = f(t) + \lambda \int_a^b H(t, \xi) f(\xi) d\xi + \sum_{i=1}^{\nu} c_i \varphi_i(t)$	Unique solution exists. $\boldsymbol{x}_i = \frac{1}{D} \sum_{j=1}^{n} D_{ji} b_j$	General solution does not exist. Condition of the existence of solution: $\boldsymbol{r}_i \circ \boldsymbol{b} = \boldsymbol{0}$ Provided $\boldsymbol{r}_i$ is the solution vector of ($\tilde{\mathbf{a}}_0$) ∞^ν of solutions exist. $\boldsymbol{x} = \boldsymbol{x}_a + \sum_{i=1}^{\nu} c_i \boldsymbol{x}_{i0}$ $\boldsymbol{x}_a$: particular solution, $\boldsymbol{x}_{i0}$: solution of ($\mathbf{a}_0$)
	$x(t) - \lambda \int_a^b K(t, \xi) x(\xi) d\xi = 0$ [F_0]		$\sum_{j=1}^{n} a_{ij} x_j = 0$, $\boldsymbol{a}_1 x_1 + \boldsymbol{a}_2 x_2 + \cdots + \boldsymbol{a}_n x_n = \boldsymbol{0}$ ($\mathbf{a}_0$)	
Homogeneous type	Unique solution exists. $x(t) \equiv 0$	There exist ν linearly independent solutions $\varphi_1(t), \varphi_2(t), \ldots, \varphi_\nu(t)$ ∞^ν of general solutions exist. $x(t) = \sum_{i=1}^{\nu} c_i \varphi_i(t)$	Unique solution exists. $\boldsymbol{x}_i = 0$	There ν exist linearly independent solutions $\boldsymbol{x}_{10}, \boldsymbol{x}_{20}, \ldots, \boldsymbol{x}_{\nu 0}$ ∞^ν of general solutions exist. $\boldsymbol{x} = \sum_{i=1}^{\nu} c_i \boldsymbol{x}_{i0}$

$$\int_a^t K(t, \xi)x(\xi)d\xi = f(t), \qquad [\mathrm{V}_1]$$

into a set of n simultaneous equations, we have

$$\left.\begin{aligned} &k_{11}x_1 = y_1,\\ &k_{21}x_1 + k_{22}x_2 = y_2,\\ &\ldots,\\ &k_{n1}x_1 + k_{n2}x_2 + \cdots + k_{nn}x_n = y_n, \end{aligned}\right\} \qquad (\mathrm{v}_1)$$

then combining each pair of consecutive equations we get

$$\left.\begin{aligned} &(k_{21} - k_{11})x_1 + k_{22}x_2 = y_2 - y_1,\\ &\ldots,\\ &(k_{i1} - k_{i-11})x_1 + (k_{i2} - k_{i-12})x_2 + \ldots + k_{ii}x_i = y_i - y_{i-1},\\ &\ldots,\\ &k_{n1} - k_{n-11})x_1 + (k_{n2} - k_{n-12})x_2 + \ldots + k_{nn}x_n = y_n - y_{n-1}. \end{aligned}\right\}$$

If

$$k_{11},\ k_{22},\ \ldots,\ k_{ii},\ \ldots,\ k_{nn} \neq 0,$$

we can express them in the same form as (v_2) in § 2.2, that is, using the symbolic expression of difference, Δ_i, the general expression becomes

$$\sum_1^{i-1}{}_j \Delta_i k_{ij} x_j + k_{ii}x_i = \Delta_i y_i. \qquad (2.36)$$

If we divide this by k_{ii} when $k_{ii} \neq 0$, and denote the result by

$$x_i + \sum_1^{i-1}{}_j (\Delta_i k_{ij}/k_{ii})x_j = (\Delta_i y_i/k_{ii}), \qquad (2.37)$$

then this has the form of (v_2). Now, if we put the equation back to the infinite dimension and make substitutions

$$i \to t, \quad j \to \xi, \sum_1^{i-1}{}_j \to \int_a^t d\xi, \quad \Delta_i \to \frac{\partial}{\partial t},$$

condition (2.35) becomes

$$K(t, t) \neq 0,$$

(2.36) becomes

$$K(t, t)x(t) + \int_a^t \frac{\partial K(t, \xi)}{\partial t} x(\xi)d\xi = \frac{\partial f(t)}{\partial t}, \qquad (2.38)$$

and (2.37) becomes

$$x(t)+\int_a^t\left\{\frac{\partial K(t,\xi)}{\partial t}\Big/K(t,t)\right\}x(\xi)d\xi=\frac{\partial f(t)}{\partial t}\Big/K(t,t), \quad (2.39)$$

which takes the form of [V_2].

The problem becomes difficult when $K(t, t)$ becomes zero somewhere in I.

Thus let us consider the case when $K(t, t) \equiv 0$. This time (2.38) becomes

$$\int_a^t \frac{\partial K(t,\xi)}{\partial t}x(\xi)d\xi = f^{(1)}(t),$$

and this is again the form of [V_1]. Then take the partial differentitaion of it in t and we get

$$\left[\frac{\partial K(t,\xi)}{\partial t}\right]_{\xi=t}x(t)+\int_a^t\frac{\partial^2 K(t,\xi)}{\partial t^2}x(\xi)d\xi=f^{(2)}(t); \quad (2.40)$$

then we can express it in the form of [V_2], if $[\partial K/\partial t]_{\xi=t} \neq 0$ holds everywhere in $t(a, b)$.

In [V_1] taking the partial integral of the left-hand side and putting

$$X(t)=\int_a^t x(t)dt, \quad (2.41)$$

we have

$$K(t,t)X(t)-\int_a^t\frac{\partial K(t,\xi)}{\partial \xi}X(\xi)d\xi=f(t); \quad (2.42)$$

hence we obtain the form of [V_2] if $K(t, t) \neq 0$.

As we have already given the uniqueness and existence theorems of the continuous solution of [V_2] in Theorem 2.2, we can introduce the same sort of theorem with regard to these cases.

THEOREM 2.9 [V_1]

If in Volterra's integral equation of the first kind:

$$\int_a^t K(t,\xi)x(\xi)d\xi=f(t), \quad [V_1]$$

the kernel $K(t, \xi)$ is continuous in the domain T and partially differentiable on t, and if $K_1 = \partial K/\partial t$ belongs to the functional space $\mathscr{R}$ in the domain T, $K(t, t)$ does not become zero in the interval I, $f(t)$ and $f^{(1)}(t)$ are continuous in the interval I, and $f(a) = 0$, *then there exists a unique continuous solution which is the solution of the integral equation of the second kind:*

$$x(t)+\int_a^t \left\{\frac{\partial K(t,\xi)}{\partial t}\Big/K(t,t)\right\}x(\xi)d\xi=\frac{df(t)}{dt}\Big/K(t,t). \quad (2.39)$$

Proof. We reduce to Theorem 2.2.

(i) *The examination of the condition of Theorem 2.4.*
From [V_1], for $x(t)$ to have a continuous solution from Lemma 1.1 it is necessary that $f(t)$ be continuous and $f(a)=0$. Under the hypothesis of differentiability, differentiate [V_1] and we can obtain (2.38), and since $K(t,t)\neq 0$, divide the equation by it and we can obtain (2.39). This is [V_2], in which the kernel is $(\partial K/\partial t)/K$, and the right-hand side is $f^{(1)}/K$.

Thus, we may see directly from the condition of the unique-existence theorem, Theorem 2.2, of the continuous solution of [V_2], that the condition of the theorem is sufficient for the existence of the solution of (2.39).

(ii) *The agreement of the solution of* [V_1] *with that of (2.39).*
In the above we have made it clear that if [V_1] has a continuous solution, it is included in (2.39). We have to show, therefore, that (2.39) does not have any solution but that of [V_1].

From (2.39) we can express

$$\frac{d}{dt}\left[f(t)-\int_a^t K(t,\xi)x(\xi)d\xi\right]=0.$$

Therefore, the continuous solution of (2.39) ought to satisfy

$$f(t)-\int_a^t K(t,\xi)x(\xi)d\xi=C \quad (2.43)$$

(where C is a constant). Since the left-hand side becomes zero if $t=a$, this constant should be zero. Hence the solution of (2.39) is at the same time the solution of [V_1].

The following theorem can be introduced quite similarly.

THEOREM 2.10 [V_1]

In Volterra's integral equation of the first kind:

$$\int_a^t K(t,\xi)x(\xi)d\xi=f(t) \quad [V_1]$$

if the kernel $K(t,\xi)$ is continuous in the domain T and partially integrable on t $(n+1)$ times successively, if $\partial^{n+1}K/\partial t^{n+1}$ belongs to the functional space $\mathscr{R}$ in the domain T, if

$$K(t, t) = \left[\frac{\partial K}{\partial t}\right]_{\xi=t} = \left[\frac{\partial^2 K}{\partial t^2}\right]_{\xi=t} = \cdots = \left[\frac{\partial^{n-1} K}{\partial t^{n-1}}\right]_{\xi=t} = 0$$

in the interval I, if $[\partial^n K/\partial t^n]_{\xi=t}$ *is non-zero in the interval I, if* $f(t), f^{(1)}(t), \ldots, f^{(n+1)}(t)$ *are continuous in the interval I, and if* $f(a) = f^{(1)}(a) = \cdots = f^{(n)}(a) = 0$; *then there exists a unique continuous solution, which is an integral equation of the second kind:*

$$x(t) + \int_a^t \left\{ \frac{\partial^{n+1} K(t, \xi)}{\partial t^{n+1}} \Big/ \left[\frac{\partial^n K}{\partial t^n}\right]_{\xi=t} \right\} x(\xi) d\xi = \frac{d^{n+1} f(t)}{dt} \Big/ \left[\frac{\partial^n K}{\partial t^n}\right]_{\xi=t}. \tag{2.44}$$

The proof is omitted

We have the following theorem.

THEOREM 2.11 [V_1]

If in Volterra's integral equation of the first kind:

$$\int_a^t K(t, \xi) x(\xi) d\xi = f(t), \qquad [V_1]$$

the kernel $K(t, \xi)$ *is continuous in the domain T, and integrable on* ξ, *if* $\partial K/\partial \xi$ *belongs to the functional space* $\mathscr{R}$ *in the domain T, if* $K(t, t)$ *does not become zero in the interval I, if* $f(t)$ *is continuous in I and if* $f(a) = 0$, *and further, if the solution of the integral equation of the second kind:*

$$X(t) - \int_a^t \left\{ \frac{\partial K(t, \xi)}{\partial \xi} \Big/ K(t, t) \right\} X(\xi) d\xi = f(t)/K(t, t), \tag{2.45}$$

is differential and its derivative is continuous, then there exists a unique continuous solution of [V_1]:

$$x(t) = X^{(1)}(t).$$

The proof is omitted (but see Question 10).

Fredholm type. In the case of Fredholm's integral equation of the first kind:

$$\int_a^b K(t, \xi) x(\xi) d\xi = f(t), \qquad [F_1]$$

divide the domain S into two triangular domains by the diagonal $\xi = t$, and set

$$\int_a^t K(t, \xi) x(\xi) d\xi + \int_t^b K(t, \xi) x(\xi) d\xi = f(t)$$

and differentiate both sides in t, then we get

$$x(t)\{K(t, t-0) - K(t, t+0)\} + \int_a^b \frac{\partial K}{\partial t} x(\xi) d\xi = f^{(1)}(t). \tag{2.46}$$

Here if

$$K(t, t-0) - K(t, t+0) \equiv L(t) \tag{2.47}$$

never vanishes in I, we can lead it to the form of $[F_2]$, dividing both sides by $L(t)$.

Now $L(t)$ is a jump on the diagonal $\xi = t$, and we may not assume that $K(t, \xi)$ always has finite jumps on such a diagonal, yet if we have a finite non-zero jump,

$$K(t, \varphi(t) - 0) - K(t, \varphi(t) + 0) = L(t)$$

everywhere on a curve in S:

$$\xi = \varphi(t), \tag{2.48}$$

which combines the two corners (a, a), (b, b) of S, intersects a line parallel to the ξ-axis only once, is continuous and changes its direction continuously; then, we can treat the problem quite similarly to the previous one. Divide S into the upper and lower domain of $\xi = \varphi(t)$ (Fig. 2.2), set

$$\int_a^{\varphi(t)} K(t, \xi) x(\xi) d\xi + \int_{\varphi(t)}^b K(t, \xi) x(\xi) d\xi = f(t).$$

and differentiate both sides in t, then we get

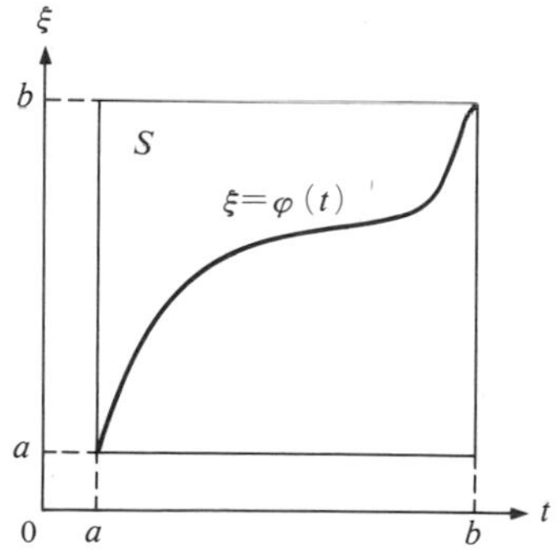

Fig. 2.2. Division of S

$$L(t)\cdot\varphi^{(1)}(t)x(t)+\int_a^b \frac{\partial K}{\partial t}x(\xi)d\xi=f^{(1)}(t). \tag{2.49}$$

We assumed that $L(t)\neq 0$ in I. If $\varphi^{(1)}(t)\neq 0$, we have

$$x(t)+\int_a^b\left\{\frac{\partial K}{\partial t}\Big/L(t)\varphi^{(1)}(t)\right\}x(\xi)d\xi=f^{(1)}(t)/L(t)\varphi^{(1)}(t).$$

Since the variable of the first term on the right-hand side is the integral variable of the regular part, take

$$t=\psi(\xi).$$

as the inverse function of (2.48), and put ξ as τ in ξ other than the integral variable, we have then

$$x(\tau)+\int_a^b\left\{\frac{\partial K}{\partial t}\Big/L(t)\varphi^{(1)}(t)\right\}_{t=\psi(\tau)}x(\xi)d\xi$$

$$=\left\{f^{(1)}(t)\Big/L(t)\varphi^{(1)}(t)\right\}_{t=\psi(\tau)}, \tag{2.50}$$

which is $[F_2]$ with the kernel

$$-\left\{\frac{\partial K}{\partial t}\Big/L(t)\varphi^{(1)}(t)\right\}_{t=\psi(\tau)}.$$

Therefore, when $f^{(1)}(t)$ is continuous (2.50) has a continuous solution if Fredholm's determinant of this kernel does not vanish. But since (2.50) is an equation derived by differentiating $[F_1]$ we can also derive (2.50) if we start from

$$f(t)+C=\int_a^b K(t,\xi)x(\xi)d\xi, \tag{2.51}$$

for C a constant, instead of $[F_1]$, so if (2.50) has a continuous solution it does not necessarily mean the solution of $[F_1]$. The continuous solution of (2.50) satisfies (2.51) for a special value of C. Hence the following theorem holds.

THEOREM 2.12 $[F_1]$

(i) *$K(t,\xi)$ is bounded in the domain S and continuous in S except on the curve $\xi=\varphi(t)$*

(ii) *$\varphi(t)$ is continuous in the interval I, and its derivative is continuous, $\varphi^{(1)}(t)$ is not zero and*

$$\varphi(a)=a,\qquad \varphi(b)=b.$$

(We denote the inverse function of φ by ψ.)

(iii) *$K(t, \varphi(t)-0)$, $K(t, \varphi(t)+0)$ exist everywhere except at both ends of the interval I, and the difference*

$$L(t) = K(t, \varphi(t)-0) - K(t, \varphi(t)+0)$$

is continuous in the interval I.

(iv) *$K(t, \xi)$ is partially differentiable on t, and $\partial K/\partial t$ belongs to the functional space $\mathscr{R}$ in the domain S.*

(v) *$f(t)$, $f^{(1)}(t)$ is continuous in I.*

Under these conditions, if the determinant or modified determinant of

$$-\left\{\frac{\partial K}{\partial t}\Big/ L(t)\varphi^{(1)}(t)\right\}_{t=\psi(\tau)}$$

is non-zero, then the integral equation of the first kind:

$$f(t) + C_0 = \int_a^b K(t, \xi)x(\xi)d\xi,$$

has a unique continuous solution for only one particular value of the parameter C. This solution becomes the unique continuous solution of the integral equation of the second kind:

$$x(\tau) + \int_a^b \left\{\frac{\partial K}{\partial t}\Big/ L(t)\varphi^{(1)}(t)\right\}_{t=\psi(\tau)} x(\xi)d\xi$$
$$= \left\{f^{(1)}(t)\Big/ L(t)\varphi^{(1)}(t)\right\}_{t=\psi(\tau)}. \tag{2.50}$$

We omit the proof of this theorem.

When $L(t)$ is always zero in I, and if

$$M(t) \equiv \frac{\partial K}{\partial t}(t, \varphi(t)-0) - \frac{\partial K}{\partial t}(t, \varphi(t)+0)$$

never vanishes, we can extend the theorem.

Example 2.7

$$\left.\begin{aligned}&\int_0^1 K(t, \xi)x(\xi)d\xi = 67t^2 - 40t^2 + 52t + 17,\\ &K(t, \xi) = \begin{cases} t\xi(t-\xi)^2 + 1, & t > \xi;\\ t\xi(t-\xi)^2, & t < \xi.\end{cases}\end{aligned}\right\} \qquad [F_1]$$

Since in this case we have

$$L(t) = 1,$$

$$\frac{\partial K}{\partial t} = 3t^2\xi - 4t\xi^2 + \xi^2,$$

$$f'(t) = 201t^2 - 80t + 52;$$

it follows that (2.46) becomes

$$x(t) + \int_0^1 (3t^2\xi - 4t\xi^2 + \xi^3)x(\xi)d\xi = 201t^2 - 80t + 52.$$

From this we get

$$x(t) = 60t^2 + 60t + 24.$$

Example 2.8

$$\left.\begin{aligned} &\int_0^1 K(t, \xi)x(\xi)d\xi = \frac{2}{3}t^2 - \frac{5}{2}t^2 - 3t + 1. \\ &K(t, \xi) = \begin{cases} t^2 + t - 2\xi + 1, & \xi < \frac{1}{2}(t^2 + t); \\ t^2 + t - 2\xi, & \xi > \frac{1}{2}(t^2 + t). \end{cases} \end{aligned}\right\} \quad [F_2]$$

In this problem, since $\varphi(t) = \frac{1}{2}(t^2 + t)$,

$$L(t) = 1, \qquad \frac{\partial K}{\partial t} = 2t + 1, \qquad \varphi^{(1)}(t) = t + \frac{1}{2},$$

$$f^{(1)}(t) = 2t^2 - 5t - 3.$$

Therefore (2.49) becomes

$$x(t) + \int_0^1 2x(\xi)d\xi = 2t - 6;$$

then

$$x(t) = 2t - 8/3.$$

Exercises

1. When $K(t, \xi)$, $G(t, \xi)$ are reciprocal functions in T, and $r(t)$ is continuous and never vanishes in I, then $(r(t)/r(\xi)K(t, \xi))$ is reciprocal to $(r(t)/r(\xi))\, G(t, \xi)$. And when $K(t, \xi)$ and $\Gamma(t, \xi)$ are reciprocal functions in S, $r(t)/r(\xi)K(t, \xi)$ and $r(t)/r(\xi)\Gamma(t, \xi)$ are reciprocal to each other. [Goursat]

2. Prove that Fredholm's determinant $\Delta(\lambda)$ is absolutely convergent.

The ratio test for the majorant series of the absolute series becomes

$$|\lambda| M(b-a)\sqrt{\left(1+\frac{1}{\nu}\right)^{\nu}}\,\frac{1}{\sqrt{\nu+1}},$$

quite similarly to the proof in Theorem 2.3 (i). The same test is adopted for the absolute convergence of Fredholm's minor.

3. Prove that where $K(t, \xi)$ is continuous, Fredholm's minor $\Delta\left(\begin{matrix}t\\ \xi\end{matrix}; \lambda\right)$ is also continuous.

[This is clear from the expansion of $\Delta\left(\begin{matrix}t\\ \xi\end{matrix}; \lambda\right)$ in § 2.2. (2.14) and Lemma 1.1.]

4. Prove that

$$\Delta\left(\begin{matrix}t\\ \xi\end{matrix}; \lambda\right) = K(t, \xi)\Delta(\lambda) - \lambda \int_a^b \Delta\left(\begin{matrix}t\\ \tau\end{matrix}; \lambda\right) K(\tau, \xi)d\tau$$

in a direct way, and then explain what significance this reciproca theorem has in the simultaneous equations (f_2).

5. Prove the following facts on the continuity of $\Delta\left(\begin{matrix}t\\ \xi\end{matrix}; \lambda\right)$. It is always continuous if $\Delta(\lambda) = 0$. Where $K(t, \xi)$ is continuous, it is continuous if $\Delta(\lambda) \neq 0$. It is discontinuous at discontinuous points but $\Delta\left(\begin{matrix}t\\ \xi\end{matrix}; \lambda\right) - K(t, \xi)\Delta(\lambda)$ is always continuous.

(Use the reciprocal theorem.)

6. What mathematical properties does the set of functions $\{E - \lambda K\}$ have with regard to the convolution of the second kind if $K(t, \xi)$ belongs to the functional space $\mathscr{R}$ in the domain S and $|\lambda| M(b-a) < 1$

7. Prove Theorem 2.5 (2).

8. Prove that $K + H - \lambda K \circ H = \sum_i K(\xi, t_i)\varphi_i'(t)$.

[Start from § 2.2 (2.26) and follow the course as we did with Theorem 2.5.]

9. Prove that if $K\left(\begin{matrix}t_1 & t_2 & \cdots & t_\nu\\ \xi_1 & \xi_2 & \cdots & \xi_\nu\end{matrix}\right)$ vanishes at $\nu = n$ then it always vanishes in $\nu \geqq n+1$.

(This expression is the general form of the minor of the νth order.)

10. Prove Theorem 2.11.

11. Expand Theorem 2.11 in I when

$$K(t, t) = \left[\frac{\partial K}{\partial \xi}\right]_{\xi=t} = \cdots = \left[\frac{\partial^{n-1} K}{\partial \xi^{n-1}}\right]_{\xi=t} = 0, \qquad \left[\frac{\partial^n K}{\partial \xi^n}\right]_{\xi=t} \neq 0,$$

following the manner of Theorem 2.10.

12. Solve

$$\int_0^{\frac{1}{2}\pi} K(t, \xi)x(\xi)d\xi = f(t), \qquad [F_1]$$

$$K(t, \xi) = \begin{cases} t(t-\xi). \\ \xi(t-\xi). \end{cases}$$

$$f(t) = at + b + c \sin t.$$

(Decide on the values of a, b, c first from the conditions for the existence of a solution.)

13. Find the condition that satisfies $f(t)$ so that

$$[F_1]\colon \int_0^1 (t^2 + \xi^2)x(\xi)d\xi = f(t)$$

may have a solution, then solve the equation.

14. Extend Theorem 2.12 in I when $L(t) \equiv 0$, $M(t) \neq 0$.

15. Prove that $\overset{*}{K}^n = \overset{*}{K}^i * \overset{*}{K}^{n-i}$, $\overset{\circ}{K}^n = \overset{\circ}{K}^n \circ \overset{\circ}{K}^{n-i}$.

16. If the reciprocal functions for the parameters λ, μ for the kernel $K(t, \xi)$ are $\Gamma(t, \xi; \lambda)$, $\Gamma(t, \xi; \mu)$, prove that

$$\Gamma(t\ \ \xi; \lambda) - \Gamma(t, \xi; \mu) = (\mu - \lambda)\int_a^b \Gamma(t, \tau; \lambda)\Gamma(\tau, \xi; \mu)d\tau.$$

[Denote the reciprocal theorem with symbols by

$$\Gamma_\lambda + K = \lambda K \circ \Gamma_\lambda = \lambda\Gamma_\lambda \circ K, \qquad \Gamma_\mu + K = \mu K \circ \Gamma_\mu = \mu\Gamma_\mu \circ K;$$

multiply the first expression by $\mu\Gamma_\mu$ from the left, the second expression by $\lambda\Gamma_\lambda$ from the right, and subtract the second from the first, then if we watch carefully the permutability of K and Γ we can arrive at

$$(\lambda - \mu)\,\Gamma_\lambda \circ \Gamma_\mu + \lambda\Gamma_\lambda \circ K - \mu\Gamma_\mu \circ K = 0.$$

Again from the above two expressions we have $\lambda\Gamma_\lambda \circ K - \mu\Gamma_\mu \circ K = \Gamma_\lambda - \Gamma_\mu$, from which we get $\Gamma_\lambda - \Gamma_\mu = (\mu - \lambda)\Gamma_\lambda \circ \Gamma_\mu$ immediately.

Or we can use the following method. Express them using iterated kernels as

$$\begin{aligned}\Gamma(t, \tau; \lambda)\Gamma(\tau, \xi; \mu) = {} & K(t, \tau)K(\tau, \xi) + \lambda K_2(t, \tau)K(\tau, \xi) \\ & + \lambda^2 K_3(t, \tau)K(\tau, \xi) + \mu K(t, \tau)K_2(\tau, \xi) \\ & + \lambda\mu K_2(t, \tau)K_2(\tau, \xi) + \mu^2 K(t, \tau)K_3(\tau, \xi) \\ & + \cdots\end{aligned}$$

and integrate over τ from a to b. Hence if we consider

$$\frac{\lambda^2 - \mu^2}{\lambda - \mu} = \lambda + \mu, \frac{\lambda^3 - \mu^3}{\lambda - \mu} = \lambda^2 + \lambda\mu + \mu^2, \ldots$$

we get

$$\Gamma_\lambda - \Gamma_\mu = (\mu - \lambda)\Gamma_\lambda \circ \Gamma_\mu$$

immediately.]

17. Find the characteristic equations for the following kernels $K(t, \xi)$, $I(a, b)$, and discuss the eigenvalue:

a. $t - \xi$, $I(0, \alpha)$; $\left[\Delta(\lambda) = 1 + \frac{\lambda^2}{12}\alpha^3\right]$;

b. $t + \xi$, $I(0, 1)$; $\left[\Delta(\lambda = 1 - \lambda - \frac{\lambda^2}{12}\right]$;

c. $t\xi$, $I(0, \alpha)$; $\left[\Delta(\lambda) = 1 - \frac{\lambda}{3}\alpha\right]$;

d. $t\xi + t^2\xi^2 + t^3\xi^3$, $I(0; 1)$;

e. $At^2 + 2Bt\xi + C\xi^2 + Dt + E\xi + F$, $I(a, b)$;

f. $e^{-t\xi}$, $I(0, 1)$;

g. $\sin t\xi$, $I(0, x)$;

h. $1 + t\xi$, $I(0, 1)$; $[\Delta(\lambda) = \lambda^2 - 16\lambda + 12]$;

i. $t\xi(t - \xi)$, $t > \xi$; $t\xi(\xi - t)$, $t < \xi$; $I(0, 1)$;

j. $\sin t \sin \xi$; $I(0, 2\pi)$; $[\Delta(\lambda) = 1 - \pi\lambda]$.

18. Calculate Fredholm's minors of the following kernels:

a. $t\xi^2$, $I(0, 1)$; $\left[\Delta(\lambda) = 1 - \frac{\lambda}{4}\right]$;

b. $t\xi + \xi^2$, $I(0, 1)$; $\left[\Delta(\lambda) = 1 - \frac{2}{3}\lambda - \frac{1}{36}\lambda^2\right]$;

c. $\begin{cases} t, & (\xi > t) \\ \xi, & (\xi < t) \end{cases}$ $I(0, 1)$; $[\Delta(\lambda) = \cos\sqrt{\lambda}]$.

19. Prove that

$$\log \Delta(\lambda) = -\left\{k_1\lambda + k_2\frac{\lambda^2}{7} + \cdots + k_n\frac{\lambda^n}{n} + \cdots\right\}$$

with regard to $K(t, \xi)$, $I(a, b)$; provided

$$k_n = \int_a^b \mathring{K}^n(t, t)dt \qquad (n = 1, 2, \ldots)$$

$$[\Delta'(\lambda)/\Delta(\lambda) = \int_a^b \Gamma(t, t, \xi)d\xi].$$

20. Prove that the number of eigenfunctions for one eigenvalue is finite.

21. Show that if $K(t, \xi) = 1$ $(t \leqq \xi)$; $K(t, \xi) = 0$ $(t > \xi)$ in I, then $\Delta(\lambda) = \exp\{-(b - a)\lambda\}$.

22. Where $K(t, \xi) = K_1(t)K_2(\xi)$ and $\int_a^b K_1(\xi)K_2(\xi)d\xi = A$ show that that $\Delta(\lambda) = 1 - A\lambda$, $\Delta\left(\begin{matrix}t\\ \xi\end{matrix}; \lambda\right) = \lambda K_1(t)K_2(\xi)$. Also, show that the solution of $[F_2]$ is

$$x(t) = f(t) + \frac{\lambda K(t)}{1 - A\lambda}\int_a^b K_2(\xi)f(\xi)d\xi.$$

(See 3.2.1.)

23. When $K(t, \xi) = a_1(t)b_1(\xi) + a_2(t)b_2(\xi)$, show that $\Delta(\lambda)$, $\Delta\left(\begin{matrix}t\\ \xi\end{matrix}; \lambda\right)$ are binomial functions of λ.

24. Show that the solutions of

$$x(t) = \lambda \int_{-\pi}^{\pi} \cos^n(t - \xi)x(\xi)d\xi$$

are $x(t) = \cos(n - 2r)t$ and $x(t) = \sin(n - 2r)t$, where i, r are positive integers smaller than $\frac{1}{2}n$.

References

Bocher, M.: (1914), *An Introduction to the Study of Integral Equations*. Cambridge Tracts in Mathematics 10 (Cambridge University Press).

3 Particular methods of solution of integral equations

In Chapter 2 we studied the uniqueness and existence theorems of the solution of integral equations and expressed these solutions in terms of iterated kernels or Fredholm's determinants. It is, however, a little inconvenient to use them as the practical methods of solution because their calculations are in most cases complicated and the convergence is not good for practical purposes. In contrast, the methods of solution of operational calculus by Laplace transform or Fourier transform are quite practical and convenient, and can be applied to the case of a singular kernel as will be described in this chapter. We shall discuss the general methods of Fourier transform and other transforms for the singular integral equation in Chapter 7.

Therefore, in this chapter we study the case where kernels belong to the functional space $\mathscr{R}$ in the bounded domain S or T as we did in the previous chapter. It goes without saying that the solution presented in this chapter agrees with that of Fredholm or Volterra since existence and uniqueness in this case have been proved. When we adopt the methods of operational calculus or series expansion of the kernel or of an unknown function, it is necessary to examine the existence of transforms and uniform convergence of series. But the cases that can be used for particular solutions are restricted to the cases of special forms, and we can easily examine the above conditions so as to make the analysis exact and strict. Therefore, from now on we shall avoid using too precise expressions for the forms of theorems and rather give the plans for solution without much formality.

3.1. Orthogonal function system

Here we give a brief explanation of the orthogonal function system (Churchill 1941), which we shall use often. If the arbitrary functions $f(t)$, $g(t)$ are integrable in the interval $I(a, b)$, the convolution of the first kind is

$$f \cdot g = \int_a^b f(t)g(t)dt.$$

When there is only one variable, we sometimes denote it as $f \cdot g$ instead of $f \circ g$. In this, if

$$f \cdot g = 0,$$

then we say that $f(t)$ and $g(t)$ intersect orthogonally. Where

$$\dot{f}^2 \equiv f \cdot f = \int_a^b f(t)f(t)dt \geqq 0,$$

we call $(f \cdot f)$ the *norm* of $f(t)$, and denote it by $||f||$. When $f(t)$ is a continuous function f^2, $||f||$ does not vanish unless f is identically zero.

Now, if

$$\varphi(t) = f(t)/||f(t)||,$$

we clearly have

$$\dot{\varphi}^2 = 1.$$

In this case we say that $\varphi(t)$ is *normalized.*

Now when a functional system

$$\varphi_1(t),\ \varphi_2(t),\ \ldots,\ \varphi_i(t),\ \ldots,$$

is given in I, if all terms are normalized and two arbitrary functions intersect orthogonally, we call the system a *normalized orthogonal function system.*. That is, for the functional space $\{\varphi_i(t)\}$, we have

$$\varphi_i \cdot \varphi_j = \begin{cases} 1, & i = j; \\ 0, & i \neq j. \end{cases}$$

We can show that $\{\varphi_i\}$ is *linearly independent.*

For, if it is not linearly independent, then

$$\sum c_i \varphi_i = 0$$

with the coefficients c_i not all zero. But if we multiply both sides by φ_k from the right, the result is

$$\sum c_i \varphi_i \cdot \varphi_k = c_k \dot{\varphi}_k{}^2 = 0,$$

and $c_k = 0$. Then all the coefficients c_i become zero for all values of k, and this is contradictory to the hypothesis.

We have the following theorem.

LEMMA 3.1

When a system of continuous functions

$$u_1(t),\ u_2(t),\ \ldots,\ u_n(t),\ \ldots,$$

is given in the interval $I(a, b)$, *we can make a normalized orthogonal function system by linear combination.*

Proof. First, make a function system

$$\Phi_n = \begin{vmatrix} a_{11} & a_{12} & \ldots a_{1n} \\ a_{21} & a_{22} & \ldots a_{2n} \\ \cdot & \cdot & \ldots \cdot \\ a_{n-1\,1} & a_{n-2\,2} & \ldots a_{n-1\,n} \\ u_1(t) & u_2(t) & \ldots u_n(t) \end{vmatrix}, \qquad (n = 1, 2, \ldots), \tag{3.1}$$

by linear combination. In this $a_{ij} = u_i \circ u_j$.

The function system $\{\Phi_n\}$ is orthogonal. For $u_i \circ \Phi_n$ is derived by replacing the last row of the above determinant by

$$a_{i1},\ a_{i2},\ \ldots,\ a_{in},$$

and this agrees with some other row if $i < n$. Therefore, Φ_n orthogonally intersects $u_1, u_2, \ldots, u_{n-1}$. Hence it orthogonally intersects all the linear combinations of these, $\Phi_1, \Phi_2, \ldots, \Phi_{n-1}$.

Next, we calculate the norm of Φ_n. Since $u_i \circ \Phi_n = 0$, we have

$$\dot{\Phi}_n{}^2 = \begin{vmatrix} a_{11} & a_{12} & \ldots a_{1n} \\ a_{21} & a_{22} & \ldots a_{2n} \\ \cdot & \cdot & \ldots \cdot \\ a_{n-1\,1} & a_{n-1\,2} & \ldots a_{n-1\,n} \\ 0 & 0 & \ldots u_n \circ \Phi_n \end{vmatrix} = u_n \cdot \Phi_n \cdot A_{n-1}. \tag{3.2}$$

Here we donote the determinants of a_{ij} by

$$A_1 = 1, \quad A_n = |a_{ij}| \quad (i, j = 1, 2, \ldots, n),$$

etc. Since from (1)

$$u_n \cdot \Phi_n = A_n,$$

we have

$$\Phi_n^2 = A_{n-1} \cdot A_n.$$

Hence, the normalized orthogonal function system is

$$\varphi_n = \frac{\Phi_n}{\sqrt{(A_{n-1}A_n)}} \qquad (n = 1, 2, \ldots). \tag{3.3}$$

Now, when a normalized orthogonal function system $\{\varphi_i(t)\}$ is given, and if for an arbitrary function $f(t)$ the expansion

$$f(t) = \sum_{i=1}^{\infty} c_i \varphi_i(t) \tag{3.4}$$

is possible in the interval $I(a, b)$ and we can take the integration of the series on the right-hand side by each term, then the coefficients are determined by

$$f \cdot \varphi_j = \sum_{i=1}^{\infty} c_i \varphi_i \cdot \varphi_j = c_j \qquad (j = 1, 2, \ldots). \tag{3.5}$$

This series is called a Fourier expansion of f by $\{\varphi\}$, and c is called a Fourier coefficient.

3.1.1. Euclidian space of infinite dimension

In Euclidian space of infinite dimension, the coordinates of one point P are represented by n coordinates $c_i (i = 1, 2, \ldots, n)$. There are n unit vectors and when they are expressed as

$$\boldsymbol{e}_i \qquad (i = 1, 2, \ldots, n),$$

the vector $\overrightarrow{\mathrm{OP}}$ is denoted by

$$\overrightarrow{\mathrm{OP}} = \sum c_i \boldsymbol{e}_i,$$

in terms of unit vectors and the components, where $c_i = \overrightarrow{\mathrm{OP}} \cdot \boldsymbol{e}_i$.

When another vector $\overrightarrow{\mathrm{OQ}}$ is expressed as

$$\overrightarrow{\mathrm{OQ}} = \sum d_i \boldsymbol{e}_i,$$

if we denote the *scalar product* of the two vectors $\overrightarrow{\mathrm{OP}}$ and $\overrightarrow{\mathrm{OQ}}$ by $\overrightarrow{\mathrm{OP}} \cdot \overrightarrow{\mathrm{OQ}}$, we have

$$\overrightarrow{\mathrm{OP}} \cdot \overrightarrow{\mathrm{OQ}} = \sum c_i d_i;$$

and if they intersect orthogonally then

$$\overrightarrow{\mathrm{OP}} \cdot \overrightarrow{\mathrm{OQ}} = 0.$$

The norm of the vector $\overrightarrow{OP}$ is

$$\overline{OP^2} = \sum c_i^2,$$

and the absolute value is

$$|\overrightarrow{OP}| = (\sum c_i^2)^{1/2}.$$

When the number of dimensions of Euclidian space n, is infinite, a point or a vector becomes a function and the set of such functions makes one functional space (called a Hilbert space). When there is a normalized orthogonal functional system $\{\varphi_i\}$ and there exists no function that intersects orthogonally with any φ_i, we say that $\{\varphi_i\}$ is complete, where φ_i has the characteristics of a unit vector, say

$$f = \sum c_i \varphi_i, \qquad c_i = f \cdot \varphi_i.$$

The scalar product of $f(t)$ and $g(t)$ is

$$f \cdot g = \int f(t) g(t) dt,$$

and if they intersect orthogonally we have $f \cdot g = 0$. And the norm of f becomes $f^2 = \sum c_i^2$ (Parseval's theorem).

3.2. Method of expansion of an orthogonal function system

Consider

$$\int_a^b K(t, \xi) x(\xi) d\xi = f(t), \qquad \text{say } K \circ x = f. \qquad [F_1]$$

Take

$$\{\varphi_i(t)\} \qquad (i = 1, 2, \ldots) \tag{3.6}$$

as a complete normalized orthogonal function system in the interval (a, b), and by denoting

$$\int_a^b K(t, \xi) \varphi_i(\xi) d\xi = K_i(t), \quad \text{say } K \circ \varphi_i = K_i, \tag{3.7}$$

we get the Fourier expansion of $K(t, \xi)$ by $\{\varphi_i(\xi)\}$ in terms of

$$K(t, \xi) = \sum_1^\infty K_i(t) \varphi_i(\xi). \tag{3.8}$$

Now, denote the expansion of the orthogonal system of the solution by

$$x(t) = \sum_1^\infty \alpha_i \varphi_i(t); \tag{3.9}$$

then from $[F_1]$

$$f(t) = \int_a^b \sum_1^\infty K_i(t)\varphi_i(\xi) \sum_1^\infty \alpha_i \varphi_i(\xi) d\xi = \sum_1^\infty \alpha_i K_i(t). \tag{3.10}$$

That is, here the coefficients in the expansion of $x(t)$ by $\varphi_i(t)$ agree with those in the expansion of $f(t)$ by $K_i(t)$. But it is difficult to find α_i with all these as they are indicated here, since $K_i(t)$ is not necessarily an orthogonal system.

From Lemma 3.1, however, we can make an orthogonal function system out of an arbitrary function sequence, then if $\{K_i(t)\}$ is linearly independent, we can make an orthogonal system $\psi_i(t)$ from its linear combination:

$$\left.\begin{aligned} \psi_1(t) &= c_{11}K_1(t), \\ \psi_2(t) &= c_{21}K_1(t) + c_{22}K_2(t), \\ \psi_3(t) &= c_{31}K_1(t) + c_{32}K_2(t) + c_{33}K_3(t), \end{aligned}\right\} \tag{3.11}$$

Expanding $f(t)$ again with this, we obtain

$$\begin{aligned} f(t) = \sum \beta_i \psi_i(t) &= \beta_1 c_{11} K_1 + \beta_2[c_{21}K_1 + c_{22}K_2] \\ &\quad + \beta_3[c_{31}K_1 + c_{32}K_2 + c_{33}K_3] + \cdots; \end{aligned} \tag{3.12}$$

then if we suppose that the right-hand side is absolutely convergent in $I(a, b)$, this equation, in which $K_i(t)$ is replaced by $\varphi_i(t)$, is $x(t)$, and

$$\begin{aligned} x(t) &= \beta_1 c_{11}\varphi_1(t) + \beta_2[c_{21}\varphi_1(t)\ c_{22}\varphi_2(t)] \\ &\quad + \beta_3[c_{31}\varphi_1(t) + c_{32}\varphi_2(t) + c_{33}\varphi_3(t)] + \cdots \end{aligned} \tag{3.13}$$

is the solution of $[F_2]$.

The substitution of $x(t)$ in the regular part of $[F_1]$ enables us to integrate each term, and we have

$$\begin{aligned} K \circ x &= \beta_1 c_{11} K \circ \varphi_1 + \beta_2[c_{21}K \circ \varphi_1 + c_{22}K \circ \varphi_2] \\ &\quad + \beta_3[c_{31}K \circ \varphi_1 + c_{32}K \circ \varphi_2 + c_{33}K \circ \varphi_3] + \cdots \\ &= \beta_1 c_{11} K_1 + \beta_2[c_{21}K_1 + c_{22}K_2] \\ &\quad + \beta_3[c_{31}K_1 + c_{32}K_2 + c_{33}K_3] + \cdots \\ &= \beta_1\psi_1 + \beta_2\psi_2 + \beta_3\psi_3 + \cdots \\ &= f. \end{aligned}$$

The plan is the same with the other types, for example, in the case of [F_2]:

$$x(t) - \lambda \int_a^b K(t, \xi)x(\xi)d\xi = f(t), \qquad (E - \lambda K) \circ x = f,$$

making use of (3.3) and (3.4) we get

$$\sum \alpha_i(\varphi_i - \lambda K_i) = f(t) \tag{3.14}$$

as the result of computation.

Here suppose that the function system $\{K_i'(t)\}$ is linearly independent with the substitution

$$\varphi_i(t) - \lambda K_i(t) \equiv K_i'(t); \tag{3.15}$$

then from its linear combination we can make the orthogonal system $\{\psi_i'(t)\}$ as in (3.11). Then, setting

$$f(t) = \sum \beta_i \psi_i'(t), \tag{3.12a}$$

as in (3.12), we may prove that (3.13), in which K_i' is replaced by φ_i', is the solution if (3.12a) is absolutely convergent.

In this method of solution, we have made the assumption that $\{K_i'\}$ is linearly independent, therefore no K_i' is identically zero, but if λ agrees with any of the eigenvalues

$$\varphi_i - \lambda_i K_i = 0 \quad (i = 1, 2, \ldots), \tag{3.16}$$

its coefficient in (3.14), α_i, becomes arbitrary and consequently we have an infinite number of solutions.

Next, in the case of [F_0]:

$$x(t) = \lambda \int_a^b K(t, \xi)x(\xi)d\xi, \qquad \text{viz. } x = \lambda K \circ x,$$

(3.14) becomes

$$\sum \alpha_i(\varphi_i - \lambda K_i) = 0. \tag{3.17}$$

Here, if $\{\varphi_i - \lambda K_i\}$ is linearly independent, and λ is different from any of the eigenvalues of (3.16), we always have

$$\alpha_i = 0,$$

and [F_0] has no other solution than $x(t) \equiv 0$, while if $\lambda = \lambda_j$, then

$$\alpha_i = \begin{cases} 0, & i \neq j, \\ \text{indeterminate}, & i = \text{j}, \end{cases}$$

and there exist an infinity of solutions $x(t) = \alpha_j \varphi_j(t)$.

When $\{\varphi_i - \lambda K_i\}$ is not linearly independent, α_i, which satisfies (3.17), need not be always zero; then there exist such solutions that are not identically zero. This time the number of eigenfunctions corresponding to the eigenvalue λ_j is not necessarily infinite.

3.2.1. $K(t, \xi) = K_1(t)K_2(\xi)$

The difficulty in this method of solution lies in the fact that $\{K_i(t)\}$ does not necessarily become an orthogonal function system if we use the normalized orthogonal function system $\{\varphi_n(t)\}$ in $t(a, b)$. This sort of difficulty, however, does not arise when the kernel takes some special forms. For example in [F_2], put

$$K(t, \xi) = K_1(t)K_2(\xi), \tag{3.18}$$

and for $\{\varphi_n(t)\}$ let each be as

$$\left.\begin{aligned} x &= \textstyle\sum \alpha_i \varphi_i, \\ K_1 &= \textstyle\sum K_{1i} \varphi_i, \\ K_2 &= \textstyle\sum K_{2i} \varphi_i, \\ f &= \textstyle\sum c_i \varphi_i, \end{aligned}\right\} \tag{3.19}$$

(for all $i = 1, 2, \ldots$) Then [F_2] becomes

$$\textstyle\sum \alpha_i \varphi_i - \lambda \sum K_{1i} \varphi_i \{\sum K_{2j} \varphi_j\} \circ \{\sum \alpha_j \varphi_j\} = \sum c_i \varphi_i, \tag{3.20}$$

and by the normal orthogonality of $\{\varphi_i\}$, the convoluted part becomes simpler as

$$\textstyle\sum \alpha_i \varphi_i - \lambda \sum K_{1i} (\sum K_{2j} \alpha_j) \cdot \varphi_i = \sum c_i \varphi_i.$$

Therefore, comparison of the coefficients of $\varphi_i(t)$ term by term produces

$$\alpha_i - \lambda K_{1i} (\textstyle\sum K_{2j} \alpha_j) = c_i, \qquad (i, j = 1, 2, \ldots). \tag{3.21}$$

Then if we write

$$K_{1i} K_{2j} = K_{ij},$$

and so obtain a determinant

$$\Delta_i(\lambda, c),$$

which arises from substituting $c_1, c_2, \ldots$ for the ith row of the determinant

$$\Delta(\lambda) = \begin{vmatrix} 1-\lambda K_{11} & -\lambda K_{12} & -\lambda K_{13} \ldots \\ -\lambda K_{21} & 1-\lambda K_{22} & -\lambda K_{23} \ldots \\ -\lambda K_{31} & -\lambda K_{32} & 1-\lambda K_{33} \ldots \\ \ldots & \ldots & \ldots \quad \ldots \end{vmatrix},$$

then we can determin the Fourier coefficient α_i of $x(t)$ as

$$\alpha_i = \Delta_i(\lambda_i c)/\Delta(\lambda), \tag{3.22}$$

with $\Delta(\lambda) \neq 0$. But $\Delta(\lambda)$ is an infinite determinant and there is no difference between this and Fredholm's method of solution. We may treat the problem as in § 2.4 when $\Delta(\lambda) = 0$. We can solve it completely when the Fourier expansion of K_2 terminates in finite terms. Minute study of this sort of finite term will be given in § 3.3.

3.2.2. *Double Fourier series expansion of the kernel*

In the integral equation of the second kind:

$$x(t) - \lambda \int_0^a K(t, \xi)x(\xi)d\xi = f(t), \tag{3.23}$$

in the interval $t(0, a)$, if $K(t, \xi)$ belongs to $\mathscr{R}$, the equation satisfies Dirichlet's test (see Question 4) and the equation is expanded in a double Fourier series and the discontinuous points are denoted by

$$K(t, \xi) = \frac{4}{a^2} \sum_{m=1}^{\infty} \sum_{n=1}^{\infty} \boldsymbol{K}_s(m, n) \sin\frac{m\pi}{a}t \sin\frac{n\pi}{a}\xi. \tag{3.24}$$

The $\boldsymbol{K}_s(m, n)$ are double Fourier coefficients such as

$$\boldsymbol{K}_s(m, n) = \int_0^a dt \int_0^a K(t, \xi) \sin\frac{m\pi}{a}t \sin\frac{n\pi}{a}\xi \, d\xi. \tag{3.25}$$

Here we suppose that we can expand $x(t), f(t)$ into a Fourier sine series and take

$$x(t) = \frac{2}{a} \sum_{n=1}^{\infty} X_s(n) \sin\frac{n\pi}{a}t; \tag{3.26}$$

$$f(t) = \frac{2}{a} \sum_{n=1}^{\infty} F_s(n) \sin\frac{n\pi}{a}t, \; F_s(n) = \int_0^a f(\pi) \sin\frac{n\pi}{a}\tau \, d\tau, \tag{3.26a}$$

$$(n = 1, 2, \ldots)$$

then putting these into the integral equations, we have

$$\sum_{n=1}^{\infty} X_s(n)\sin\frac{n\pi}{a}t - \lambda\sum_{m=1}^{\infty}\sum_{n=1}^{\infty} \boldsymbol{K}_s(m, n)X_s(n)\sin\frac{m\pi}{a}t$$
$$= \sum_{n=1}^{\infty} F_s(n)\sin\frac{n\pi}{a}t.$$

Hence the solution of the simultaneous equations of $X_s(n)$ is obtained by solving the simultaneous equations of the infinite dimension:

$$X_s(n) - \lambda\sum_{m=1}^{\infty} \boldsymbol{K}_s(n, m)X_s(m) = F_s(n), \quad (n = 1, 2, \ldots) \qquad (3.27)$$

3.2.3. $\boldsymbol{K(t, \xi) = K(mt + n\xi), K(t, \xi) = K(t + \xi), K(t, \xi) = K(t - \xi)}$

For example, in

$$x(t) - \lambda\int_{-h}^{h} K(t - \xi)x(\xi)d\xi = f(t), \qquad (3.28)$$

denote the coefficients of $K(t), f(t)$ by $\boldsymbol{K}_c(n)$, $\boldsymbol{K}_s(n)$; $F_c(n)$, $F_s(n)$ in the interval $(-h, h)$, and set†

$$\boldsymbol{K}_c(n) = \int_{-h}^{h} K(\tau)\cos\frac{n\pi}{h}\tau\, d\tau, \qquad \boldsymbol{K}_s(n) = \int_{-h}^{h} K(\tau)\sin\frac{n\pi}{h}\tau\, d\tau; \qquad (3.29)$$

$$F_c(n) = \int_{-h}^{h} F(\tau)\cos\frac{n\pi}{h}\tau\, d\tau, \qquad F_s(n) = \int_{-h}^{h} F(\tau)\sin\frac{n\pi}{h}\tau\, d\tau. \qquad (3.30)$$

then we have

$$K(t - \xi) = \frac{1}{h}\left\{\boldsymbol{K}_c(0)\cdot\frac{1}{2} + \sum \boldsymbol{K}_c(n)\cos\frac{n\pi}{h}(t - \xi)\right.$$
$$\left. + \sum \boldsymbol{K}_s(n)\sin\frac{n\pi}{h}(t - \xi)\right\},$$

$$f(t) = \frac{1}{h}\left\{F_c(0)\cdot\frac{1}{2} + \sum F_c(n)\cos\frac{n\pi}{h}t + \sum F_s(n)\sin\frac{n\pi}{h}t\right\}.††$$

† $\boldsymbol{K}_c(n)$, $F_c(n)$ are called the finite cosine transform of the nth order of $K(t)$, $f(t)$, and $\boldsymbol{K}_s(n)$, $F_s(n)$ the finite sine transform of the nth order.

†† In this case, in order to facilitate computation, it is better not to use the normalized orthogonal function system expansion.

Now suppose that $x(t)$ can be expanded in a Fourier series also, and put the series into the original integral equation provided

$$x(t) = \frac{1}{h}\left\{X_c(0)\frac{1}{2} + \sum X_c(n)\cos\frac{n\pi}{h}t + \sum X_s(n)\sin\frac{n\pi}{h}t\right\} \quad (3.31)$$

Then the result of computation runs as

$$\begin{aligned}
&\frac{1}{h}\{X_c(0) - \lambda \boldsymbol{K}_c(0)X_c(0)\}\frac{1}{2} + \frac{1}{h}\sum\{X_c(n) - \lambda \boldsymbol{K}_c(n)X_c(n) \\
&\qquad + \lambda \boldsymbol{K}_s(n)X_s(n)\}\cos\frac{n\pi}{h}t \\
&\qquad + \frac{1}{h}\sum\{X_s(n) - \lambda \boldsymbol{K}_c(n)X_s(n) - \lambda \boldsymbol{K}_c(n)X_c(n)\}\sin\frac{n\pi}{h}t \\
&\qquad = \frac{1}{h}\left\{F_c(0)\frac{1}{2} + \sum F_c(n)\cos\frac{n\pi}{h}t + \sum F_s(n)\sin\frac{n\pi}{h}t\right\}.
\end{aligned}$$

Hence, comparing the coefficients, we get the solutions of X_c, X_s as

$$\left.\begin{aligned}
X_c(0) &= F_c(0)/\{1 - \lambda \boldsymbol{K}_c(0)\}, \\
X_c(n) &= \{[1 - \lambda \boldsymbol{K}_c(n)]F_c(n) - \lambda \boldsymbol{K}_s(n)F_s(n)\}/\Delta(\lambda), \\
X_s(n) &= \{[1 - \lambda \boldsymbol{K}_c(n)]F_s(n) + \lambda \boldsymbol{K}_s(n)F_c(n)\}/\Delta(\lambda).
\end{aligned}\right\} \quad (3.32)$$

Here $\Delta(\lambda)$ is

$$\Delta(\lambda) = [1 - \lambda \boldsymbol{K}_c(n)]^2 + [\lambda \boldsymbol{K}_s(n)]^2,$$

and when $\boldsymbol{K}_c(n) \neq 0$, $\boldsymbol{K}_s(n) = 0$,

$$\lambda_n = 1/\boldsymbol{K}_c(n)$$

becomes the eigenvalue, but otherwise $\Delta(\lambda) \neq 0$.

Now take the values of $\Gamma_c(0)$, $\Gamma_c(n)$, $\Gamma_s(n)$ so that

$$\left.\begin{aligned}
&\{1 - \lambda \boldsymbol{K}_c(0)\}\{1 - \lambda \Gamma_c(0)\} = 1, \\
&\Delta(\lambda)\{1 - \lambda \Gamma_c(n)\} = 1 - \lambda \boldsymbol{K}_c(n), \\
&\Delta(\lambda)\Gamma_s(n) = -\boldsymbol{K}_s(n);
\end{aligned}\right\} \quad (3.33)$$

then we have

$$\left.\begin{aligned}
X_c(0) &= F_c(0) - \lambda \Gamma_c(0)F_c(0), \\
X_c(n) &= F_c(n) - \lambda \Gamma_c(n)F_c(n) + \lambda \Gamma_s(n)F_s(n), \\
X_s(n) &= F_s(n) - \lambda \Gamma_c(n)F_s(n) - \lambda \Gamma_s(n)F_c(n),
\end{aligned}\right\} \quad (3.34)$$

and we may write

$$x(t) = f(t) - \lambda \int_{-h}^{h} \Gamma(t - \xi, \lambda) f(\xi) d\xi.$$

Here the kernel of the solution $\Gamma(t - \xi, \lambda)$ is a Fourier series with $\Gamma_c(0)$, $\Gamma_c(n)$, $\Gamma_s(n)$ as its coefficients, which is a function expressed as

$$\Gamma(t) = \frac{1}{h}\left\{\Gamma_c(0)\frac{1}{2} + \sum \Gamma_c(n)\cos\frac{n\pi}{h}t + \sum \Gamma_s(n)\sin\frac{n\pi}{h}t\right\}.$$

Here we can now easily introduce the reciprocal theorem:

$$K(t) + \Gamma(t) = \lambda \int_{-h}^{h} K(t - \xi)\Gamma(\xi)d\xi = \lambda \int_{-h}^{h} K(\xi)\Gamma(t - \xi)d\xi,$$

since we have

$$\left.\begin{aligned} \boldsymbol{K}_c(0) + \Gamma_c(0) &= \lambda \boldsymbol{K}_c(0) \cdot \Gamma_c(0), \\ \boldsymbol{K}_c(n) + \Gamma_c(n) &= \lambda\{\boldsymbol{K}_c(n)\Gamma_c(n) - \boldsymbol{K}_s(n)\Gamma_s(n)\}, \\ \boldsymbol{K}_s(n) + \Gamma_s(n) &= \lambda\{\boldsymbol{K}_s(n)\Gamma_c(n) + \boldsymbol{K}_c(n)\Gamma_s(n)\} \end{aligned}\right\}$$

from (3.33).

We can proceed in the same manner when we compute in more general cases, as m, n are integers and the kernel takes the form $K(mt + nt)$.

3.3. The method of undetermined coefficients†

As one of the ways to solve a linear differential equation, there is a method of forming a power-series expansion of an unknown function hypothetically and determining the unknown coefficients successively from the given equation. In the case of an integral equation, if the kernel $K(t, \xi)$ and a known function $f(t)$ are analytic and the Taylor expansions in the neighbourhood of $t = a$ are expressed as

$$K(t, \xi) = K(a, \xi) + (t - a)\frac{1}{1!}\left(\frac{\partial K}{\partial t}\right)_{t=a} + (t - a)^2\frac{1}{2!}\left(\frac{\partial^2 K}{\partial t^2}\right)_{t=a}$$
$$+ \cdots,$$

† This is also called the method of series expansion. See, for example, (Forsyth 1941).

$$f(t) = f(a) + (t-a)\frac{1}{1!}f^{(1)}(a) + (t-a)^2\frac{1}{2!}f^{(2)}(a) + \ldots,$$

then suppose that the solution $x(t)$ is also expressed as

$$x(t) = x(a) + (t-a)\frac{1}{1!}x^{(1)}(a) + (t-a)^2\frac{1}{2!}x^{(2)}(a) + \ldots,$$

and we have only to proceed to the integral equation. After computing, the integration of the regular part leads to the simultaneous equations in

$$x(a),\ x^{(1)}(a),\ x^{(2)}(a),\ \ldots.$$

Yet this method is not practical, except for a special case, since the dimension in the method is infinite.

In a special case, we can express a solution in terms of simultaneous equations of $n+1$ unknown coefficients $\alpha_0, \alpha_1, \ldots, \alpha_n$, under the assumption that the solution is expressed in the form of a power series of finite terms:

$$x(t) = \alpha_0 + \alpha_1 t + \alpha_2 t^2 + \cdots + \alpha_n t^n. \tag{3.35}$$

For example, suppose we have

$$K(t, \xi) = K_0(\xi) + tK_1(\xi) + t^2K_2(\xi) + \cdots + t^nK_n(\xi), \tag{3.36}$$

$$f(t) = \beta_0 + \beta_1 t + \beta_2 t^2 + \cdots + \beta_n t^n. \tag{3.37}$$

Write the form of the solution as in (3.35) and determine the coefficient.

Let us put

$$\int_a^b \xi^j K_i(\xi)d\xi \equiv c_{ij} \qquad (i, j = 1, 2, 3, \ldots, n); \tag{3.38}$$

Then (noting that the numbers of terms in (3.35) to (3.37) need not agree) the integral equation becomes

$$\textstyle\sum_i \alpha_i t^i - \lambda \sum_i t^i(\sum_j \alpha_j c_{ij}) = \sum_i \beta_i t^i.$$

On comparing the coefficients of t^i, we get simultaneous equations in α_i:

$$\alpha_i - \lambda \textstyle\sum_j \alpha_j c_{ij} = \beta_i \qquad (i, j = 1, 2, \ldots, n). \tag{3.39}$$

Therefore, if we set the determinant of the coefficients as

$$\Delta(\lambda) = \begin{vmatrix} 1-\lambda c_{11} & -\lambda c_{12} & \dots & -\lambda c_{1n} \\ -\lambda c_{21} & 1-\lambda c_{22} & \dots & -\lambda c_{2n} \\ \dots & \dots & \dots & \dots \\ -\lambda c_{n1} & -\lambda c_{n2} & \dots & 1-\lambda c_{nn} \end{vmatrix}, \tag{3.40}$$

we know that this is the Fredholm's determinant by the calculation of (2.10). Here we get the following results.

(1) *The case* $\Delta(\lambda) \neq 0$. The simultaneous equations (3.39) are solved for α_i and the group of the solutions is unique. Hence, [F_2] has a unique solution (3.35). Here the root α_i is

$$\begin{aligned} \alpha_i &= -\frac{1}{\lambda}\left(\frac{\partial\Delta}{\partial c_{1i}}\beta_1 + \frac{\partial\Delta}{\partial c_{2i}}\beta_2 + \cdots + \frac{\partial\Delta}{\partial c_{ni}}\beta_n\right)\Big/\Delta(\lambda) \\ &= -\frac{1}{\lambda\Delta(\lambda)}\frac{\partial\Delta}{\partial c_{ii}}\beta_i - \frac{1}{\lambda\Delta(\lambda)}\sum{}'\frac{\partial\Delta}{\partial c_{ij}}\beta_j, \end{aligned} \tag{3.41}$$

where $\sum'$ denotes the sum over j except $j = i$.

(2) *The case* $\Delta(\lambda) = 0$. Here $\Delta(\lambda)$ has generally n roots since $\Delta(\lambda)$ is of the nth degree with regard to λ. If for $\lambda = \lambda_0$ the rank of $\Delta(\lambda)$ is $n - \nu$, that is, if ν is the largest degree of a non-zero minor of the determinant, then the general solution of the homogeneous equation:

$$\alpha_i - \lambda \textstyle\sum_j \alpha_j c_{ij} = 0, \tag{3.39a}$$

is expressed as

$$\alpha_i = \gamma_1 a_{1i} + \gamma_2 a_{2i} + \cdots + \gamma_\nu a_{\nu i} \qquad (i = 1, 2, \dots, n),$$

where $\gamma_1, \gamma_2, \dots, \gamma_\nu$ are arbitrary constants. Putting these values in (3.35) and setting

$$\varphi_r(t) = a_{r1} + a_{r2}t + \cdots + a_{rn}t^n \qquad (r = 1, 2, \dots, \nu),$$

we get

$$u(t) = \gamma_1\varphi_1 + \gamma_2\varphi_2 + \cdots + \gamma_\nu\varphi_\nu$$

as the solution of the homogeneous integral equation for $\lambda = \lambda_0$. Here ν denotes the number of eigenfunctions

$$\varphi_1(t), \varphi_2(t), \dots, \varphi_\nu(t)$$

which are linearly independent, and the solution of [F_0] is expressed by a linear combination of these.

3.3.1. $K(t, \xi) = K(t^m\xi^n)$, $K(t, \xi) = K(t\xi)$, $K(t, \xi) = K(t/\xi)$

For example, in

$$x(t) - \lambda \int_a^b K(t\xi)x(\xi)d\xi = f(t), \tag{3.42}$$

where

$$K(t\xi) = K_0 + K_1 t\xi + K_2 t^2\xi^2 + \dots + K_n t^n \xi^n,$$

if we put

$$\int_a^b K_i \xi^{i+j} d\xi = \frac{(b-a)^{i+j+1}}{i+j+1} K_i \equiv c_{ij}, \quad (i, j - 1, 2, 3, \dots, n) \tag{3.43}$$

this is the same as we had before.

We can make a similar calculation in the case of $K(t/\xi)$ when $t(a, b)$ does not contain zero.

Example 3.1 [F₂]

$$x(t) - \lambda \int_0^1 t\xi \, x(\xi)d\xi = at. \tag{3.44}$$

Put $x(t) = At$ and we have

$$At - \frac{\lambda}{3} At = at;$$

then we have $A = 3a/(3 - \lambda)$ if $\lambda \neq 3$. Hence, the solution is

$$x(t) = \frac{3at}{3 - \lambda}$$

(see Chapter 2, Example 1).

Example 3.2

$$6\int_0^1 (1 + \xi + t\xi)x(\xi)d\xi = -t + 1. \tag{3.45}$$

Put $x(t) = \alpha_0 + \alpha_1 t$ and we have

$$(9\alpha_0 + 5\alpha_1) + (3\alpha_0 + 2\alpha_1)t = 1 - t;$$

then we get the simultaneous equations

$$9\alpha_0 + 5\alpha_1 = 1, \qquad 3\alpha_0 + 2\alpha_1 = -1.$$

From these come $\alpha_0 = 7/3$, $\alpha_1 = -4$. Hence the solution is

$$x(t) = 7/3 - 4t.$$

But, in addition, $5/3 - 4t^2$, etc., also serve as solutions. This means that solutions are not unique.

Example 3.3 [F_0]

$$x(t) = \lambda \int_0^1 (1 + t\xi)x(\xi)d\xi. \tag{3.46}$$

Put $x(t) = \alpha_0 + \alpha_1 t$, and we get the simultaneous equations

$$(1 - \lambda)\alpha_0 - \frac{\lambda}{2}\alpha_1 = 0, \qquad \frac{\lambda}{2}\alpha_0 - \left(1 - \frac{\lambda}{3}\right)\alpha_1 = 0.$$

Hence the eigenvalues are $8 \pm 2\sqrt{13}$.

Example 3.4 [F_2]

$$x(t) - 6\int_0^1 t\xi^2 x(\xi)d\xi = 2e^t - t + 1. \tag{3.47}$$

Since the regular part is the term of t of the first order, we have only to assume the solution as

$$x(t) = 2e^t + kt + 1$$

from the first and determine k. The equation to determine k is

$$k - 6\int_0^1 \xi^2(2e\xi + k\xi + 1)d\xi = -1,$$

from which we get $k = 22$; therefore, the solution is

$$x(t) = 2e^t + 22t + 1.$$

3.4. Special cases

Next we have some simpler special methods of solution when a kernel takes some special form. In this section we study a few of the comparatively interesting cases:

$$K(t, \xi) = \sum_{i=1}^{n} a_i(t)b_i(\xi), \tag{3.48}$$

$$K(t, \xi) = \sum_{i=0}^{\infty} a_{i+1}(t)\frac{(t - \xi)^i}{i!}. \tag{3.49}$$

We have already treated

$$\left.\begin{aligned} K(t, \xi) &= K_1(t)K_2(\xi), \\ K(t, \xi) &= K(mt + n\xi),\ K(t + \xi),\ K(t - \xi), \\ K(t, \xi) &= K(t^m\xi^n),\ K(t\xi),\ K(t/\xi). \end{aligned}\right\} \qquad (3.50)$$

Now, in the cases we are going to treat in the next section, the kernel is also of the form $K(t - \xi)$.

3.4.1. $K(t, \xi) = a_1(t)b_1(\xi) + a_2(t)b_2(\xi) + \cdots + a_n(t)b_n(\xi) = \sum a_i(t)b_i(\xi)$

In [F_2] where the kernel is

$$K(t, \xi) = \sum_{i=1}^{n} a_i(t)b_i(\xi), \qquad (3.51)$$

we express [F_2] as

$$x = f + \lambda K \circ x,$$

and set the right-hand side as

$$b_j \circ x \equiv \alpha_j \qquad (j = 1, 2, \ldots, n); \qquad (3.52)$$

by carrying out the regular part, then we can express it as

$$x = f + \lambda \textstyle\sum_j \alpha_j a_j. \qquad (3.53)$$

Take the convolution of b_i with both sides of (3.53). Here if

$$b_i \circ f \equiv \beta_i, \quad a_i \circ b_j \equiv c_{ij}, \qquad (3.54)$$

we obtain

$$\alpha_i = \beta_i + \lambda \textstyle\sum_j c_{ji}\alpha_j. \qquad (3.55)$$

This is the same as we had in (3.39), and the succeeding computation proceeds similarly. Yet, in this case, we can compute the kernel of the solution since the equation is set as (3.53). For example, when $\Delta(\lambda) = 0$ we have

$$\alpha_i = -\frac{1}{\lambda\Delta(\lambda)} \sum_j \frac{\partial\Delta}{\partial c_{ji}}\beta_j, \qquad (3.56)$$

then the solution follows:

$$x(t) = f(t) - \frac{1}{\Delta(\lambda)}\left\{a_1(t)\sum_j \frac{\partial\Delta}{\partial c_{j1}}\beta_j + a_2(t)\sum_j \frac{\partial\Delta}{\partial c_{j2}}\beta_j + \cdots \right.$$
$$\left. + a_n(t)\sum \frac{\partial\Delta}{\partial c_{jn}}\beta_j\right\}$$

$$= f(t) - \frac{1}{\Delta(\lambda)}\left\{\Sigma_i a_i(t)\frac{\partial\Delta}{\partial c_{1i}}\beta_1 + \Sigma_i a_i(t)\frac{\partial\Delta}{\partial c_{2i}}\beta_2 + \cdots\right.$$

$$\left. + \Sigma_i a_i(t)\frac{\partial\Delta}{\partial c_{ni}}\beta_n\right\}$$

$$= f(t) - \frac{1}{\Delta(\lambda)}\Sigma_j\left\{\Sigma_i a_i(t)\frac{\partial\Delta}{\partial c_{ji}}\beta_j\right\}$$

$$= f(t) - \frac{1}{\Delta(\lambda)}\Sigma_j\Sigma_i a_i(t)\frac{\partial\Delta}{\partial c_{ji}}\int_a^b b_j(\xi)f(\xi)d\xi. \tag{3.57}$$

Hence, the kernel of the solution can be expressed as

$$\Gamma(t, \xi, \lambda) = \frac{1}{\lambda\Delta(\lambda)}\Sigma_i\Sigma_j\frac{\partial\Delta}{\partial c_{ji}}a_i(t)b_j(\xi). \tag{3.58}$$

Further, the kernel of the simultaneous equations $[\tilde{\mathrm{F}}_2]$ is

$$K(\xi, t) = \sum_{i=1}^{n} a_i(\xi)b_i(t).$$

If we then substitute

$$a_j \circ x \equiv \alpha_j, \quad b_i \circ f \equiv \beta_i, \quad a_j \circ b_i \equiv c_{ji},$$

the following simultaneous equations hold:

$$\alpha_i = \beta_i + \lambda\,\Sigma_j c_{ij}\alpha_j,$$

and since $c_{ij} = c_{ji}$ we get (3.55). Hence the determinant agrees with the former $\Delta(\lambda)$ and we have

$$\tilde{\Gamma}(\xi, t;\, \lambda) = \frac{1}{\lambda\Delta(\lambda)}\Sigma_i\Sigma_j\frac{\partial\Delta}{\partial c_{ij}}a_i(\xi)b_j(t).$$

If $\Delta(\lambda) = 0$ and the rank of $\Delta(\lambda)$ is $n - \nu$, there exist ν eigenfunctions. Now if the eigenfunctions of $[\tilde{\mathrm{F}}_0]$ are

$$\tilde{\varphi}_1(t),\ \tilde{\varphi}_2(t),\ \ldots,\ \tilde{\varphi}_\nu(t),$$

we need

$$f \circ \tilde{\varphi}_i = 0 \qquad (i = 1, 2, \ldots, \nu)$$

for the solution of $[\mathrm{F}_2]$.

3.4.2. $K(t, \xi) = a_1(t) + a_2(t)(t-\xi) + a_3(t)(t-\xi)^2/2! + a_4(t)(t-\xi)^3/3! + \ldots$

In $[\mathrm{V}_2]$ if the kernel is denoted by

$$K = a_1 + a_2\overset{*}{1}{}^2 + a_3\overset{*}{1}{}^3 + a_4\overset{*}{1}{}^4 + \ldots,$$

using the notation of § 1.4, if we set the kernel of the solution as

$$G = b_1 + b_2\overset{*}{1}{}^2 + b_3\overset{*}{1}{}^3 + a_4\overset{*}{1}{}^4 + \dots,$$

we can deduce a reciprocal theorem

$$K + G = K * G = G * K.$$

By integration by parts we get

$$\begin{aligned}
\int_\xi^t a_1(\tau)d\tau &= \Big[(\tau-\xi)a_1(\tau)\Big]_\xi^t - \int_\xi^t (\tau-\xi)a^{(1)}(\tau)d\tau \\
&= (t-\xi)a_1(t) - \left[\frac{(\tau-\xi)^2}{2!}a^{(1)}(\tau)\right]_\xi^t \\
&\quad + \int_\xi^t \frac{(\tau-\xi)^2}{2!}a^{(2)}(\tau)d\tau \\
&\quad - \dots \\
&= a_1\overset{*}{1}{}^2 - a_1{}^{(1)}\overset{*}{1}{}^3 + a_1{}^{(2)}\overset{*}{1}{}^4 - \cdots.
\end{aligned}$$

The reciprocal theorem becomes

$$\begin{aligned}
&(a_1+b_1) + (a_2+b_2)\overset{*}{1}{}^2 + (a_3+b_3)\overset{*}{1}{}^3 + (a_4+b_4)\overset{*}{1}{}^4 \\
&\quad + (a_5+b_5)\overset{*}{1}{}^5 + \cdots \\
&\quad = (b_1 + b_2\overset{*}{1}{}^2 + b_3\overset{*}{1}{}^3 + \cdots) * (a_1 + a_2\overset{*}{1}{}^2 + a_3\overset{*}{1}{}^3 + \cdots),
\end{aligned}$$

and by the above integral formula the right-hand side becomes

$$\begin{aligned}
a_1b_1\overset{*}{1}{}^2 - a_1{}^{(1)}b_1\overset{*}{1}{}^3 + a_1{}^{(2)}b_1\overset{*}{1}{}^4 - a_1{}^{(3)}b_1\overset{*}{1}{}^5 &+ \cdots \\
+ a_2b_1\overset{*}{1}{}^3 - a_2{}^{(1)}b_1\overset{*}{1}{}^4 + a_2{}^{(2)}b_1\overset{*}{1}{}^5 &- \cdots \\
+ a_1b_2\overset{*}{1}{}^3 - a_1{}^{(1)}b_2\overset{*}{1}{}^4 + a_1{}^{(2)}b_2\overset{*}{1}{}^5 &- \cdots \\
+ a_2b_2\overset{*}{1}{}^4 - 2a_2{}^{(1)}b_2\overset{*}{1}{}^5 &+ \cdots \\
+ a_3b_1\overset{*}{1}{}^4 - a_3{}^{(1)}b_1\overset{*}{1}{}^5 &+ \cdots \\
+ a_1b_3\overset{*}{1}{}^4 - a_1{}^{(1)}b_3\overset{*}{1}{}^5 &+ \cdots \\
+ a_1b_4\overset{*}{1}{}^5 &- \cdots \\
+ a_4b_1\overset{*}{1}{}^5 &- \cdots \\
+ a_2b_3\overset{*}{1}{}^5 &- \cdots \\
+ a_3b_2\overset{*}{1}{}^5 &- \cdots \\
&+ \cdots;
\end{aligned}$$

then by comparing the coefficients of $\overset{*}{1}$ we get

$$a_1 + b_1 = 0,$$
$$a_2 - a_1 b_1 + b_2 = 0,$$
$$a_3 - a_2 b_1 - a_1 b_2 + b_3 + a_1{}^{(1)} b_1 = 0,$$
$$a_4 - a_3 b_1 - a_2 b_2 - a_1 b_3 + b_4 + a_1{}^{(2)} b_2 + a_2{}^{(1)} b_1 = 0,$$
$$a_5 - a_4 b_1 - a_3 b_2 - a_2 b_3 - a_1 b_4 + b_5 + a_1{}^{(3)} b_1 - a_2{}^{(2)} b_1$$
$$- a_1{}^{(2)} b_2 + 2a_2{}^{(1)} b_2 + a_3{}^{(1)} b_1 + a_1{}^{(1)} b_3 = 0,$$
$$\cdots\cdot$$

For example, with regard to

$$K = t^2(t - \xi) + t(t - \xi)^2,$$

by computing a few terms of G, we get

$$G = -t^2(t - \xi) - t(t - \xi)^2 + t^2(t^2 - 4)(t - \xi)^3/3! + \cdots.$$

3.5. Methods of solution by operational calculus

When a function $f(t)$ is single-valued in $t > 0$ and there exists an integral

$$\int_0^\infty e^{-st} f(t)dt,$$

we consider this to be a function of s and express it as

$$\int_0^\infty e^{-st} f(t)dt = F(s), \tag{3.59}$$

where $F(s)$ is called the Laplace image function (or Laplace transform) of $f(t)$, and $f(t)$ is the Laplace original function of $F(s)$.

When $f(t)$ is defined in $0 < t < \infty$ and is piecewise continuous,† and we have

$$\int_0^\infty e^{-ct} |f(t)| dt < \infty$$

for a positive value c, then the set of functions $\{f(t)\}$ is called the Laplace original space. We denote this by $\mathscr{L}_D$. Then when $F(s)$ is an

† Piecewise continuous means that the number of discontinuous points in any finite interval in $(0, \infty)$ is finite, and the limits of the functions at those points both from the right and the left are finite.

analytic function of a complex number s and it has order s^{-k} in $R(s) \geqq c$ for the constants $c > 0$, $k > 1$, we call the set of this sort of analytic functions $\{F(s)\}$ the Laplace image space and denote it by $\mathscr{L}_R$. Here $R(s)$ is the real part of s.

The following lemma describe the correspondence between the original function and the image function.

LEMMA 3.2

When $f(t)$ belongs to $\mathscr{L}_D$,

$$f(t) = \frac{1}{2\pi i}\int_{\gamma-i\infty}^{\gamma+i\infty} e^{st}F(s)ds, \ \gamma > c. \tag{3.60}$$

This kind of integration in the complex domain is called a Bromwith integral with sign $\int_{Br}$.

LEMMA 3.3

When $F(s)$ belongs to $\mathscr{L}_R$, the integration (3.60) in the complex domain is convergent, and if its value is $f(t)$, (3.59) *holds; in addition, $f(t)$ is continuous in $t \geq 0$, and $f(0) = 0$ and has the order $e^{\gamma t}$ in $t \geqq 0$.*

Therefore, in a theorem like this, the image function of all functions of $\mathscr{L}_D$ is included in $\mathscr{L}_R$, and although the original function of all functions of $\mathscr{L}_R$ is not included in $\mathscr{L}_D$ inversely, yet the original function and the image function are combined to each other by (3.59) *and* (3.60) *at least when $f(t) \in \mathscr{L}_D$ or $F(s) \in \mathscr{L}_R$. We express such a case as*

$$f(t) \supset F(s). \tag{3.61}$$

From this point on, image functions of original functions printed in small letters are printed in capital letters and image functions of original functions printed in capital letters are printed in bold type; the variable of the original space is denoted by t and the variable of the image space by s. For example, we denote the image function of $f(t)$, $F(t)$ by $F(s)$, $\mathbf{F}(s)$, respectively. If the original function is restricted to the continuous function belonging to $\mathscr{L}_D$, the original function and the image function are in one-to-one correspondence.

Then we have the characteristics that will be stated in the following. Using these characteristics we can convert a functional equation in the original space to an equation adaptable for a comparatively easier treatment in the image space, and we may take the original function of the solution as the solution of a given functional equation. Methods of this sort belong to the operational calculus. We can solve an integral equation by the operational calculus only when the regular part is of convolution type, and we call this sort of equation *convolution type* or *faltungs type* of Volterra type. Where the kernel $K(t)$ and $f(t)$ belong to $\mathscr{L}_D$, we look for the solution that is continuous and belongs to $\mathscr{L}_D$.

(1) *Convolution type—Integral equations of the first kind.*

$$\int_0^t K(t-\xi)x(\xi)d\xi = f(t). \tag{A}$$

This is also called *Abel type*. If

$$K(t) \supset \boldsymbol{K}(s), \quad x(t) \supset X(s), \quad f(t) \supset F(s)$$

in the image space, we have

$$\boldsymbol{K}(s)X(s) = F(s); \tag{3.62}$$

then

$$X(s) = F(s)/\boldsymbol{K}(s),$$

and, with regard to $G(s)$, if

$$s\boldsymbol{G}(s) = 1/\boldsymbol{K}(s), \tag{3.63}$$

we get

$$X(s) = s\boldsymbol{G}(s)F(s). \tag{3.64}$$

Now since

$$s\boldsymbol{G}(s) - G(+0) \subset G^{(1)}(t)$$

or

$$sF(s) - f(+0) \subset f^{(1)}(t),$$

then transform (3.64) to the original space, expressing it as

$$X(s) = \{s\boldsymbol{G}(s) - G(+0)\}F(s) + G(+0)F(s);$$

then we get

$$\begin{aligned} x(t) &= G(+0)f(t) + \int_0^t G^{(1)}(t-\xi)f(\xi)d\xi \\ &= G(+0)f(t) + \int_0^t G^{(1)}(\xi)f(t,\xi)d\xi; \end{aligned} \tag{3.65}$$

similarly we have

$$x(t) = G(t)f(+0) + \int_0^t G(t-\xi)f^{(1)}(\xi)d\xi$$
$$= G(t)f(+0) + \int_0^t G(\xi)f^{(1)}(t-\xi)d\xi. \qquad (3.66)$$

We call these Duhamel's formulae.

Of course $\boldsymbol{G}(s) \subset G(t)$ and $\boldsymbol{K}(s)\boldsymbol{G}(s) = 1/s$, therefore we have

$$\int_0^t K(t-\xi)G(\xi)d\xi = \int_0^t K(\xi)G(t-\xi) = 1, \qquad (3.67)$$

where $G(t)$ is the kernel of the solution.

Example 3.5 [V_1]

$$\int_0^t (1-t+\xi)x(\xi)d\xi = t \qquad [V_1] \qquad (3.68)$$

is of convolution type, and since

$$K(t) = 1-t, \qquad f(t) = t,$$

we have

$$\boldsymbol{K}(s) = \frac{1}{s} - \frac{1}{s^2}, \qquad F(s) = \frac{1}{s^2}$$

and (3.68) becomes

$$\left(\frac{1}{s} - \frac{1}{s^2}\right)X(s) = \frac{1}{s^2}$$

in the image space, therefore,

$$X(s) = \frac{1}{s-1},$$

and since $s = 1$ is a single pole, we get the solution

$$x(t) = e^t \qquad (3.69)$$

directly by integration (3.60) in the complex domain.

(2) *Convolution type—Integral equations of the second kind.*
Now consider

$$x(t) - \int_0^t K(t-\xi)x(\xi)d\xi = f(t). \qquad [P]$$

This is an often applied equation, which is also called *Poisson type*. In the image space it takes the form

$$X(s) - \boldsymbol{K}(s)X(s) = F(s); \tag{3.70}$$

we solve this in $X(s)$ and get

$$X(s) = \frac{F(s)}{1 - \boldsymbol{K}(s)}.$$

Now set

$$\boldsymbol{G}(s) = -\frac{\boldsymbol{K}(s)}{1 - \boldsymbol{K}(s)} \tag{3.71}$$

and it follows that

$$X(s) = F(s) - \boldsymbol{G}(s)F(s); \tag{3.72}$$

hence, in the original space,

$$x(t) = f(t) - \int_0^t G(t - \xi)f(\xi)d\xi. \tag{3.73}$$

Further, we get

$$\boldsymbol{K}(s) + \boldsymbol{G}(s) = \boldsymbol{K}(s)\boldsymbol{G}(s) = \boldsymbol{G}(s)\boldsymbol{K}(s)$$

from (3.71), from which we derive the reciprocal theorem

$$K(t) + G(t) = \int_0^t K(t - \xi)G(\xi)d\xi = \int_0^t G(t - \xi)K(\xi)d\xi. \tag{3.74}$$

Therefore, we can get the solution by finding the reciprocal function $G(t)$, for which we have only to carry out the integration in the complex domain:

$$\frac{1}{2\pi i}\int_{Br} e^{st}\boldsymbol{G}(s)ds.$$

For example, if $\boldsymbol{G}(s)$ is a rational function and is expressible as the ratio of integral functions $g_1(s)$ and $g_2(s)$ as

$$\boldsymbol{G}(s) = g_1(s)/g_2(s), \tag{3.75}$$

then $G(t)$ becomes

$$G(t) = \sum_i \frac{g_1(\lambda_i)}{g_2{}'(\lambda_i)}\exp(\lambda_i t) \tag{3.76}$$

by the residue theorem, provided λ_i is a simple root of the characteristic equation

$$g_2(\lambda) = 0, \tag{3.77}$$

and $g_2'(\lambda_i)$ represents $[dg_2(s)/ds]_{s=\lambda_i}$. When the characteristic equation (3.77) has a multiple root, for example, suppose one of the k-tupl roots is κ, the equation becomes

$$G(t) = \sum_i \frac{g_1(\lambda_i)}{g_2'(\lambda_i)} \exp(\lambda_i t) + \sum' \frac{1}{(\kappa - 1)!} \left\{ \frac{d^{k-1}}{ds^{k-1}} \left[(s - t)^k \frac{g_1(s)}{g_2(s)} \exp(st) \right] \right\}_{s=\kappa}, \tag{3.78}$$

where the first sum $\sum$ shows the sum over the residues of the simple roots and the second, $\sum'$, those of the multiple roots.

Example 3.6 [V_2]

$$x(t) - \int_0^t [3 + 6(t - \xi) - 4(t - \xi)^2] x(\xi) d\xi = 1 - 2t - 4t^2 \tag{3.79}$$

is of convolution type and

$$K(t) = 3 + 6t - 4t^2, \qquad f(t) = 1 - 2t - 4t^2.$$

Therefore, we have

$$\boldsymbol{K}(s) = \frac{3}{s} + \frac{6}{s^2} - \frac{8}{s^3}, \qquad F(s) = \frac{1}{s} - \frac{2}{s^2} - \frac{8}{s^3},$$

which, in the image space, becomes

$$X(s) - \left(\frac{3}{s} + \frac{6}{s^2} - \frac{8}{s^3}\right) X(s) = \frac{1}{s} - \frac{2}{s^2} - \frac{8}{s^3}.$$

Then

$$X(s) = \frac{1}{s - 1},$$

hence we have

$$x(t) = e^t. \tag{3.80}$$

Furthermore, when we look for the kernel of the solution we get

$$\boldsymbol{G}(s) = \frac{-(3s^2 + 6s - 8)}{s^3 - 3s^2 - 6s + 8}$$

from (3.71). Therefore, all the roots of the characteristic equation

$$\lambda^3 - 3\lambda^2 - 6\lambda + 8 = 0$$

are simple roots and

$$\lambda = -2, \quad \lambda = 1, \quad \lambda = 4;$$

then we have

$$G(t) = \frac{1}{9}(4e^{-2t} + e^t + 32e^{4t}) \tag{3.81}$$

by (3.76). Therefore, we may express $x(t)$ as

$$x(t) = 1 - 2t - 4t^2 - \frac{1}{9}\int_0^t [4e^{-2(t-\xi)} + e^{t-\xi} + 32e^{4(t-\xi)}]$$
$$(1 - 2\xi - 4\xi^2)d\xi,$$

using the kernel of the solution.

Computation of this gives us

$$x(t) = e^t.$$

Furthermore, we can prove the reciprocal theorem.

3.5.1. $K(t) = k_1e^{a_1t} + k_2e^{a_2t} + \cdots + k_ne^{a_nt}$.

In [P], when the kernel takes the form

$$K(t) = \sum_{i=1}^{n} k_ie^{a_it}, \tag{3.82}$$

we have

$$\boldsymbol{K}(s) = \frac{k_1}{s - a_1} + \frac{k_2}{s - a_2} + \cdots + \frac{k_n}{s - a_n};$$

then from (3.71) we obtain

$$\boldsymbol{G}(s) = -\frac{\boldsymbol{K}(s)}{1 - \boldsymbol{K}(s)}.$$

Cancel the denominator of $\boldsymbol{K}(s)$, and we can determine the equation in the form of $\boldsymbol{G} = g_1/g_2$ (3.75). Now let

$$\lambda_1, \lambda_2, \ldots, \lambda_n$$

be n simple roots of the equation

$$\frac{k_1}{\lambda - a_1} + \frac{k_2}{\lambda - a_2} + \cdots + \frac{k_n}{\lambda - a_n} - 1 = 0; \tag{3.83}$$

then

$$\frac{k_1}{\lambda_i - a_1} + \frac{k_2}{\lambda_i - a_2} + \cdots + \frac{k_n}{\lambda_i - a_n} = 1;$$

therefore, it is followed by

$$g_1(\lambda_i) = (\lambda_i - a_1)(\lambda_i - a_2) \ldots (\lambda_i - a_n).$$

Next, computing $g_2'(\lambda_i)$ we find that

$$g_2'(\lambda_i) = (\lambda_i - \lambda_1)(\lambda_i - \lambda_2) \cdots (\lambda_i - \lambda_{i-1}) \times (\lambda_i - \lambda_{i+1}) \cdots (\lambda_i - \lambda_n);$$

then we can obtain

$$\begin{aligned} G(t) &= \sum \frac{g_1(\lambda_i)}{g_2'(\lambda_i)} \exp(\lambda_i t) \\ &= \sum \frac{(\lambda_i - a_1)(\lambda_i - a_2) \cdots (\lambda_i - a_n)}{(\lambda_i - \lambda_1) \cdots (\lambda_i - \lambda_n)} \exp(\lambda_i t) \end{aligned} \tag{3.84}$$

from (3.76) (see Question 15).

3.5.2. $\boldsymbol{K(t) = k_0 + k_1 t + \frac{k_2}{2!} t_2 + \cdots + \frac{k_{n-1}}{(n-1)!} t^{n-1}}$.

In [P], where the kernel is given by the following polynomial expression of the *n*th order:

$$K(t) = \sum_{i=0}^{n-1} \frac{k_i}{i!} t^i, \tag{3.85}$$

suppose that the simple roots of the equation

$$\lambda^n + k_0 \lambda^{n-1} + \cdots + k_{n-1} = 0 \tag{3.86}$$

are $\lambda_1, \lambda_2, \ldots, \lambda_n$; then it can be proved that the kernel of the solution is given by

$$G(t) = \sum \frac{\lambda_i^n}{(\lambda_i - \lambda_1) \ldots (\lambda_i - \lambda_n)} \exp(\lambda_i t). \tag{3.87}$$

The denominator of this equation is the product of all factors except $\lambda_i - \lambda_i$.

In this case, we have

$$K(s) = \frac{k_0}{s} + \frac{k_1}{s^2} + \cdots + \frac{k_{n-1}}{s^n}.$$

All of the computational procedures that follow are similar to those of 3.5.1.

In this section we have studied the cases when (3.83) and (3.86) have only simple roots, but for the case of multiple roots, the formula (3.78) is applied.

3.6. Integral equation of Fredholm type with Pincherle–Goursat kernel

When the kernel can be expressed as

$$K(t, \xi) = \sum_{i=1}^{n} a_i(t) b_i(\xi) \tag{3.88}$$

we call this a Pincherle–Goursat kernel.

If we put

$$\int_a^b b_i(\xi) x(\xi) d\xi = \alpha_i \tag{3.84}$$

[F_2] is reduced to

$$x(t) - \lambda \sum_{i=1}^{n} \alpha_i a_i(t) = f(t). \tag{3.85}$$

This indicates that $x(t) - f(t)$ can be expressed as a linear combination of $a_i(t)$.

Integrating (3.85) from a to b, after multiplying by $b_j(t)$ on both sides, we have

$$\alpha_j - \lambda \sum_{i=1}^{n} \alpha_i \int_a^b a_i(t) b_j(t) dt = \int_a^b f(t) b_j(t) dt \qquad (i, j = 1, 2, \ldots, n). \tag{3.86}$$

Then, we put

$$\int_a^b a_i(t) b_j(t) dt = k_{ji}, \qquad \int_a^b f(t) b_j(t) dt = c_j \tag{3.87}$$

We obtain simultaneous equations for $\alpha_1, \alpha_2, \ldots, \alpha_n$:

$$\left.\begin{array}{l} (1 - \lambda k_{11})\alpha_1 - \lambda k_{12}\alpha_2 - \cdots - \lambda k_{1n}\alpha_n = c_1, \\ -\lambda k_{21}\alpha_1 + (1 - \lambda k_{22})\alpha_2 - \cdots - \lambda k_{2n}\alpha_n = c_2, \\ \ldots, \\ -\lambda k_{n1}\alpha_1 - \lambda k_{n2}\alpha_2 - \cdots + (1 - \lambda k_{nn})\alpha_n = c_n. \end{array}\right\} \tag{3.88}$$

Then, the solution yields

$$x(t) = f(t) + \lambda \sum_i \alpha_i a_i(t).$$ (3.89)

The determinant of the simultaneous equations is

$$\Delta(\lambda) \equiv \begin{vmatrix} 1 - \lambda k_{11} & - \lambda k_{12} \ldots & - \lambda k_{1n} \\ - \lambda k_{21} & 1 - \lambda k_{22} \ldots & - \lambda k_{2n} \\ \cdots & \cdots & \cdots & \cdots \\ - \lambda k_{n1} & - \lambda k_{n2} \ldots & 1 - \lambda k_{nn} \end{vmatrix}$$ (3.90)

which is a polynomial in λ of nth degree.

(1) $\Delta(\lambda) \neq 0$

If we write $\Delta(\lambda, c)$ when the elements of the determinant $\Delta(\lambda)$ are replaced by $c_1, c_2, \ldots, c_n$, the solutions of (3.87) are

$$\alpha_i = \frac{\Delta_i(\lambda, c)}{\Delta(\lambda)},$$

or

$$\alpha_i = \frac{1}{\Delta(\lambda)} \sum_j \Delta_{ji} c_j,$$

where Δ_{ji} is the cofactor of the jith element of the determinant $\Delta(\lambda)$. Finally, we have

$$x(t) = f(t) + \frac{\lambda}{\Delta(\lambda)} \sum_i \left\{ \left(\sum_j \Delta_{ji} \int_a^b (f(\xi) b_j(\xi) d\xi \right) a_i(t) \right\}$$

$$= f(t) + \frac{\lambda}{\Delta(\lambda)} \int_a^b \{ \sum_i [\Delta_{1i} b_1(\xi) + \Delta_{2i} b_2(\xi) + \cdots$$

$$+ \Delta_{ni} b_n(\xi)] a_i(t) \} f(\xi) d\xi.$$ (3.91)

However, if we put

$$\Delta\left(\begin{matrix} t \\ \xi \end{matrix}; \lambda\right) \equiv \begin{vmatrix} 0 & - a_1(t) & - a_2(t) \ldots & - a_n(t) \\ b_1(\xi) & 1 - \lambda k_{11} & - \lambda k_{12} \ldots & - \lambda k_{1n} \\ b_2(\xi) & - \lambda k_{21} & 1 - \lambda k_{22} \ldots & - \lambda k_{2n} \\ \cdots & \cdots & \cdots & \cdots & \cdots \\ b_n(\xi) & - \lambda k_{n1} & - \lambda k_{n2} \ldots & 1 - \lambda k_{nn} \end{vmatrix}$$ (3.92)

then the solution becomes

$$x(t) = f(t) + \frac{\lambda}{\Delta(\lambda)} \int_a^b \Delta\binom{t}{\xi}; \lambda\Big) f(\xi) d\xi. \tag{3.93}$$

Consequently if we put

$$\Gamma(t, \xi; \lambda) = -\Delta\Big(\begin{matrix} t \\ \xi \end{matrix}; \lambda\Big) \Big/ \Delta(\lambda), \tag{3.94}$$

then we have

$$x(t) = f(t) - \lambda \int_a^b \Gamma(t, \xi; \lambda) f(\xi) d\xi, \tag{3.95}$$

where the kernel Γ of the solution is a quotient of nth degree polynomials.

If $\Delta(\lambda) \neq 0$, the solution of $[F_0]$ is

$$x(t) \equiv 0. \tag{3.96}$$

(2) $\Delta(\lambda) = 0$.

In this case, there is no solution for the simultaneous equations (3.87). However, when $\Delta(\lambda_0) = 0$, λ_0 is a characteristic root of (3.87). If the rank of the determinant (3.88) is r $(1 \leqq r \leqq n-1)$, then there are ν fundamental solutions; linearly independent, they are

$$\begin{aligned}
&\alpha_1{}^{(1)}, \alpha_2{}^{(1)}, \ldots, \alpha_n{}^{(1)} \\
&\alpha_1{}^{(2)}, \alpha_2{}^{(2)}, \ldots, \alpha_n{}^{(2)} \\
&\ldots, \\
&\alpha_1{}^{(\nu)}, \alpha_2{}^{(\nu)}, \ldots, \alpha_n{}^{(\nu)},
\end{aligned}$$

where $\nu = n - r$ and superior symbols denote the group number, not the order of differentiation for the moment.

Then there exist an infinite number of solutions of the simultaneous equations (3.88) which are linear combinations of fundamental solutions with ν arbitrary constants:

$$\alpha_i = A_1\alpha_i{}^{(1)} + A_2\alpha_i{}^{(2)} + \cdots + A_\nu\alpha_i{}^{(\nu)}. \tag{3.97}$$

Hence, when λ_0 is not an eigenvalue, there exist $\nu = n - r$ linearly independent fundamental solutions

$$\begin{aligned}
\phi^{(1)}(t) &= \alpha_1{}^{(1)} a_1(t) + \alpha_2{}^{(1)} a_2(t) + \cdots + \alpha_n{}^{(1)} a_n(t), \\
\phi^{(2)}(t) &= \alpha_1{}^{(2)} a_1(t) + \alpha_2{}^{(2)} a_2(t) + \cdots + \alpha_n{}^{(2)} a_n(t), \\
&\ldots, \\
\phi^{(\nu)}(t) &= \alpha_1{}^{(\nu)} a_1(t) + \alpha_2{}^{(\nu)} a_2(t) + \cdots + \alpha_n{}^{(\nu)} a_n(t),
\end{aligned}$$

and the linear combination of fundamental solutions

$$x(t) = A_1\phi^{(1)}(t) + A_2\phi^{(2)}(t) + \cdots + A_\nu\phi^{(\nu)}(t) \qquad (3.98)$$

yields the solution of the homogeneous integral equation [F_0], where $A_1, A_2, \ldots, A_\nu$ are arbitrary constants.

For the associated integral equation of (3.84); that is,

$$x(t) - \lambda\int_a^b \tilde{K}(\xi, t)x(\xi)d\xi = f(t), \qquad [\tilde{F}_2] \qquad (3.99)$$

we have

$$\tilde{K}(\xi, t) = \sum_{i=1}^{n} a_i(\xi)b_i(t). \qquad (3.100)$$

If we put

$$\left.\begin{aligned} &\int_a^b a_i(\xi)x(\xi)d\xi = \tilde{\alpha}_i, \\ &\int_a^b a_j(t)b_i(t)dt = \tilde{k}_{ji}, \quad \int_a^b f(t)a_j(t)dt = \tilde{c}_j, \end{aligned}\right\} \qquad (3.101)$$

then we have the simultaneous equations

$$\left.\begin{aligned} &(1-\lambda\tilde{k}_{11})\alpha_1 - \lambda\tilde{k}_{12}\alpha_2 - \cdots - \lambda\tilde{k}_{1n}\alpha_n = \tilde{c}_1, \\ &-\lambda\tilde{k}_{21}\alpha_1 + (1-\lambda\tilde{k}_{22})\alpha_2 - \cdots - \lambda\tilde{k}_{2n}\alpha_n = \tilde{c}_2, \\ &\ldots, \\ &-\lambda\tilde{k}_{n1}\alpha_1 - \lambda\tilde{k}_{n2}\alpha_2 - \cdots + (1-\lambda\tilde{k}_{nn})\alpha_n = \tilde{c}_n \end{aligned}\right\} \qquad (3.102)$$

for $\alpha_1, \alpha_2, \ldots, \alpha_n$.

If we compare with (3.97) and (3.101), then

$$\tilde{k}_{ji} = k_{ij} \qquad (3.103)$$

holds, hence we have

$$\left.\begin{aligned} &(1-\lambda k_{11})\tilde{\alpha}_1 - \lambda k_{21}\tilde{\alpha}_2 - \cdots - \lambda k_{n1}\tilde{\alpha}_n = \tilde{c}_1, \\ &-\lambda k_{12}\tilde{\alpha}_1 + (1-\lambda k_{22})\tilde{\alpha}_2 \cdots - \lambda k_{n2}\tilde{\alpha}_n = \tilde{c}_2, \\ &\ldots, \\ &-\lambda k_{1n}\tilde{\alpha}_1 - \lambda k_{2n}\tilde{\alpha}_2 - \cdots + (1-\lambda k_{nn})\tilde{\alpha}_n = \tilde{c}_n, \end{aligned}\right\} \qquad (3.104)$$

corresponding to (3.102). The determinant of coefficients is the transposed determinant of (3.89). Consequently eigenvalues of [F_0] coincide with those of [$\tilde{F}_0$]. When the rank of the determinant is r there are ν $(= n - r)$ linearly independent fundamental solutions. They are

$$\tilde{\alpha}_1{}^{(1)},\ \tilde{\alpha}_2{}^{(1)},\ \ldots,\ \tilde{\alpha}_n{}^{(1)},$$
$$\tilde{\alpha}_1{}^{(2)},\ \tilde{\alpha}_2{}^{(2)},\ \ldots,\ \tilde{\alpha}_n{}^{(2)},$$
$$\ldots$$
$$\tilde{\alpha}_1{}^{(\nu)},\ \tilde{\alpha}_2{}^{(\nu)},\ \ldots,\ \tilde{\alpha}_n{}^{(\nu)}.$$

These are the solutions of homogeneous simultaneous equations, obtained by putting 0 for $\tilde{c}_i$ on the right-hand side.

The homogeneous integral equation $[\tilde{\mathrm{F}}_0]$ has r linearly independent solutions. They are

$$\tilde{\varphi}^{(1)}(t) = \tilde{\alpha}_1{}^{(1)} b_1(t) + \tilde{\alpha}_2{}^{(1)} b_2(t) + \cdots + \tilde{\alpha}_n{}^{(1)} b_n(t),$$
$$\tilde{\varphi}^{(2)}(t) = \tilde{\alpha}_1{}^{(2)} b_1(t) + \tilde{\alpha}_2{}^{(2)} b_2(t) + \cdots + \tilde{\alpha}_n{}^{(2)} b_n(t),$$
$$\ldots,$$
$$\tilde{\varphi}^{(\nu)}(t) = \tilde{\alpha}_1{}^{(\nu)} b_1(t) + \tilde{\alpha}_2{}^{(\nu)} b_2(t) + \cdots + \tilde{\alpha}_n{}^{(\nu)} b_n(t).$$

A linear combination of these furnishes a solution of $[\tilde{\mathrm{F}}_0]$; this is

$$\tilde{x}(t) = A_1\tilde{\varphi}^{(1)}(t) + A_2\tilde{\varphi}^{(2)}(t) + \cdots + A_\nu\tilde{\varphi}^{(\nu)}(t). \tag{3.104}$$

If we put the eigenfunction of $[\mathrm{F}_0]$ or $[\tilde{\mathrm{F}}_0]$ as $\phi_1(t)$, $\tilde{\phi}_2(t)$, corresponding to λ_1 and λ_2 respectively, we have

$$\varphi_1(t) = \lambda_1 \int_a^b K(t, \xi)\varphi_1(\xi)d\xi, \qquad \tilde{\varphi}_2(t) = \lambda_2 \int_a^b K(\xi, t)\tilde{\varphi}_2(\xi)d\xi.$$

Then, we obtain

$$\begin{aligned} I = \int_a^b \varphi_1(t)\tilde{\varphi}_2(t)dt &= \lambda_1 \int_a^b \tilde{\varphi}_2(t)dt \int_a^b K(t, \xi)\varphi_1(\xi)d\xi \\ &= \lambda_1 \int_a^b \varphi_1(\xi)d\xi \int_a^b K(t, \xi)\tilde{\varphi}_2(t)dt \\ &= \frac{\lambda_1}{\lambda_2} \int_a^b \varphi_1(\xi)\tilde{\varphi}_2(\xi)d\xi \\ &= \frac{\lambda_1}{\lambda_2} I. \end{aligned} \tag{3.105}$$

Therefore, we must have $I = 0$ in order this equation holds. We assumed that $\lambda_1 \neq \lambda_2$. This means that eigenfunctions for $[\mathrm{F}_0]$ and $[\tilde{\mathrm{F}}_0]$ are orthogonal.

(3) $\Delta(\lambda_0) = 0$

For any eigenfunction of $[\tilde{\mathrm{F}}_0]$ we have

$$\int_a^b f(t)\tilde{\phi}_h(t)dt = 0 \qquad (h = 1, 2, \ldots, \nu). \tag{3.106}$$

This may be proved in the following way. If we assume that $[F_2]$ has a solution, then we have

$$\int_a^b f(t)\tilde{\varphi}_h(t)dt = \int_a^b x(t)\tilde{\varphi}_h(t)dt - \lambda_0 \int_a^b \tilde{\varphi}_h(t)dt \int_a^b K(t, \xi)x(\xi)d\xi$$
$$= \int_a^b x(t)\tilde{\varphi}_h(t)dt - \lambda_0 \int_a^b x(\xi)d\xi \int_a^b K(t, \xi)\tilde{\varphi}_h(t)dt.$$

But since $\tilde{\varphi}_h(t)$ is a solution of an associated homogeneous integral equation $[\tilde{F}_0]$, we have

$$\tilde{\varphi}_h(\xi) = \lambda_0 \int_a^b K(t, \xi)\tilde{\varphi}_h(t)dt;$$

This means that (3.106) is a necessary condition for $[F_2]$ to have a solution when $\lambda = \lambda_0$. Now if the integral equation has a solution, the simultaneous equations (3.88) hold, and ν is the number of linearly independent solutions of (3.88). If we add the simultaneous solution (3.88), after multiplying by $\tilde{\alpha}_1{}^{(h)}$, $\tilde{\alpha}_2{}^{(h)}$, ..., and $\tilde{\alpha}_n{}^{(h)}$ respectively, we obtain

$$\begin{aligned}
&[(1 - \lambda k_{11})\tilde{\alpha}_1{}^{(h)} - \lambda k_{21}\tilde{\alpha}_2{}^{(h)} - \cdots - \lambda k_{n1}\tilde{\alpha}_n{}^{(h)}]\alpha_1 \\
&\quad + [- \lambda k_{12}\tilde{\alpha}_1{}^{(h)} + (1 - \lambda k_{22})\tilde{\alpha}_2{}^{(h)} - \cdots - \lambda k_{n2}\tilde{\alpha}_n{}^{(h)}]\alpha_2 \\
&\quad + \cdots \\
&\quad + [- \lambda k_{1n}\tilde{\alpha}_1{}^{(h)} - \lambda k_{2n}\tilde{\alpha}_2{}^{(h)} - \cdots + (1 - \lambda k_{nn})\tilde{\alpha}_n{}^{(h)}]\alpha_n,
\end{aligned}$$

for the left-hand side. Since $\tilde{\alpha}^{(h)}$ is a solution of the homogeneous equations which we obtain by putting zero on the right-hand sides, it follows that all the coefficients of α will disappear and we just have

$$\sum_{i=1}^{n} c_i\tilde{\alpha}_i{}^{(h)} = 0,$$

for the right-hand side.

From (3.87), we have

$$\sum_{i=1}^{n} c_i\tilde{\alpha}_i{}^{(h)} = \int_a^b f(t)[\sum \tilde{\alpha}_i{}^{(h)} b_i(t)]dt,$$

where the summation in parenthesis is $\tilde{\varphi}^{(h)}(t)$. Hence, we have

$$\int_a^b f(t)\tilde{\varphi}^{(h)}(t)dt = 0. \tag{3.107}$$

This means that (3.106) is a sufficient condition for $[F_2]$ to have a solution. When λ_0 is a eigenvalue, and condition (3.106) holds, then we have

$$x(t) = h(t) + \sum_{h=1}^{\nu} A_h \varphi_h(t) \tag{3.108}$$

as the solution, where $h(t)$ is a particular solution and φ_r is an eigenfunction corresponding to the eigenvalue λ_0.

Example 3.7 $[F_2]$

For the integral equation

$$x(t) - \lambda \int_0^1 [t\xi(t + \xi + 1) - t + \xi - 1]x(\xi)d\xi = -2t^2 + \frac{1}{5}t + \frac{10}{3}$$

the kernel is

$$K(t, \xi) = t^2\xi + t\xi^2 + (t + 1)(\xi - 1).$$

This is the Pincherle–Goursat kernel. If we compare this with (3.88) we have

$$a_1(t) = t^2, \quad a_1(t) = t, \quad a_3(t) = t + 1;$$
$$b_1(\xi) = \xi, \quad b_2(\xi) = \xi^2, \quad b_3(\xi) = \xi - 1.$$

Now, we put

$$\int_0^1 \xi x(\xi)d\xi = \alpha_1, \qquad \int_0^1 \xi^2 x(\xi)d\xi = \alpha_2, \qquad \int_0^1 (\xi - 1)x(\xi)d\xi = \alpha_3.$$

Then by (3.87)

$$k_{11} = \int_0^1 a_1(\xi)b_1(\xi)d\xi = \int_0^1 \xi^3 d\xi = \frac{1}{4}, \; k_{12} = \int_0^1 a_2(\xi)b_1(\xi)d\xi = \frac{1}{3},$$
$$k_{13} = \int_0^1 a_3(\xi)b_1(\xi)d\xi = \frac{5}{6}.$$

Similarly,

$$k_{21} = \frac{1}{5}, \qquad k_{22} = \frac{1}{4}, \qquad k_{23} = \frac{7}{12};$$
$$k_{31} = -\frac{1}{12}, \; k_{32} = -\frac{1}{6}, \; k_{33} = -\frac{2}{3}$$

and

$$c_1 = \int_0^1 f(t)b_1(t)dt = -\int_0^1 \left(2t^2 - \frac{1}{5}t^2 - \frac{10}{3}t\right)dt$$

$$= -\left(\frac{2}{4} - \frac{1}{15} - \frac{5}{3}\right) = \frac{37}{30}.$$

$$c_2 = \frac{137}{180}, \qquad c_3 = \frac{26}{30}.$$

We have simultaneous equations for α_1, α_2 and α_3:

$$\left(1 - \frac{\lambda}{4}\right)\alpha_1 - \frac{\lambda}{3}\alpha_2 - \frac{5}{6}\lambda\alpha_3 = \frac{37}{30},$$

$$-\frac{\lambda}{5}\alpha_1 + \left(1 - \frac{\lambda}{4}\right)\alpha_2 - \frac{7}{12}\lambda\alpha_3 = \frac{137}{180}.$$

$$\frac{\lambda}{12}\alpha_1 + \frac{\lambda}{6}\alpha_2 + \left(1 + \frac{2}{3}\lambda\right)\alpha_3 = \frac{26}{30}.$$

The determinant of coefficients is

$$\Delta(\lambda) = \begin{vmatrix} 1 - \frac{\lambda}{4} & -\frac{\lambda}{3} & -\frac{5}{6}\lambda \\ -\frac{\lambda}{5} & 1 - \frac{\lambda}{4} & -\frac{7}{12}\lambda \\ \frac{\lambda}{12} & \frac{\lambda}{6} & 1 + \frac{2}{3}\lambda \end{vmatrix}$$

The equation $\Delta(\lambda) = 0$ is

$$\frac{1}{2160}\lambda^3 + \frac{41}{240}\lambda^2 - \frac{1}{6}\lambda - 1 = 0.$$

Then we have

$$\lambda^2 - \lambda - 6 = 0$$

dropping the coefficient of λ^3, and the eigenvalues are

$$\lambda = -2, +3$$

By Newton's method, we have

$$\lambda = -1.9848, \quad 2.9901, \quad -364.0251.$$

If we put $\lambda = 4$ which is not coincident with an eigenvalue, then we have

$$-\frac{4}{3}\alpha_2 - \frac{10}{3}\alpha_3 = \frac{37}{30}.$$

$$-\frac{4}{5}\alpha_1 \qquad -\frac{7}{3}\alpha_3 = \frac{137}{180},$$

$$\frac{1}{3}\alpha_1 + \frac{2}{3}\alpha_2 + \frac{11}{3}\alpha_3 = \frac{26}{30}$$

to determine α_1, α_2 and α_3. The results are

$$\alpha_1 = -6.0608, \quad \alpha_2 = -5.3045, \quad \alpha_3 = 1.7518$$

Consequently the solution is

$$x(t) = f(t) + 4[\alpha_1 t^2 + \alpha_2 t + \alpha_3(t+1)] = t^2 + 1.$$

3.7. Solution by the Monte-Carlo method

Random motion of a particle on a line. Consider a particle which moves on a straight line t at random on every second. If the position of the particle is t_n after n seconds and $\mathscr{U}$ is the length of the step at that time, then we have

$$t_n = t_{n-1} + \mathscr{U}_n,$$

where we assume that $\mathscr{U}_1$, $\mathscr{U}_2$, ... are independent and the probability density function is $K(\mathscr{U})$.

Now $x(t)$ is the mean time for a particle at t passing through the interval (a, b). Consider the particle at t in (a, b); if the step is large, it will step out the interval in 1 second, while if the step is small, it will come to the position ξ in (a, b); however, this particle will step out of the interval in $1 + x(\xi)$ seconds. Consequently, we have

$$x(t) = 1 \times \int_b^a K(t-\xi)d\xi + \int_b^\infty \{K(t-\xi)d\xi\}$$
$$+ \int_a^b \{1 + x(\xi)\} K(t-\xi)d\xi.$$

The Monte-Carlo method can be applied to solve the problem. Several particles can be chosen at random from the initial distribution $f(t)$. Then we trace the movement of the particle. Some particles may jump outside of the interval (a, b) and then the computation is stopped, while for the other particles the computation will carry on until the movements are invisible. We can calculate the probability $\varphi(t)$ of a particle jumping out of the interval (a, b).

If the motion of a particle on a line t is similar to that of neutrons in an absorbing wall (a, b), the probability of passing through the protection wall (a, b) can be obtained as

$$\int_a^b \varphi(t)x(t)dt,$$

which yields

$$x(t) = \int_{-\infty}^{\infty} K(t-\xi)d\xi + \int_a^b K(t-\xi)x(\xi)d\xi.$$

The first integral on the right is 1, since $K(t)$ is a probability density function. Finally we have an integral equation

$$x(t) - \int_a^b K(t-\xi)x(\xi)d\xi = 1.$$

If we assume that

$$K(t) = \frac{1}{\sqrt{(2\pi)}}\exp\left(-\frac{t^2}{2}\right), \qquad a = -1, b = 1,$$

we can calculate the solution of the equation by counting the number of steps for a sample particle to jump out of the interval $(-1, 1)$ by means of the normal random numbers.

Exercises

1. Prove that a function system

$$\sin t,\ \sin 2t,\ \sin 3t,\ \ldots,\ 1,\ \cos t,\ \cos 2t,\ \ldots$$

in the interval $(-\pi, \pi)$ forms a complete orthogonal system, then construct the normalized orthogonal function system.

2. Define the distance between two elements $f(t)$ and $g(t)$ in Hilbert space.

3. Connect § 3.1.1 to Euclidian space by the principle of passage from discontinuity to continuity.

4. If the kernel belongs to k in a regular square domain S with its boundary, show that it satisfies Dirichlet's condition and is expansible in a double Fourier series.

5. In $[F_2]$, $I(0, a)$, when the kernel is given by

$$K(t, \xi) = \frac{4}{a^2}\sum_{m=1}^{4}\sum_{n=1}^{4} K_s(m, n)\sin\frac{m\pi}{a}t\cos\frac{n\pi}{a}\xi,$$

suppose that

$$x(t) = \frac{2}{a}\sum_{n=1}^{\infty} X_s(n)\sin\frac{n\pi}{a}t,$$

and establish the Fourier coefficients, provided

$$f(t) = \frac{2}{a}\left(\sin\frac{\pi}{a}t + \sin\frac{2\pi}{a}t\right).$$

6. In $[F_2]$ obtain the solution if

$$K(t, \xi) = a + bt + c\xi + dt^2 + et\xi + f\xi^2,$$
$$f(t) = \beta_0 + \beta_1 t + \beta_2 t^2.$$

7. Obtain the solution of $[F_2]$: $x(t) - \lambda\int_a^b K(t-\xi)x(\xi)d\xi = f(t)$ when the kernel is given by

$$K(t) = \sum_{i=1}^{n} k_i e^{a_i t}.$$

8. Show that § 3.3 (3.40) agrees with the Fredholm's determinant of the given equation by computing the Fredholm's determinant.
9. Compute Fredholm's determinant when the kernel is given by

$$K(t, \xi) = a + bt + c\xi + et^2 + ft\xi + g\xi^2.$$

10. If

$$K(t) = \sum_{0}^{n} b_n \sin nt$$

in $[V_2]$:

$$x(t) - \int_0^t K(t-\xi)x(\xi)d\xi = f(t),$$

find a method of solution.
11. Solve the following integral equations.

(a) $x(t) - \lambda\int_0^1 t\xi\, x(\xi)d\xi = 3t^2 + 2t + 1,$

$$\left[x(t) = 3t^2 + \left(2 + \frac{5}{4}\lambda\right)\left(1 - \frac{\lambda}{3}\right)^{-1} t + 1\right];$$

(b) $x(t) - \lambda\int_0^1 (t-\xi)x(\xi)d\xi = at + b, \qquad [(\Delta(\lambda) = \lambda^3 + 12)];$

(c) $x(t) - \lambda\int_0^{\pi} \sin(t-\xi)x(\xi)d\xi = a_1 \sin t + a_2 \sin 2t,$

$$\left[\Delta(\lambda) = 1 + \frac{\pi^2}{4}\lambda^2\right];$$

(d) $x(t) - \lambda \int_0^{\pi} \sin^2(t - \xi)x(\xi)d\xi = 3\cos t, \qquad [x(t) = 3\cos t];$

(e) $x(t) - \frac{1}{2}\int_0^1 x(\xi)d\xi = e^t - \frac{t}{2} + \frac{1}{2},$

$$\left[x(t) = e^t - \frac{t}{2} + \left(e - \frac{1}{4}\right)\right];$$

(f) $x(t) - \lambda \int_0^1 (t^2\xi^2 + t\xi)x(\xi)d\xi = at + b,$

$$[\Delta(t) = \lambda^2 - 128\lambda + 240];$$

(g) $x(t) - \frac{1}{2}\int_0^1 t\xi\, x(\xi)d\xi = \frac{5}{6}t, \qquad [x(t) = t];$

(h) $x(t) - \frac{1}{3}\int_0^1 (t + \xi)x(\xi)d\xi = \frac{5}{6}t \quad \frac{1}{9}, \qquad [x(t) - t)];$

(i) $x(t) - \frac{1}{2}\int_0^1 x(\xi)d\xi = e^t - \frac{e}{2} + \frac{1}{2}, \qquad [x(t) = e^t];$

(j) $x(t) - \frac{1}{4}\int_0^{\pi/2} t\xi\, x(\xi)d\xi = \sin t - \frac{t}{4}, \qquad [x(t) = \sin t];$

(k) $x(t) - \int_0^{1/2} x(\xi)d\xi = t, \qquad \left[x(t) = t + \frac{1}{4}\right];$

(l) $x(t) - \int_0^t (t\xi + 2t + 3\xi)x(\xi)d\xi = te^t - e^t + te - 3e - t + 9,$

$$[x(t) = e^t(t - 1)];$$

(m) $x(t) - \frac{1}{2}\int_0^1 \xi\, x(\xi)d\xi = \frac{3}{2}e^t - \frac{t}{2}e^t - \frac{1}{2}, \qquad [x(t) = e^t];$

(n) $x(t) - \lambda \int_1^2 \frac{t}{3}x(\xi) = at + b, \quad [x(t) = (a - b)(1 - \lambda/2)^{-1}t + b];$

(o) $x(t) - 2\int_0^1 (t - \xi)x(\xi)d\xi = t^2 + 4t, \quad \left[x(t) = t^2 + \frac{25}{4}t - \frac{7}{3}\right];$

(p) $x(t) = \lambda \int_0^{\pi} \sin(t - \xi)x(\xi)d\xi, \qquad [\text{no solution, } \Delta(\lambda) = \lambda^2 + 1];$

(q) $x(t) + \int_0^{10} t\xi\, x(\xi)d\xi = \sin t + 4t + 1, \quad [\text{Put } x(t) = \sin t + at + 1];$

(r) $x(t) - 2\int_0^{\pi/2} \{\sin(t - \xi) + \sin 2(t - \xi)\}x(\xi)d\xi = \cos t$

[express $x(t)$ in the form of a Fourier cosine series];

(s) $x(t) - \int_0^{\pi} \sin t \, x(\xi)d\xi = \cos t, \qquad [x(t) = \cos t];$

(t) $x(t) + \int_0^1 x(\xi)d\xi = \sec^2 t, \qquad \left[x(t) = \sec^2 t - \frac{1}{2}\tan 1\right];$

(u) $x(t) - \lambda \int_0^1 (t - \xi)^2 x(\xi)d\xi = t, \qquad [\text{put } x(t) = at^2 + bt + c];$

(v) $x(t) - \lambda \int_0^1 (1 + t\xi)x(\xi)d\xi = t^2, \qquad [\text{put } x(t) = t^2 + bt + c];$

(w) $x(t) = \frac{3}{125} \int_0^5 t\xi \, x(\xi)d\xi, \qquad [x(t) = Ct]:$

(x) $x(t) = \frac{1}{e^2 - 1} \int_0^1 2e^t e^{\xi} x(\xi)d\xi,$

[One of the solutions is given in the form of $x(t) = Ce^t$];

(y) $x(t) = \lambda \int_{2a}^{2b} t \, x(\xi)d\xi, \ (a \neq b)$

$$\left[\text{when } \lambda = \frac{1}{2(b^2 - a^2)}, \ x(t) = Ct; \text{ when } \lambda \neq \frac{1}{2(b^2 - a^2)}, \ x(t) = 0\right].$$

12. Solve the following integral equations using the operational calculus.

(a) $x(t) + \int_0^t \sin(t - \xi)x(\xi)d\xi = f(t),$

$$\left[x(t) = f(t) + \int_0^t \sin(t - \xi)f(\xi)d\xi\right];$$

(b) $x(t) + \int_0^t \cos(t - \xi)x(\xi)d\xi = f(t),$

(c) $\int_0^t (1 + t - \xi)x(\xi)d\xi = t, \qquad [x(t) = e^{-t}];$

(d) $\int_0^t e^{2(t-\xi)} x(\xi)d\xi = at + be^{2t}, \ b \neq 0,$

[there is no continuous solution];

(e) $x(t) - \int_0^t \sin(t - \xi)x(\xi)d\xi = a \sin t, \qquad [x(t) = at];$

(f) $x(t) - \lambda \int_0^t \sin k(t - \xi)x(\xi)d\xi = a \sin kt,$

$$\left[x(t) = \frac{ak}{\sqrt{(k^2 - k\lambda)}} \sin t \sqrt{(k^2 - k\lambda)}\right];$$

(g) $x(t) - \int_0^t (t - \xi)x(\xi)d\xi = t, \qquad [x(t) = \sin t];$

(h) $x(t) - \int_0^t (t - \xi)x(\xi)d\xi = 1, \qquad [x(t) = \cos t];$

(i) $x(t) - \int_0^t x(\xi)d\xi = 1, \qquad [x(t) = e^t];$

(j) $x(t) - \int_0^t (6t - 6\xi + 5)x(\xi)d\xi = 29 + 6t, \qquad [x(t) = e^{2t} - e^{3t}];$

(k) $x(t) - \int_0^t (t - \xi)x(\xi)d\xi = \cos t - t - 2,$

$[x(t) = \sin t + t \sin t].$

13. Show that an integral equation:

$$x(t) - \int_0^t [c + d(t - \xi)]x(\xi)d\xi = a + bt$$

has the solution of the type

$$x(t) = a_1 e^{\lambda_1 t} + a_2 e^{\lambda_2 t},$$

and write a_1, a_2, λ_1, λ_2 in the symbols a, b, c, d.
14. In [V_1], [V_2], when the kernel is given by Dirichlet's series, prove that the kernels of the solution can also be given by this series. [Kondo]
15. When the characteristic equations have multiple roots in § 3.5 [A], [P], find in each case the form of the kernel of the solution.
16. In [V_2] if

$$K(t) = A^n \frac{t^{n-1}}{(n-1)!},$$

show that

$$G(t) = \frac{A}{n}\Big\{\exp(At) + \omega \exp(\omega At) + \omega^2 \exp(\omega^2 At) + \cdots$$
$$+ \omega^{n-1}\exp(\omega^{n-1}At)\Big\}.$$

provided ω is the nth power root of 1. [Evans]
17. Show that when

$$K(t) = A^2 t$$

in [V_2], then

$$G(t) = A \sinh At.$$

18. In $[V_2]$, if $K(t)$ is expansible in a Taylor series:

$$K(t) = k_0 + k_1 t + \frac{k_2}{2!}t^2 + \cdots,$$

prove that the kernel of the solution is expressed as

$$G(t) = k_0 - \begin{vmatrix} k_0 & 1 \\ k_1 & k_0 \end{vmatrix} t + \begin{vmatrix} k_0 & 1 & 0 \\ k_1 & k_0 & 1 \\ k_2 & k_1 & k_0 \end{vmatrix} \frac{t^2}{2!} + \cdots.$$

[Whittaker, G. Prasad, Kondo]

19. Prove that the result obtained in the previous question and the solution of § 3.5.2, [P] agree.

20. In $[V_2]$, when the kernel is given by

$$K(t) = a\cos t + b\sin t,$$

show that the kernel of the solution is

$$G(t) = \frac{b + a^2 - \lambda_2 a}{\lambda_1 - \lambda_2}\exp(\lambda_1 t) + \frac{b + a^2 - \lambda_1 a}{\lambda_2 - \lambda_1}\exp(\lambda_2 t),$$

provided λ_1, λ_2 are the roots of

$$\lambda^2 + a\lambda + (1 - b) = 0.$$

21. Make an orthogonal functional system in $t(-1, 1)$ out of a functional series: $1, t, t^2, \ldots$.

[Whittaker; $1, t\ t^2 - {}^1/_3, t^3 - {}^3/_5 t, t^4 - {}^6/_7 t^2 + {}^3/_{35}, \ldots$

22. Make an orthogonal function system: $f_0(t)$, $f_1(t)$, $f_2(t)$, ... in $t(a, b)$ out of a function series; $1, t\ t^2, \ldots$.

Suppose that $f_n(t) = d^n\{(t-a)^n(t-b)^n\}/dt^n$. [Whittaker]

References

Churchill, R.V. (1941) *Fourier Series and Boundary Value Problems* (McGraw–Hill, New York) pp. 37–40.

Forsyth, A.R. (1941) *A Treatise on Differential Equations* (Macmillan, London) p. 148.

4 Method of approximate solutions of integral equations

The iterated kernel and the Fredholm's determinant that we studied in Chapter 2 are rather inconvenient as practical methods of solution since their computation is generally troublesome and their convergence is not always good. Moreover, as we explained in Chapter 3, it is only when a kernel takes a special form that we can easily solve an integral equation analytically. Even when the form of a kernel is not so complicated, we can scarcely denote the exact solution of an integral equation in a simple expression. Thus it is required that we find an approximate solution correct enough for practical applications. This is the method of approximate solutions of integral equations, in which a kernel or a disturbance function takes a special form. The method of numerical solution is also a special one to obtain an approximate solution.

The methods of approximate solutions of integral equations are classified into two main groups according to the way the integration of an unknown function is treated.

(a) There is the method of using the formula of numerical integration, in decomposing its regular part into the form of a sum and then reducing it to a set of simultaneous linear equations with the values of unknown functions at some points for their unknown quantities.

(b) The second method is to get a kernel or an unknown function to approximate to a special form applicable for an easy solution as shown in the previous chapter, and then to determine an exact solution or unknown coefficients of the approximate solution for this approximate kernel. Of the two, method (a) is a most fundamental way to solve an integral equation, since reducing it to simultaneous equations, as has been already explained in §§ 1.4 and 2.2, and the precision of the result arrived at by (a) depends on the accuracy of numerical integration, while in (b) it depends on the exactness of the approximate kernel.

In this chapter we first study the formula of numerical integration and the approximation of a kernel, and then proceed to explain the methods for solving various kinds of integral equations in comparison with the exact solutions. The method (b) is also treated in § 4.3. At the end of the chapter there will be a brief explanation of how to solve an integral equation using a computer.

We adopt the form of Dwyer (1951) for the numerical solution of simultaneous equations. It must be added that many of the methods of approximate solution considered in this chapter are applied similarly in the cases of a singular kernel and the nonlinear type which will be treated in Part II, as the special theories.

4.1. Approximation formulae

There are two main methods for solving an integral equation approximately: (a) to re-form its regular part into a sum, or (b) to denote the kernel in an approximation. In preparation for this, we briefly set forth here the formulae of numerical integration for (a), and the various kinds of interpolation and the methods of Fourier analysis for (b). In the following pages we shall treat
(i) numerical integration formulae,
(ii) Newton–Lagrange interpolation formula—to approximate a function by polynomials,
(iii) Prony's interpolation formula—to represent a function using finite terms of Dirichlet series,
(iv) Fourier analysis—to represent a function using finite terms of a Fourier series,
(v) other methods—in particular an expansion in an orthogonal function system.

4.1.1. Numerical integration formulae

There is a method to approximate the value of a finite integral

$$L = \int_{-\frac{1}{2}h}^{\frac{1}{2}h} f(\xi)d\xi$$

by weighted means of the value $f(\xi_i)$ of an integrand at some representative points $\xi = \xi_i$ $(i = 1, 2, \ldots, n)$ in the interval $\xi\,(-\dfrac{1}{2}h, \dfrac{1}{2}h)$. In this if we put $\xi_i/h \equiv t_i$, then we have $t_i\,(-1/2, 1/2)$, and these are independent of the integral interval. If the weight to be given to the ith representative value is w_i, then

$$L = \sum_{i=1}^{n} w_i f(\xi_i/h) \cdot h + \varepsilon$$

is obtained, where ε is an error which depends upon the representative coordinates (places) and their number (n). These formulae are called numerical integration formulae by the method of representative points (or weighted means).

Many formulae have been found with regard to the method of representative points. By classifying these according to their characteristics, and when the number n of coordinate points are given, we have the following methods.

(a) The method of taking the weight w_i as constant and selecting ξ_i/h so that the error ε becomes a minimum—the Newton–Cotes method, Maclaurin's method, Tschebyscheff's method, etc.

(b) The method of putting representative points at regular intervals and setting the value of w_i so that ε becomes a minimum—Maclaurin's method, etc.

(c) The method of selecting the values of ξ_i and w_i so that the error ε becomes minimal—Gauss's method, the Gauss–Moor method, the Gauss–Lobato method, etc.

Of course ε depends upon the number n of representative coordinates, and the absolute error decreases as n increases for the same kind of approximate integration formulae, but the relative error does not always decrease in constant proportion to the increase of n.

The Newton–Cotes method, which is the simplest of all, is to put ξ_i at regular intervals and take the weight w_i uniformly, but it cannot be avoided that when n is constant, ε takes a rather large value compared with the other methods. It is widely used when a computer is available.

4.1.2. *Formula to approximate a function by a polynomial*

When the values of a function $K(t)$ for $n + 1$ values of an independent variable t are given as in Table 4.1, we may use Newton's interpolation formula and Lagrange's interpolation formula to apply a polynomial of the nth order

$$a_n t^n + a_{n-1} t^{n-1} + \cdots + a_1 t + a_0$$

to $K(t)$.

TABLE 4.1 Value of Function $K(t)$

t	$K(t)$
t_0	k_0
t_1	k_1
t_2	k_2
$\vdots$	$\vdots$
t_n	k_n

TABLE 4.2 Numerical Integration Formulae.

(Method of weighted means; method of representative coordinates)

$$L = \int_{-\frac{1}{2}h}^{\frac{1}{2}h} f(x)dx = \sum w_i f(\xi_i/h) \cdot h$$

Formulae		Newton–Cotes		Maclaurin		Tschebyscheff		Gauss
n	i	ξ_i/h	w_i	ξ_i/h	w_i	ξ_i/h	w_i	ξ_i/h
2	1	− 1/2	1/2	− 1/2	1/2	− 0.28867513	1/2	− 0.28867513
	2	1/2	1/2	1/2	1/2	0.28867513	1/2	0.28867513
	1	− 1/2	1/6	− 1/3	3/8	− 0.35355339	1/3	− 0.38729833
3	2	0	4/6	0	2/8	0	1/3	0
	3	1/2	1/6	1/3	3/8	− 0.35355339	1/3	0.38729833
	1	− 1/2	1/8	− 3/8	13/48	− 0.39732724	1/4	− 0.430568155797
4	2	− 1/6	3/8	− 1/8	11/48	− 0.09379624	1/4	− 0.169990521792
	3	1/6	3/8	1/8	11/48	0.09379624	1/4	0.169990521792
	4	1/2	1/8	3/8	13/48	0.39732724	1/4	0.430568155797
	1	− 1/2	7/90	− 4/10	275/1152	− 0.41624874	1/5	− 0.4530899229693320
	2	1/4	32/90	− 2/10	100/1152	− 0.18727072	1/5	− 0.2692346550528415
5	3	0	12/90	0	402/1152	0	1/5	0
	4	1/4	32/90	2/10	100/1152	0.18727072	1/5	0.2692346550528415
	5	1/2	7/90	4/10	275/1152	0.41624874	1/5	0.4530899229693320
	1	− 5/10	19/288	− 5/12	247/1280	− 0.43312341	1/6	− 0.4662347571015760
	2	− 3/10	75/288	− 3/12	139/1280	− 0.21125933	1/6	− 0.3306046932331323
6	3	− 1/10	50/288	− 1/12	254/1280	− 0.13331770	1/6	− 0.1193095930415985
	4	1/10	50/288	1/12	254/1280	0.13331770	1/6	0.1193095930415985
	5	3/10	75/288	3/12	139/1280	0.21125933	1/6	0.3306046932331323
	6	5/10	19/288	5/12	247/1280	0.43312341	1/6	0.4662347571015760
		$[V_2]$,	$[F_1]$	$[V_1]$		$[F_2]$		$[F_2]$

In Newton's formula, if independent variables are distributed at equal intervals and

$$t_n - t_{n-1} = t_{n-1} - t_{n-2} = \cdots = t_1 - t_0 \, (\equiv h),$$

then, using the differences Δk_0, $\Delta^2 k_0$, ..., $\Delta^n k_0$,† we approximate $K(t)$ making it

$$K(t) = k_0 + \frac{t - t_0}{1!} \frac{\Delta k_0}{h} + \frac{(t - t_0)(t - t_1)}{2!} \frac{\Delta^2 k_0}{h^2} + \cdots$$

$$+ \frac{(t - t_0)(t - t_1) \cdots (t - t_{n-1})}{n!} \frac{\Delta^n k_0}{h^n}.$$

Lagrange's formula is used when the interval is irregular, and $K(t)$

† $\Delta k_0 = k_1 - k_0$, $\Delta k_1 = k_2 - k_1$, ..., $\Delta^2 k_0 = \Delta k_1 - \Delta k_0$, $\Delta^2 k_1 = \Delta k_2 - \Delta k_1 \ldots$; $\Delta^3 k_0 = \Delta^2 k_1 - \Delta^2 k_0$, ...

TABLE 4.2—*Continued*

	Gauss–Moor		Gauss–Lobato	
w_i	ξ_i/h	w_i	ξ_i/h	w_i
1/2				
1/2				
5/18	− 0.3872983	0.277777827439		
8/18	0	0.444444345122		
5/18	0.3872983	0.277777827439		
0.173927422569	− 0.43056816	0.173927419816		
0.326072577431	− 0.16999052	0.326072580184		
0.326072577431	0.16999052	0.326072580184		
0.173927422569	0.43056816	0.173927419816		
0.1184634425280945	− 0.45308992	0.118463448212	− 0.5	0.049999911974
0.2393143352496832	− 0.26923465	0.239314332532	0.327327	0.272222153758
64/225	0	0.284444438512	0	0.355555868534
0.2393143352496832	0.26923465	0.239314332532	0.327327	0.272222153758
0.1184634425280945	0.45308992	0.118463448212	0.5	0.049999911974
0.0856622461895852	− 0.46623476	0.085662243842		
0.1803807865240693	− 0.33060469	0.180380794947		
0.2339569672863455	0.11930959	0.233956961211		
0.2339569672863455	− 0.11930959	0.233956961211		
0.1803807865240693	0.33060469	0.180380794947		
0.0856622461895852	0.46623476	0.085662243842		
	[F_2]		[F_2]	

is to be taken as

$$K(t) = \sum_{i=0}^{n} k_i \frac{(t - t_0)(t - t_1) \dots (t - t_n)}{(t_i - t_0)(t_i - t_1) \dots (t_i - t_n)},$$

provided the denominator of this expression shows the product of all the factors except $(t_i - t_i)$, and the numerator that of all the factors except $(t - t_i)$.

Besides these we have Bessel's formula obtained from Newton's formula and others.

We shall leave the discussion of these formulae and the error evaluations to later works.

4.1.3. Formula to approximate a function by a Dirichlet series

When given the value (Table 4.1) of the function $K(t)$ for $2n$ values of an independent value t, which are equally distributed in the intervals of h, we have Prony's interpolation formula for the approximation of $K(t)$ with a Dirichlet series:

$$k_1 e^{a_1 t} + k_2 e^{a_2 t} + \cdots + k_n e^{a_n t}.$$

If we denote a determinant of the mth order by

$$\begin{vmatrix} K(t_m) & K(t_{m-1}) & \ldots K(t_2) & K(t_1) \\ K(t_{m+1}) & K(t_m) & \ldots K(t_3) & K(t_2) \\ \ldots & \ldots & \ldots \;\ldots & \ldots \\ K(t_{2m-1}) & K(t_{2m-2}) & \ldots K(t_{m+1}) & K(t_m) \end{vmatrix} \equiv \mathrm{D},$$

and write D_i for D with the $(m - i + 1)$th column replaced by $K(t_{m+1})$, $K(t_{m+2})$, ..., $K(t_{2m})$, then if the roots of the algebraic equation

$$Dx^m - D_1 x^{m-1} + D_2 x^{m-2} - \cdots - (-1)^m D_{m-1}\, x + (-1)^m D_m = 0$$

are d_1, d_2, ..., d_m, the index a_i is fixed as

$$a_i = \frac{1}{h} \log \alpha_i \qquad (i = 1, 2, \ldots, m).$$

Next, from the simultaneous equations

$$\left.\begin{aligned} K(t_1) &= Q_1 + Q_2 + \cdots + Q_m, \\ K(t_2) &= Q_1 d_1 + Q_2 d_2 + \cdots + Q_m d_m, \\ &\ldots, \\ K(t_{i+1}) &= Q_1 d_1{}^i + Q_2 d_2{}^i + \cdots + Q_m d_m{}^i, \\ &\ldots, \\ K(t_{m+1}) &= Q_1 d_1{}^m + Q_2 d_2{}^m + \cdots + Q_m d_m{}^m, \end{aligned}\right\}$$

we find Q_1, Q_2, ..., Q_m, and then determine the coefficients with

$$k_i = Q_i \exp(-\, a_i x_1).$$

4.1.4. *Fourier analysis*

To denote a function $K(t)$ with finite terms of a Fourier series

$$K(t) \sim a_0 + a_1 \cos t + \cdots + a_n \cos nt + b_1 \sin t + \cdots + b_n \sin nt$$

and to determine the coefficients a_i, b_i numerically, we have only to take a numerical integration of

$$a_0 = \frac{1}{2\pi}\int_0^{2\pi} K(t)dt, \qquad a_i = \frac{1}{\pi}\int_0^{2\pi} K(t)\cos it\, dt,$$

$$b_i = \frac{1}{\pi}\int_0^{2\pi} K(t)\sin it\, dt.$$

For example, the interval between 0 and 2π is divided into $2m$ equal parts, and putting

$$h = \frac{2\pi}{2m} = \frac{\pi}{m}$$

we may compute the coefficients as

$$a_0 = \frac{1}{2m}\sum_{j=1}^{2m} K(jh), \qquad a_i = \frac{1}{m}\sum_{j=1}^{2m} K(jh)\cos(jih),$$

$$b_i = \frac{1}{m}\sum_{j=1}^{2m-1} K(jh)\sin(jih).$$

In order to carry out practical numerical computations, a mechanical method is adopted using a special sort of computation format. It is also easily carried out on a computer.

4.1.5. Other methods

In addition to the methods mentioned above, we may use change of variables and the method of least squares for the transformation or the approximation of the kernel $K(t, \xi)$ into some form suitable for computation. With these methods, accuracy of the result is estimated since the error of an approximate solution of an integral equation depends upon the accuracy of these approximate expressions.

In approximating $K(t)$ by an orthogonal step function

$$H_i(t) \qquad (i = 1, 2, 3, \ldots, n)$$

as shown in Fig. 4.1, if

$$K(t) = C_1H_1 + C_2H_2 + 2C_3H_3 + 2C_4H_4 + 4C_5H_5 + 4C_6H_6 + 4C_7H_7 + 4C_8H_8 + 8C_9H_9 + \cdots,$$

then it is determined by

$$C_1 = \int_0^1 K(t)dt, \qquad C_2 = \int_0^{\frac{1}{2}} K(t)dt - \int_{\frac{1}{2}}^1 K(t)dt,$$

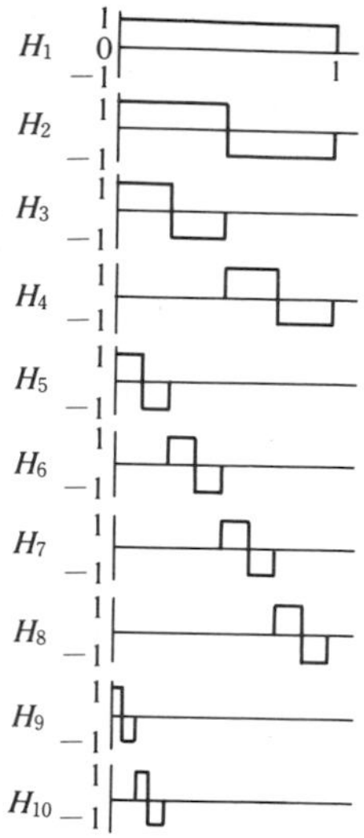

Fig. 4.1. Orthogonal step function

$$C_2 = \int_0^{\frac{1}{4}} K(t)dt - \int_{\frac{1}{4}}^{\frac{1}{2}} K(t)dt, \ldots .$$

4.2. Methods of numerical solution

The regular part of an integral equation is denoted in the form of a sum by the numerical integral formula explained in the previous section, by estimating the values of an unknown function at a few representative points. In this case we may derive simultaneous linear equations with the values of an unknown function as the unknown quantities. Here, it is required to have values at some points symmetrical for $t = \xi$ for the kernel $K(t, \xi)$, with a similar constraint on the values of a disturbance function at representative points. In the case of Fredholm's type, the increased number of representative points prevents us from arriving at a solution because of the increase in dimensions; therefore, we have to use Gauss's high precision approximation formula. When a computer is available, the increase of unknown variables in a set of simultaneous equations causes longer computation time. In the case of Volterra type, however, we may adopt Newton's approximate formula, which is convenient for calculation since simultaneous equations are solved successively.

For the numerical computation of an integral equation, the number n of representative points and the numerical integration formulae are selected so as to best fit the requirement of precision of the solution, and the form of simultaneous equations which are to be derived. The following methods are used for the computation. The Newton–Cotes method for [F_1], the Tschebyscheff–Gauss method for [F_2], Maclaurin's method for [V_1], the Newton–Cotes method for [V_2]. We apply the most suitable integration formula depending on the type of integral equation.

After the values $x(t_i) \equiv x_i$ of an unknown function for n values of t are given, to find the value for an arbitrary value of t we may use Newton's interpolation formula if f_i is given at regular intervals, and Lagrange's interpolation formula in the case of irregular intervals.

In the following, we show the ways of assigning representative points and the methods of numerical computation in various kinds of integral equations using the following figures and tables, and, by comparing the results with exact solutions, examine the accuracy of the methods of solution in examples. We are not going to introduce a computer program; instead we shall explain the principles of numerical solutions. This will help in the programming.

4.2.1. Fredholm-type integral equations of the first kind

For

$$\int_a^b K(t, \xi)x(\xi)d\xi = f(t), \qquad [F_1]$$

we have only to solve the simultaneous equations

$$\sum_1^n w_j K_{ij} x_j = f_i \quad \left(\begin{matrix} i \\ j \end{matrix} = 1, 2, \ldots, n\right),$$

making use of the values of $f(t)$ and those of the kernel at the representative points shown in Table 4.3 and Fig. 4.2. When the number of dimensions of the simultaneous equations increases in general, it suddenly becomes difficult to solve the equations. So, in such a case, a highly efficient approximate integration formula is adopted to raise the precision of the approximate solution. The formulae of Tschebyscheff, Gauss, Gauss and Moor, and Gauss and Lobato are often adopted for such purposes.

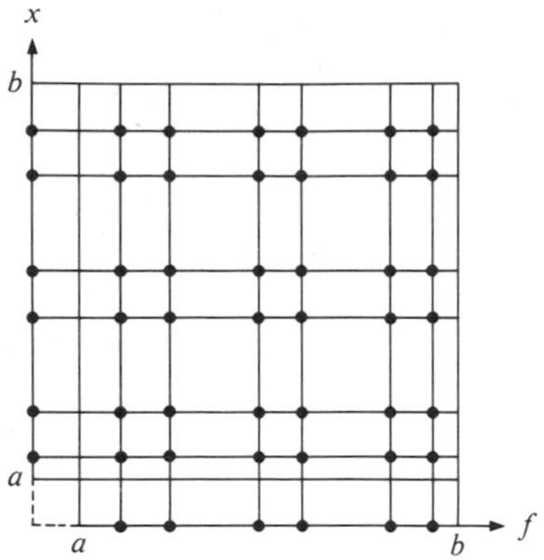

Fig. 4.2. Representative points

TABLE 4.3 Values of $K(t, \xi)$, $f(t)$, $x(t)$

x_n	K_{1n}	K_{2n}	:	K_{in}	:	K_{nn}
x_j	K_{1j}	K_{2j}	:	K_{ij}	:	K_{nj}
x_2	K_{12}	K_{22}	·	K_{i2}	·	K_{n2}
x_1	K_{11}	K_{21}	·	K_{i1}	·	K_{n1}
	f_1	f_2	·	f_i	·	f_n

Example 4.1 [F_1]

$$\int_0^1 (1 + t\xi)x(\xi)d\xi = 2t + 3. \qquad [F_1] \tag{4.1}$$

This equation has a solution such as

$$x(t) = 6t; \tag{4.2}$$

however, this solution is not unique as has been observed in § 3.4.

Given four as the number of representative points, let us solve this

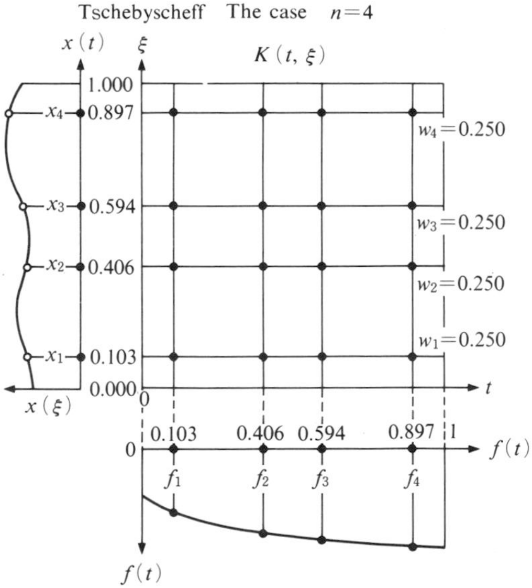

Fig. 4.3. Representative points (Tschebyscheff)

TABLE 4.4. Values of $K(t, \xi)$ and $f(t)$ at representative points

0.250 000	**0.897 327**	1.092 131	1.364 498	1.532 829	1.805 196
0.250 000	**0.593 796**	1.060 967	1.241 202	1.352 594	1.532 829
0.250 000	**0.406 204**	1.041 706	1.165 002	1.241 202	1.364 498
0.250 000	**0.102 673**	1.010 542	1.041 706	1.060 967	1.092 131
Weight	ξ / t	**0.102 673**	**0.406 204**	**0.593 796**	**0.897 327**
	$f(t)$	3.205 346	3.812 408	4.187 592	4.794 654
	$4f(t)$	12.821 384	15.249 632	16.750 368	19.178 616

case numerically by the numerical integration formulae of Tscheby-scheff and Gauss (see Question 3).

By Tschebyscheff's formula, the values of $K(t, \xi)$ and $f(t)$ at the representative points are as shown in Table 4.4 and Fig. 4.3. As the weight is uniform in Tschebyscheff's formula, we make use of the values of $K(t)$ in the table of coefficients just as they are, and the simultaneous equations are derived directly from Table 4.4 if the value of the right-hand side $f(t)$ is multiplied by the inverse of the weight.

The simultaneous equations become

$$\left.\begin{aligned}
&1.010\,542\,x_1 + 1.041\,706\,x_2 + 1.060\,967\,x_3 + 1.092\,131\,x_4\\
&\quad = 12.821\,384\\
&1.041\,706\,x_1 + 1.165\,002\,x_2 + 1.241\,202\,x_3 + 1.364\,498\,x_4\\
&\quad = 15.249\,632\\
&1.060\,967\,x_1 + 1.241\,202\,x_2 + 1.352\,594\,x_3 + 1.532\,829\,x_4\\
&\quad = 16.750\,368\\
&1.092\,131\,x_1 + 1.364\,498\,x_2 + 1.532\,829\,x_3 + 1.805\,196\,x_4\\
&\quad = 19.198\,616
\end{aligned}\right\}, \tag{4.3}$$

Note the difference in the order of the coefficients from that of Table 4.4.

Now generally in solving simultaneous equations:

$$\left.\begin{aligned}
a_{11}x_1 + a_{12}x_2 + a_{13}x_3 + a_{14}x_4 &= a_{15},\\
a_{21}x_1 + a_{22}x_2 + a_{23}x_3 + a_{24}x_4 &= a_{25},\\
a_{31}x_1 + a_{32}x_2 + a_{33}x_3 + a_{34}x_4 &= a_{35},\\
a_{41}x_1 + a_{42}x_2 + a_{43}x_3 + a_{44}x_4 &= a_{45},
\end{aligned}\right\} \tag{4.4}$$

we combine the first equation with the others in order to eliminate x_1, then we get

$$\left.\begin{aligned} m_{22-1}x_2 + m_{23-1}x_3 + m_{24-1}x_4 &= m_{25-1}, \\ m_{32-1}x_2 + m_{33-1}x_3 + m_{34-1}x_4 &= m_{35-1}, \\ m_{42-1}x_2 + m_{43-1}x_3 + m_{44-1}x_4 &= m_{45-1}. \end{aligned}\right\} \tag{4.5}$$

Here, the coefficients are obtained by a calculation like

$$m_{ij-1} = a_{11}a_{ij} - a_{i1}a_{1j}, \tag{4.5a}$$

which can also be carried out on a computer.

Next, evaluate x_2 and we have

$$m_{33-12}x_3 + m_{34-12}x_4 = m_{35-12}, \quad m_{43-12}x_3 + m_{44-12}x_4 = m_{45-12}, \tag{4.6}$$

where the coefficients are

$$m_{ij-12} = m_{22-1}m_{ij-1} - m_{i2-1}m_{3j-1}. \tag{4.6a}$$

Further, evaluating x_2 we get

$$m_{44-123}x_4 = m_{43-123}. \tag{4.7}$$

In this,

$$m_{ij-123} = m_{33-12}m_{ij-22} - m_{i2-12}m_{2j-12}. \tag{4.7a}$$

Thus, we get

$$x_4 = m_{45-123}/m_{44-123}. \tag{4.8}$$

Substitute the values in the first equation of (4.6) and x_3 is determined first, then x_2 by the first equation of (4.5), and lastly x_1 is determined by the first equation of (4.4).

This calculation is carried out through the form of Table 4.5 (A). The last column in this table is headed 'Checksum'. This is used to make sure of the congruence of the row total m_{iT-1} and the result of

TABLE 4.5(A). The methods of solution of simultaneous equations (4.4)

x_1	x_2	x_3	x_4		Row total	Checksum
a_{11}	a_{12}	a_{13}	a_{14}	a_{15}	a_{1T}	a_{15}'
a_{21}	a_{22}	a_{23}	a_{24}	a_{25}	a_{2T}	a_{25}'
a_{31}	a_{32}	a_{33}	a_{34}	a_{35}	a_{3T}	a_{35}'
a_{41}	a_{42}	a_{43}	a_{44}	a_{45}	a_{4T}	a_{45}'
	m_{22-1}	m_{23-1}	m_{24-1}	m_{25-1}	m_{2T-1}	m_{2T-1}'
	m_{32-1}	m_{33-1}	$m_{34=1}$	m_{35-1}	m_{3T-1}	m_{3T-1}'
	m_{42-1}	m_{43-1}	m_{44-1}	$m_{45=1}$	m_{4T-1}	m_{4T-1}'
		m_{33-12}	m_{34-12}	m_{35-12}	m_{3T-12}	m_{3T-12}'
		m_{43-12}	m_{44-12}	m_{45-12}	m_{4T-12}	m_{4T-12}'
			m_{44-123}	m_{45-123}	m_{4T-123}	m_{4T-123}'
x_1	x_2	x_3	x_4			

TABLE 4.5.(B) The example of solutions of simultaneous equations (4.3)

x_1	x_2	x_3	x_4	a_{i5}
1.010 542	1.041 706	1.060 967	1.092 131	12.821 584
1.041 706	1.165 002	1.241 202	1.364 498	15.249 632
1.060 967	1.241 202	1.532 594	1.532 829	16.750 368
1.092 131	1.364 498	1.532 829	1.805 196	19.178 656
	$0.092\ 132_{041}$	$0.149\ 071_{042}$	$0.241\ 203_{122}$	$2.054\ 072_{638}$
	$0.149\ 071_{062}$	$0.241\ 202_{071}$	$0.390\ 273_{123}$	$3.323\ 672_{868}$
	$0.241\ 203_{122}$	$0.390\ 273_{133}$	$0.631\ 476_{255}$	$5.377\ 988_{036}$
		$0.000\ 002\ 596_{23}$	$0.000\ 002\ 596_{22}$	$0.000\ 138\ 325_{21}$
		$0.000\ 002\ 596_{23}$	$0.000\ 002\ 596_{23}$	$0.000\ 362\ 205_{97}$
			0.000 656	0.000 777
... ...	... ...	... ...	... ...	

$$m'_{iT-1} = a_{11}a_{iT} - a_{i1}a_{1T}, \tag{4.5b}$$

etc., carried through in accordance with (4.5a), (4.6a), (4.7a), and thus, to discover mistakes made in the course of computation. The last column is also used to compare a_{i5}' with a_{i5} by putting the determined solutions x_1, x_2, x_3, x_4 into the left side of (4.4), thus determining the right-hand side a_{i5}'. This method is similarly applied in the case of higher orders.

We give here the course of solution of (4.3) as an example of the method of using Table 4.5 (B). No solution is obtained in this example. By Gauss's formula, the table of values of the kernel $K(t, \xi)$, $w_i\ K(t, \xi)$ and $f(t)$ at the representative points given in Fig. 4.4 becomes as shown in Table 4.6. In this case, we may compute in the manner of Table 4.5, but we have another method for it as given in Table 4.7. The rows (1′), (2′), etc., are the results of dividing each element of rows (1), (2), etc., by its first row element; (5), (6), (7) are the results of subtracting successively from each row element of (1′) the row elements of (2′),

TABLE 4.6 Values of $K(t, \xi)$, $w_i K(t, \xi)$ and $f(t)$ at representative points (Gauss)

	0.930 568	1.064 611	1.307 096	1.623 472	1.865 957
0.173 927		0.185 165	0.227 339	0.282 366	0.324 540
	0.669 991	1.046 519	1.221 103	1.448 888	1.623 472
0.326 073		0.341 242	0.398 169	0.472 443	0.529 370
	0.330 009	1.002 913	1.108 906	1.221 103	1.307 096
0.326 073		0.333 543	0.361 584	0.398 169	0.426 209
	0.069 432	1.004 821	1.022 913	1.046 519	1.064 611
0.173 927		0.174 766	0.177 912	0.182 018	0.185 165
Weight	ξ t	**0.069 432**	**0.330 009**	**0.669 991**	**0.930 568**
	$f(t)$	3.138 864	3.660 018	4.339 982	4.861 136

TABLE 4.7. Examples of solutions of simultaneous equations

	x_1	x_2	x_3	x_4	a_{is}	Row total	Checksum
(1)	0.174 766	0.333 543	0.341 242	0.185 165	3.138 864	4.173 580	3.138 871
(2)	0.177 912	0.361 584	0.398 169	0.227 339	3.660 018	4.825 022	3.659 767
(3)	0.182 018	0.398 169	0.472 443	0.282 366	4.339 982	5.674 978	4.339 833
(4)	0.185 165	0.426 209	0.529 370	0.324 540	4.861 136	6.326 420	4.861 114
(1′)	1.000 000	1.908 512	1.952 565	1.059 502	17.960 381	24.880 960	23.880 961
(2′)	1.000 000	2.032 376	2.238 011	1.277 817	20.572 069	27.120 273	27.120 273
(3′)	1.000 000	2.187 525	2.595 584	1.551 308	23.843 697	31.178 114	31.178 114
(4′)	1.000 000	2.301 779	2.858 910	1.752 707	26.252 996	34.166 392	34.166 392
(5)	(1′)–(2′)	− 0.123 864	− 0.285 446	− 0.218 315	− 2.611 688	− 3.239 313	− 3.239 312
(6)	(1′)–(3′)	− 0.279 013	− 0.643 019	− 0.491 806	− 5.883 316	− 7.297 154	− 7.297 153
(7)	(1′)–(4′)	− 0.393 267	− 0.906 345	− 0.693 205	− 8.292 615	− 10.285 432	− 10.285 432
(5′)		1.000 000	2.304 511	1.762 538	21.085 126	26.152 175	26.152 167
(6′)		1.000 000	2.304 620	1.762 663	21.086 172	26.153 455	26.153 452
(7′)		1.000 000	2.304 656	1.762 683	21.086 476	26.153 815	26.153 812
(8)		(5′)–(6′)	− 0.000 109	− 0.000 125	− 0.001 046	− 0.001 280	− 0.001 285
(9)		(5′)–(9′)	− 0.000 145	− 0.000 145	− 0.001 350	− 0.001 640	− 0.001 645
(8′)			1.000 000	1.146 789	9.596 3303	11.743 119	11.788 991
(9′)			1.000 000	1.000 000	9.310 3448	11.310 345	11.344 276
(10)			(8′)–(9′)	(11) 0.146 789	(12) 0.285 9855	0.432 774	0.444 715
	202.685 303	− 182.925 814	73.620 687	(12) ÷ (11) 19.482 761			

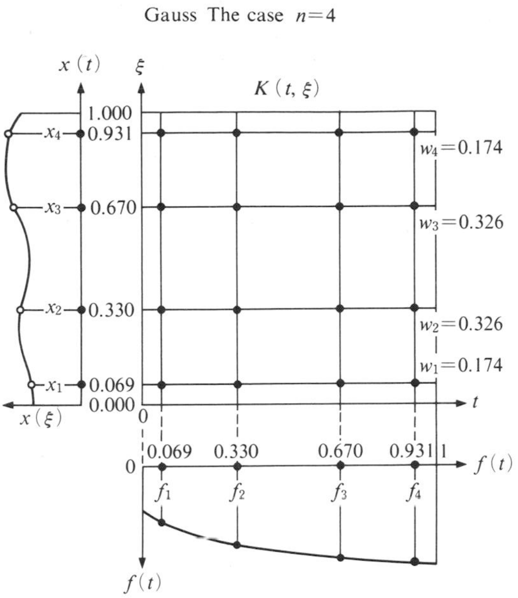

Fig. 4.4. Representative points (Gauss)

(3′), (4′); x_4 is the quotient of the (10)th row element, that is, the result of (12) ÷ (11).

The kernel of the integral equation (4.1) is symmetric. Then the coefficient determinant of the simultaneous equations (4.3) derived by Tschebyscheff's formula proves to be symmetric and its value becomes zero. But it must be noticed that this symmetry does not come out in Gauss's formula.

4.2.2. *Fredholm type integral equations of the second kind*

$$x(t) - \lambda \int_a^b K(t, \xi)x(\xi)d\xi = f(t). \qquad [F_2]$$

We have only to solve the simultaneous equations

$$x_i - \lambda \sum_1^n w_j K_{ij} x_j = f_i, \quad \left(\begin{matrix} i \\ j \end{matrix} = 1, 2, \ldots, n\right);$$

using the value of $f(t)$ and that of the kernel at representative points, such as those given in Fig. 4.2 and Table 4.3. When the number of dimensions increases it becomes rather difficult to solve this set of simultaneous equations. Then, in order to raise the accuracy of an approximate solution, we should use good approximate integration formulae.

In such a case, the formulae of Tschebyscheff, Gauss, Gauss and Moor, Gauss and Lobato, etc., are used. In simultaneous equations derived from [F_2], the orthogonal elements of the coefficient determinant are comparatively larger than the other coefficients; it is then possible for us to use this characteristic in finding an approximate solution successively.

Example 4.2 Numerical solution of [F_2]

$$x(t) + 2\int_0^1 (1 + t\xi)x(\xi)d\xi = t^2. \qquad [F_2] \tag{4.9}$$

The exact solution of this equation is

$$x(t) = t^2 - \frac{5}{24}t - \frac{11}{72}. \tag{4.10}$$

According to Tschebyscheff's formula, the values of $K(t, \xi)$ and $f(t)$ at representative points are as in Table 4.4. Then, noting the uniformity of weight we may obtain the simultaneous equations

$$\left.\begin{aligned}
&6.021\,084\, x_1 + 2.083\,412\, x_2 + 2.121\,934\, x_3 + 2.184\,262\, x_4 \\
&\quad = 0.042\,168, \\
&2.083\,412\, x_1 + 6.330\,004\, x_2 + 2.482\,404\, x_3 + 2.728\,996\, x_4 \\
&\quad = 0.660\,008, \\
&2.121\,934\, x_1 + 2.482\,404\, x_2 + 6.705\,188\, x_3 + 3.065\,658\, x_4 \\
&\quad = 1.410\,376, \\
&2.184\,262\, x_1 + 2.728\,996\, x_2 + 3.065\,658\, x_3 + 7.610\,392\, x_4 \\
&\quad = 3.220\,784,
\end{aligned}\right\} \tag{4.11}$$

TABLE 4.8 Values of $K(t, \xi)$ and $f(t)$ at representative points (Tschebyscheff)

Weight	ξ \ t	**0.102 673**	**0.406 204**	**0.593 796**	**0.897 327**
0.250 000	**0.897 327**	2.184 262	2.728 996	3.065 658	3.610 392
0.250 000	**0.593 796**	2.121 934	2.482 404	2.705 188	3.065 658
0.250 000	**0.406 204**	2.083 412	2.330 004	2.482 404	2.728 996
0.250 000	**0.102 673**	2.021 084	2.083 412	2.121 934	2.184 262
	$f(t)$	0.010 542	0.165 002	0.352 594	0.805 196
	$4f(t)$	0.042 168	0.660 008	1.410 376	3.220 784

TABLE 4.9. Numerical solution of the integral Equation (4.9) (Tschebyscheff, $n = 4$)

Representative points t	Solution $x(t)$		Error	Relative error (%)
	Approximate solution	Exact solution		
0.102 673	– 0.163 000	– 0.163 626	– 0.000 626	0.111
0.406 201	– 0.072 000	– 0.072 401	– 0.000 401	0.011
0.593 796	0.076 000	0.076 109	0.000 109	0.011
0.897 327	0.456 475	0.465 475	0.000 475	0.111

TABLE 4.10. Numerical solution of the integral equations (4.9) (Gauss, $n = 4$)

Representative points t	Solution $x(t)$		Error	Relative error (%)
	Approximate solution	Exact solution		
0.069 432	– 0.162 565	– 0.162 422	– 0.000 123	0.076
0.330 009	– 0.112 605	– 0.112 625	0.000 020	0.018
0.669 991	0.156 544	0.156 529	0.000 015	0.010
0.930 568	0.519 324	0.519 511	– 0.000 187	0.036

where care is taken with the exceptional part and 4 is added to the orthoogonal element. If we solve this by the method of Table 4.7, we can obtain the values of unknown functions at the representative points.

Making use of Gauss's formula, we can calculate $K(t, \xi)$, $w_i K(t, \xi)$ and $f(t)$ in the manner of Table 4.6, and then solve the simultaneous equations

$$\left.\begin{aligned} &1.348\,532\, x_1 + 0.667\,086\, x_2 + 0.682\,484\, x_3 + 0.370\,330\, x_4 \\ &\quad = 0.004\,821, \\ &0.355\,824\, x_1 + 1.723\,168\, x_2 + 0.796\,338\, x_3 + 0.454\,678\, x_4 \\ &\quad = 0.108\,906, \\ &0.364\,036\, x_1 + 0.796\,338\, x_2 + 1.944\,886\, x_3 + 0.564\,732\, x_4 \\ &\quad = 0.448\,888, \\ &0.370\,330\, x_1 + 0.852\,418\, x_2 + 1.058\,740\, x_3 + 1.649\,080\, x_4 \\ &\quad = 0.865\,957. \end{aligned}\right\} \quad (4.12)$$

In the case of Gauss's formula we can obtain highly exact solutions, yet we have to carry out troublesome calculations since the coefficient determinant has no symmetry.

4.2.3. *Fredholm type homogeneous integral equations*

$$x(t) = \lambda \int_a^b K(t, \xi) x(\xi) d\xi \qquad [F_0]$$

Again, in this case we have only to select such representative points as were given in Fig. 4.2 and Table 4.3, and derermine the eigenvalue of λ as the root of the coefficient determinant, which is set to be zero, and of the simultaneous equations:

$$x_i - \lambda \textstyle\sum_j w_j K_{ij} x_j = 0 \qquad (i, j = 1, 2, \ldots, n).$$

Here the values of λ are counted as n. Therefore, the number of eigenvalues sought by the method of numerical solution agrees with that of the representative points and does not agree with the number of original eigenvalues of [F_0]. Next, after eigenvalues have been determined, we have to numerically compute (2.14) in § 2.2 to settle the eigenfunctions. In this process, the method of numerical solution serves only for the determination of eigenvalues, yet in applications there are many cases where the forms of eigenfunctions are already known and it is only necessary to determine the eigenvalues which are contained as parameters in the eigenfunctions.

Example 4.3 Numerical solution of [F_0]

$$x(t) = \lambda \int_0^1 (t + \xi) x(\xi) d\xi. \qquad [F_0] \tag{4.13}$$

In this case the eigenvalue is

$$\lambda = -6 - \sqrt{48}, \qquad \lambda = -6 + \sqrt{48}, \tag{4.14}$$

by Question 17 in Chapter 2.

Put four as the number of representative points and by using Gauss's formula we can get the simultaneous equations:

$$\left.\begin{aligned}
x_1 &= 0.02415\,\lambda\, x_1 + 0.06947\,\lambda\, x_2 + 0.12861\,\lambda\, x_3 + 0.17393\,\lambda\, x_4,\\
x_2 &= 0.13025\,\lambda\, x_1 + 0.21521\,\lambda\, x_2 + 0.32607\,\lambda\, x_3 + 0.41104\,\lambda\, x_4,\\
x_3 &= 0.24110\,\lambda\, x_1 + 0.32607\,\lambda\, x_2 + 0.43693\,\lambda\, x_3 + 0.52189\,\lambda\, x_4,\\
x_4 &= 0.17393\,\lambda\, x_1 + 0.21925\,\lambda\, x_2 + 0.27839\,\lambda\, x_3 + 0.32371\,\lambda\, x_4.
\end{aligned}\right\} \tag{4.15}$$

Generally, for example, as in the condition that

$$\left.\begin{aligned}
a_{11}x_1 + a_{12}x_2 + a_{13}x_3 + a_{14}x_4 &= \lambda\, x_1,\\
a_{21}x_1 + a_{22}x_2 + a_{23}x_3 + a_{24}x_4 &= \lambda\, x_2,\\
a_{31}x_1 + a_{32}x_2 + a_{33}x_3 + a_{34}x_4 &= \lambda\, x_3,\\
a_{41}x_1 + a_{42}x_2 + a_{43}x_3 + a_{44}x_4 &= \lambda\, x_4
\end{aligned}\right\} \tag{4.16}$$

have roots other than $x_1 = x_2 = x_3 = x_4 = 0$, we have, the coefficient determinant

$$\begin{vmatrix} a_{11}-\lambda & a_{12} & a_{13} & a_{14} \\ a_{21} & a_{22}-\lambda & a_{23} & a_{24} \\ a_{31} & a_{32} & a_{33}-\lambda & a_{34} \\ a_{41} & a_{42} & a_{42} & a_{44}-\lambda \end{vmatrix} = 0. \tag{4.17}$$

This is the integral equation of the fourth order in λ and we denote it by

$$\lambda^4 - c_3\lambda^3 + c_2\lambda^2 - c_1\lambda + c_0 = 0; \tag{4.18}$$

then c_3, c_2, c_1, c_0 are the functions of the elements of a determinant a_{ij}. For example, we have

$$c_3 = a_{11} + a_{22} + a_{33} + a_{44}, \ldots, c_0 = |a_{ij}|. \tag{4.19}$$

To calculate thc cocfficients, we have only to compute them in the form given in Table 4.11 (A).

Here d is given as

$$\left.\begin{aligned} d_{ij-1} &= m_{ij-1}, \\ d_{ij-12} &= m_{ij-12}/a_{11}, \\ d_{ij-123} &= m_{ij-123}/d_{22-1}, \text{ etc.,} \end{aligned}\right| \tag{4.20}$$

for m in (4.5).

In this case, we have, for example,

$$\begin{aligned} c_0 \equiv |a_{ij}| &= \frac{1}{a_{11}{}^2} \begin{vmatrix} a_{11} & a_{12} & a_{13} & a_{14} \\ 0 & d_{22-1} & d_{23-1} & d_{24-1} \\ 0 & d_{32-1} & d_{33-1} & d_{34-1} \\ 0 & d_{42-1} & d_{43-1} & d_{44-1} \end{vmatrix} \\ &= \frac{1}{a_{11}{}^2} \begin{vmatrix} d_{22-1} & d_{23-1} & d_{24-1} \\ d_{32-1} & d_{33-1} & d_{34-1} \\ d_{42-1} & d_{43-1} & d_{44-1} \end{vmatrix} \\ &= \frac{1}{d_{22-1}} \begin{vmatrix} d_{33-12} & d_{34-12} \\ d_{43-12} & d_{44-12} \end{vmatrix} = d_{44-123}. \end{aligned} \tag{4.21}$$

Thus, the coefficient c is the sum of the underlined elements in Table 4.11 (A).

TABLE 4.11. (A) Calculation forms of the main determinants

$\underline{a_{11}}$	a_{12}	a_{13}	a_{14}	
a_{21}	$\underline{a_{22}}$	a_{23}	d_{24}	c_3
a_{31}	a_{32}	$\underline{a_{33}}$	d_{34}	
a_{41}	a_{42}	a_{43}	$\underline{d_{44}}$	
(a_{11})	$\underline{d_{22-1}}$	d_{23-1}	d_{24-1}	
	d_{32-1}	$\underline{d_{33-1}}$	d_{34-1}	
	d_{42-1}	d_{43-1}	$\underline{d_{44-1}}$	
(a_{22})		$\underline{d_{33-2}}$	d_{34-2}	c_2
		d_{43-2}	$\underline{d_{44-2}}$	
(a_{33})			$\underline{d_{44-3}}$	
(d_{22-1})		$\underline{d_{33-12}}$	d_{34-12}	
		d_{43-12}	$\underline{d_{44-12}}$	
(d_{33-1})			$\underline{d_{44-13}}$	c_1
(d_{33-2})			$\underline{d_{44-23}}$	
(d_{33-12})			$\underline{d_{44-123}}$	c_0

In the case of (4.15), if we adopt the form of calculation of Table 4.11 (B), we get

$$(1/\lambda)^4 - (1/\lambda)^3 - 0.08333639(1/\lambda)^2 + 0\lambda + 0 = 0 \qquad (4.22)$$

as a characteristic equation. This enables us to have the almost exact result

$$1 - \lambda - \frac{1}{12}\lambda^2 = 0.$$

TABLE 4.11.(B) Examples of calculations of the main determinants

0.02415	0.06947	0.12861	0.17393	
0.13025	**0.21521**	0.32607	0.41104	1.00000000
0.24110	0.32607	**0.43693**	0.52189	
0.17393	0.21925	0.27839	**0.32371**	
(0.02415)	− **0.00385114**	− 0.00887686	− 0.01272777	
	− 0.00887462	− **0.02045602**	− 0.02933088	
	− 0.00678803	− 0.01564602	− **0.02243405**	
(0.21521)		− **0.01228994**	− 0.02171187	− 0.08333639
		− 0.01157853	− **0.02045490**	
(0.43693)			− **0.00385034**	
(−0.00385114)		**0.00000000**	0.00000013	
		− 0.00000005	**0.00000000**	
(−0.02045602)			**0.00000000**	
(−0.01228994)			**0.00000000**	0.00000000
(0.00000000)			**0.00000000**	0.00000000

4.2.4. *Volterra type integral equations of the first kind*

$$\int_a^t K(t, \xi)x(\xi)d\xi = f(t). \qquad [V_1]$$

Here we show the values of the kernel at the lattice points distributed at regular intervals of $2h$, and the values of $f(t)$ and $x(\xi)$ at regular intervals of $2h$ in the form of Table 4.12. Making use of Maclaurin's formula and entering the weights as in Fig. 4.5, the simultaneous equations are solved as follows:

$$K_{21} \cdot x_1 \cdot (2h) = f_2,$$

$$\left(\frac{1}{2}K_{41} \cdot x_1 + \frac{1}{2}K_{43} \cdot x_3\right)(4h) = f_4,$$

TABLE 4.12. Values of $K(t, \xi), f(t), x(\xi)$

↑					
x_9					K_{109}
x_7				K_{87}	K_{107}
x_5			K_{65}	K_{85}	K_{105}
x_3		K_{43}	K_{63}	K_{83}	K_{103}
x_1	K_{21}	K_{41}	K_{61}	K_{81}	K_{101}
	f_2	f_4	f_6	f_8	f_{10}

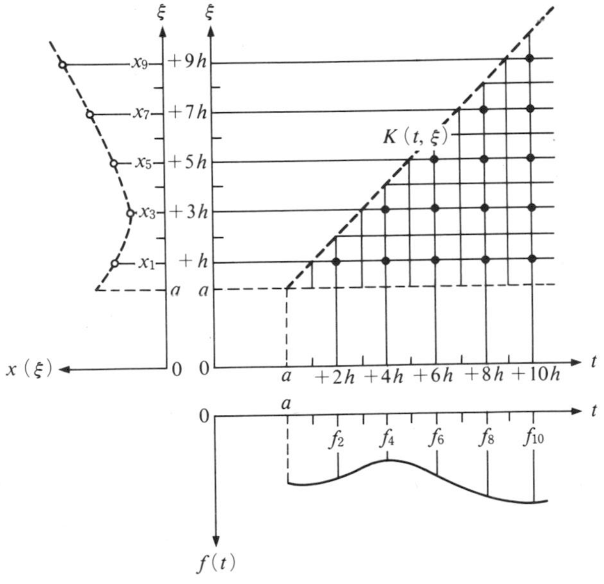

Fig. 4.5. Representative points

$$\left(\frac{3}{8}K_{61}\cdot x_1+\frac{2}{8}K_{63}\cdot x_3+\frac{3}{8}K_{65}\cdot x_5\right)(6h)=f_6,$$

$$\left(\frac{13}{48}K_{81}\cdot x_1+\frac{11}{48}K_{83}\cdot x_3+\frac{11}{48}K_{85}\cdot x_5+\frac{13}{48}K_{87}\cdot x_7\right)(8h)=f_8,$$

$$\left(\frac{275}{1152}K_{101}x_1+\frac{100}{1152}K_{103}x_3+\frac{402}{1152}K_{105}x_5+\frac{100}{1152}K_{107}x_7\right.$$
$$\left.+\frac{275}{1152}K_{109}x_9\right)(10h)=f_{10},$$

....

Example 4.4 Numerical solution of [V_1]

$$\int_0^t (1-t+\xi)x(\xi)d\xi = t. \qquad [V_1]$$

The exact solution of this equation is

$$x(t)=e^t,$$

from Example 3.5 of Chapter 3. Let $h=0.05$ and calculate the kernel and the disturbance function, and then the result is as shown in Table 4.13.

Thus, take the weight as given in Table 4.2 and the simultaneous equation is solved successively as follows:

$$0.95\times x_1\times 0.1 = 0.10000,\ x_1 = 1.05264,$$

$$\left(\frac{1}{2}\times 0.85\times x_1+\frac{1}{2}\times 0.95\times x_3\right)\times 0.2 = 0.20000,$$

$$x_3 = 1.16343,$$

TABLE 4.13. Values of $K(t, \xi)$ and $f(t)$ at representative points (Maclaurin's formula)

↑	↑ **0.55**							
1.56959	**0.45**						0.95	→
1.41863	**0.35**					0.95	0.85	→
1.28201	**0.25**				0.95	0.85	0.75	→
1.16343	**0.15**			0.95	0.85	0.75	0.65	→
1.05264	**0.05**		0.95	0.85	0.75	0.65	0.55	→
$x(\xi)$	ξ	t	**0.10**	**0.20**	**0.30**	**0.40**	**0.50**	→
	$f(t)$		0.10000	0.20000	0.30000	0.40000	0.50000	→

TABLE 4.14. Errors in the numerical solution by Maclaurin's formula

Representative points t	Solution $x(t)$		Error	Relative error (%)
	Approximate solution	Exact solution		
0.05	1.05264	1.10517	−0.05253	4.75311
0.15	1.16343	1.16183	0.00160	0.13771
0.25	1.28201	1.28403	−0.00202	0.15732
0.35	1.41863	1.41907	−0.00044	0.03101
0.45	1.56959	1.56831	0.00128	0.08162
0.55		1.73325		

$$\left(\frac{3}{8} \times 0.75 \times x_1 + \frac{2}{8} \times 0.85 \times x_3 + \frac{3}{8} \times 0.95 \times x_5\right) \times 0.3$$
$$= 0.30000,\ x_5 = 1.28201,$$

$$\left(\frac{13}{48} \times 0.65 \times x_1 + \frac{11}{48} \times 0.75 \times x_3 + \frac{11}{48} \times 0.85 \times x_5\right.$$
$$\left. + \frac{13}{48} \times 0.95 \times x_7\right) \times 0.4 = 0.40000,\ x_2 = 1.41863;$$

$$\left(\frac{275}{1152} \times 0.55 \times x_1 + \frac{100}{1152} \times 0.65 \times x_3 + \frac{402}{1152} \times 0.75 \times x_5\right.$$
$$\left. + \frac{100}{1152} \times 0.85 \times x_1 + \frac{275}{1152} \times 0.95 \times x_9\right) \times 0.5$$
$$= 0.50000,\ x_9 = 1.56959:$$

....

The errors are shown in Table 4.14.

4.2.5. *Volterra integral equations of the second kind*

Consider

$$x(t) - \int_a^t K(t, \xi)x(\xi)d\xi = f(t). \qquad [V_2]$$

We use the values of $f(t)$ and $x(\xi)$ at the same regular intervals with the values of the kernel at the lattice points regularly distributed in the interval, and write these in the form of Table 4.15. Then putting a, $a + h$, $a + 2h$, ... into t in order, we obtain the weights as given in Fig 4.6, and the simultaneous equations are solved as follows:

$$x_0 = f_0,$$

$$x_1 = f_1 + \left(\frac{1}{2}K_{10}x_0 + \frac{1}{2}K_{11}x_1\right)h,$$

TABLE 4.15 $K(t, \xi), f(t), x(\xi)$

↑					
x_4					K_{44}
x_3				K_{33}	K_{43}
x_2			K_{22}	K_{32}	K_{42}
x_1		K_{11}	K_{21}	K_{31}	K_{41}
x_0	K_{00}	K_{10}	K_{20}	K_{30}	K_{40}
	f_0	f_1	f_2	f_3	f_4

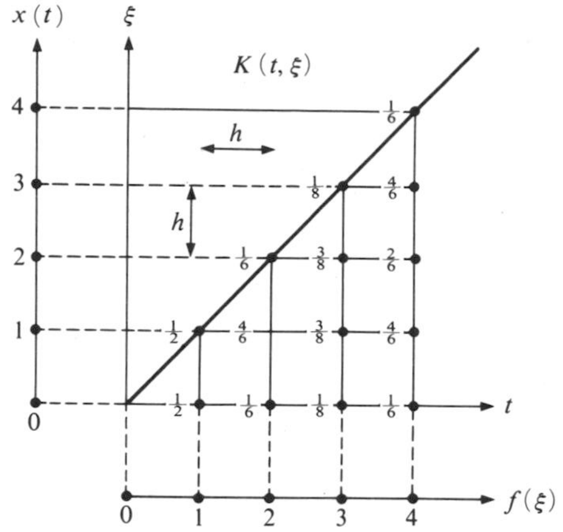

Fig. 4.6. Representative points

$$x_2 = f_2 + \left(\frac{1}{6}K_{20}x_0 + \frac{4}{6}K_{21}x_1 + \frac{1}{6}K_{22}x_2\right)2h,$$

$$x_3 = f_3 + \left(\frac{1}{8}K_{30}x_0 + \frac{3}{8}K_{31}x_1 + \frac{3}{8}K_{32}x_2 + \frac{1}{8}K_{33}x_3\right)3h,$$

$$x_4 = f_4 + \left(\frac{1}{6}K_{40}x_0 + \frac{4}{6}K_{41}x_1 + \frac{2}{6}K_{42}x_2 + \frac{4}{6}K_{43}x_3 + \frac{1}{6}K_{44}x_4\right)4h,$$

....

The weights are determined by combining Simpson's one-third law and three-eighths law. Since these may be solved successively, the value x_n of an unknown function in $t = a + nh$ depends on the accuracy of $x_{n-1}, x_{n-2}, \ldots, x_1, x_0$ and on h. But we may have the exact value of x_0; therefore, it is desirable to allow only a small error in x_1. Yet the precision of the second equation in the above simultaneous equations is very low because of the trapezoidal rule. Then we have another way of carrying out the calculation to increase the accuracy of x_1; that is, to subdivide the interval $t(a, a + h)$, and obtain a good approximation of the x_1–value before we proceed to the third equation.

Example 4.5

$$x(t) - \int_0^t \{3 + 6(t - \xi) - 4(t - \xi)^2\}x(\xi)d\xi = 1 - 2t - 4t^2. \qquad [V_2]$$

From example (3.79), the exact solution of this equation is

$$x(t) = e^t.$$

TABLE 4.16 Values of $K(t, \xi)$ and $f(t)$ at representative points (NewtongCotes' formula)

1.22140	0.20					3.00
1.16183	0.15				3.00	3.29
1.10517	0.10			3.00	3.29	3.56
1.05127	0.05		4.00	3.29	3.56	3.81
1.00000	0.00	3.00	3.29	3.56	3.81	4.04
$x(\xi)$	ξ / t	0.00	0.05	0.10	0.15	0.20
	$f(t)$	1.00	0.89	0.76	0.61	0.44

Put $h = 0.05$ and calculate the kernel and the disturbance function, the results of which appear in Table 4.16. Hence, take the weights as indicated in Fig. 4.7 and we may solve the simultaneous equations successively as follows.

$$x_0 = 1.00000;$$

$$x_1 - \left(\frac{1}{2} \times 3.29 \times x_0 + \frac{1}{2} \times 3.00 \times x_1\right) \times 0.05 = 0.89000,$$

$$x_1 = 1.05108;$$

$$x_2 - \left(\frac{1}{6} \times 3.56 \times x_0 + \frac{4}{6} \times 3.29 \times x_1 + \frac{1}{6} \times 3.00 \times x_2\right)$$

$$\times 0.10 = 0.76000, \; x_2 = 1.10513;$$

$$x_3 - \left(\frac{1}{8} \times 3.81 \times x_0 + \frac{3}{8} \times 3.56 \times x_1 + \frac{3}{8} \times 3.29 \times x_2\right.$$

$$\left. + \frac{1}{8} \times 3.00 \times x_3\right) \times 0.15 = 0.61000, \; x_3 = 1.16178;$$

$$x_4 - \left(\frac{1}{6} \times 4.04 \times x_0 + \frac{4}{6} \times 3.81 \times x_1 + \frac{2}{6} \times 3.56 \times x_2\right.$$

$$\left. + \frac{4}{6} \times 3.29 \times x_3 + \frac{1}{6} \times 3.00 \times x_4\right) \times 0.10 = 0.44000,$$

$$x_4 = 1.10513;$$

....

The results of the comparison between the numerical and the exact solutions are shown in Table 4.17. The reason for the peculiarly large relative error of x_1 is that the approximate integration formula of the first step is a trapezoidal rule and the error caused by it is large. That error of x_1 exerts a bad influence on the succeeding approximate

TABLE 4.17 Errors in the numerical solution of [V_2] by the Newton–Cotes formula

Representative points t	Solution $x(t)$ Approximate solution	Exact solution	Error	Relative error (%)
0.00	1.00000	1.00000	− 0.00000	0.000
0.05	1.05108	1.05127	− 0.00019	0.019
0.10	1.10513	1.10517	− 0.00004	0.004
0.15	1.16178	1.16183	− 0.00005	0.005
0.20	1.22133	1.22140	− 0.00007	0.007

TABLE 4.18 Errors in the method of numerical solution (improved values)

Representative points t	Solution $x(t)$ Approximate solution	Exact solution	Error	Relative error (%)
0.00	1.00000	1.00000	0.00000	0.000
0.025	1.02568	1.02531	0.00037	0.037
0.05	1.05131	1.05127	0.00004	0.004
0.10	1.10512	1.10517	0.00001	0.001
0.15	1.16183	1.16183	0.00000	0.000
0.20	1.22141	1.22140	0.00001	0.001

values. To improve the value at $t = 0.05$, perform the calculation with $h = 0.025$, and we get

$$x(0.05) = 1.05131$$

as far as the second step. Employ this value as x_1 and carry out the calculations as before; then we get the results shown in Table 4.18, in which the precision of approximation is seen to be remarkably improved.

If we calculate the problem with $h = 0.1$, we get the results shown in Table 4.19 in which the errors are seen to be gradually increasing in value.

TABLE 4.19 Errors in numerical solution

Representative points t	Solution $x(t)$ Approximate solution	Exact solution	Error	Relative error (%)
0.00	1.000 000	1.000 000	0.000 000	0.000
0.10	1.105 126	1.105 171	− 0.000 045	0.004
0.20	1.221 370	1.221 403	− 0.000 033	0.003
0.30	1.349 785	1.349 859	− 0.000 074	0.006
0.40	1.491 729	1.491 825	− 0.000 096	0.007
0.50	1.648 562	1.648 721	− 0.000 159	0.010
0.60	1.821 882	1.822 119	− 0.000 237	0.013
0.70	2.013 392	2.013 753	− 0.000 361	0.018
0.80	2.224 987	2.225 541	− 0.000 554	0.025

4.3. Methods for using approximate formulae

In Chapter 3 we explained how to construct a special method of solution, and in the first section of this chapter the approximation of a function. Now it is also possible to adopt approximate forms of kernels and unknown functions that are capable of being handled by the analytic methods explained in the previous chapter and thus to find the precise solution corresponding to the approximate solution obtained in this way.

For example, if a kernel takes the form $K(t-\xi)$, we may express it as a polynomial in $t-\xi$ by Newton's or Lagrange's method, and then solve it by the method explained in § 3.3 in the case of Fredholm type, or by the method adopted in § 3.5.2 for the Volterra type. Or we may express $K(t-\xi)$ in a Fourier series of finite terms using Fourier analysis, and solve it by the methods employed in § 3.2.3 or § 3.5.1.

These approximate methods of solution are available in cases when $K(t, \xi)$ takes the form of separated variables and can be expressed in the form $K_1(t)K_2(\xi)$. Even when $K(t, \xi)$ takes a fairly complicated form, we can solve it by the methods treated in the previous chapter if we approximate it by

$$\sum_{i=1}^{n} a_i(t)b_i(\xi), \quad \sum a_i(t)\frac{(t-\xi)^i}{i!}, \quad \sum K_s(m, n)\sin\frac{m\pi}{a}t\sin\frac{n\pi}{a}\xi.$$

In this sort of approximation, we may employ Taylor expansion, Fourier analysis, Newton's and Lagrange's interpolation formulae, etc.

4.3.1. Batemann's method (Batemann 1921)

We may easily solve a Fredholm homogeneous integral equation

$$x(t) = \lambda\int_0^1 K(t, \xi)x(\xi)d\xi \qquad [F_0]$$

if the kernel is given in the form of $\sum_{m=1}^{m} \alpha_m(t)\beta_m(\xi)$ (see § 3.4). A special technique is required to approximate a general function $K(t, \xi)$ by the sum of the products of the functions of the two variables. (It is a very difficult problem to select appropriate functions for α and β. Schmidt says that the first n terms of the orthogonal function system expansion of K will do, but from the practical point of view most of the time, these do not serve for numerical computation. With all these considerations, Batemann derived a more practical form of the problem. We shall present here just a brief introduction to his method, without going far into its detailed proof. The method is to approximate the given $K(t, \xi)$ by $\sum\alpha\beta$ so that they coincide on $n \times n$ lattice points.

Suppose that $\Gamma'(t, \xi; \lambda)$ is the approximation of the kernel of the solution of $[F_0]$; then it is given in the following way. Take n as an odd number, and

$$\begin{vmatrix} \Gamma'(t, \xi; \lambda) & K(t, \xi_1) & K(t, \xi_2) & \cdot & K(t, \xi_n) \\ K(t_1, \xi) & A_{11}(\lambda) & A_{12}(\lambda) & \cdot & A_{1n}(\lambda) \\ K(t_2, \xi) & A_{21}(\lambda) & A_{22}(\lambda) & \cdot & A_{2n}(\lambda) \\ \cdot & \cdot & \cdot & \cdot & \cdot \\ K(t_n, \xi) & A_{n1}(\lambda) & A_{n2}(\lambda) & \cdot & A_{nn}(\lambda) \end{vmatrix} = 0,$$

where $$A_{ij} = K(t_i, \xi_j) - \lambda \int_a^b K(t_i, \tau)K(\tau, \xi_j)d\tau.$$

The characteristic function is given in terms of the cofactors of the above determinant $\Gamma'(t, \xi; \lambda)$:

$$|A_{ij}(\lambda)| = 0.$$

Example 4.6 Eigenvalues

In the case of

$$K(t, \xi) = \begin{cases} t(1 - \xi), & t \leqq \xi; \\ \xi(1 - t), & \xi \leqq t; \end{cases}$$

the eigenvalues are π^2, $(2\pi^2)$, $(3\pi)^2$, ... (see Chapter 5).

Now

$$\int_0^1 K(t, \tau)\, K(\tau, \xi)d\tau = \frac{1}{6}t(1 - \xi)(2\xi - t^2 - \xi^2),\ t \leqq \xi,$$

or

$$\frac{1}{6}\xi(1 - t)(2t - \xi^2 - t^2),\ \xi \leqq t.$$

Let $n = 5$, put the lattice points at regular integrals, and put

$$t_1 = \xi_1 = \frac{1}{6}, \quad t_2 = \xi_2 = \frac{2}{6}, \quad \ldots, \quad t_3 = \xi_3 = \frac{5}{6}.$$

Furthermore, put $\lambda/216$ as a new unknown quantity x to simplify the calculation; then the characteristic function in x, namely $|A_{ij}(\lambda)|$, becomes

$$\begin{vmatrix} 5(1-10\,x) & 4(1-19\,x) & 3(1-26\,x) & 2(1-31\,x) & 1-34\,x \\ 4(1-19\,x) & 8(1-16\,x) & 6(1-23\,x) & 4(1-28\,x) & 2(1-31\,x) \\ 3(1-26\,x) & 6(1-23\,x) & 9(1-18\,x) & 6(1-23\,x) & 3(1-26\,x) \\ 2(1-31\,x) & 4(1-28\,x) & 6(1-23\,x) & 8(1-16\,x) & 4(1-19\,x) \\ 1-34\,x & 2(1-31\,x) & 3(1-26\,x) & 4(1-19\,x) & 5(1-10\,x) \end{vmatrix} = 0.$$

Then we calculate

$$130x^5 - 441x^4 + 488x^3 - 206x^2 + 30x - 1 = 0,$$

$$(x-1)(2x-1)(5x-1)(13x^2 - 22x + 1) = 0.$$

From this find x and re-express it by λ, then we have

$$\lambda_1 = 10.09, \quad \lambda_2 = 43.2, \quad \lambda_3 = 108, \quad \lambda_4 = 216, \quad \lambda_5 = 355.2$$

as the eigenvalues, where the error in the first root is about 2%, and that in the second 9%.

4.4. Historical methods for solving integral equations by machine

The development of the electric computer and the analogue computer in the late 1940s, made it possible to rapidly solve successive approximation and simultaneous equations by means of these machines, yet still at that time there was no suitable equipment applicable to solving the integral equation, like the differential analyser used for the ordinary differential equation.

Then Wallman (1947) invented an electric integral transform computer. This machine consisted of a battery for the disturbance function $f(t)$, a television battery and a valve multiplier for the kernel $K(t, \xi)$, a valve average calculator, and a synchronized cathode-ray tube system.

By dividing the kernel $K(t, \xi)$ into $n \times n$ small squares in a square domain of definition $S(0 \leq t, \ \xi \leq 1)$ and by supposing that the kernel takes a constant value within each square, the kernel K could be approximated by charging each square to a constant level. Techniques used in television enabled charging the lattice board (the surface of the Brown tube), which is mentioned above, by more than a million microcondensers. Thus, it was possible to approximate $K(t, \xi)$ in a sufficiently precise manner.

Then, in an integral transformer, the multiplier was used to get $K(t, \xi)f(\xi)$, and the average calculator was used to calculate an integration: $\int_0^1 K(t, \xi)f(\xi)d\xi$. These machines gave the result of the transform $K \circ f$ in only 0.01 of a second, if the charges corresponding to the value of $f(t)$ and $K(t, \xi)$ were input and all the preparatory arrangements were done beforehand.

Now to solve Fredholm's integral equation of the second kind,

$$x(t) - \int_0^1 K(t, \xi)x(\xi)d\xi = f(t),$$

making use of this integral transformer, we take f as input and $K \circ f$ as output. Feedback a part of the output and retransform it, then we have $\mathring{K}^2 \circ f$. In this way we get the solution:

$$f(t) + \sum_1^{\infty} \mathring{K}^i \circ f.$$

This is the very method of successive approximation which we studied in § 2.1.1. Since this expression converges, the output $f + \sum K \circ f$ is stabilized in a short time.

In the transform $\int_0^1 K(t, \xi)f(\xi)d\xi$, make t its parameter and we can calculate it using the integrator. In such a case, when the kernel takes a special form such as $K_1(t) + K_2(\xi)$, we may employ a mechanical integrator, but it takes a longer time to get the result than by using the valve one; besides the precision is not good. Yet for the case where the form of the kernel is restricted, as in the potential problem in § 5.5.1, special equipment was conceived. Föttinger (1928) later conceived the potential computer with sixteen wheels.

As for Volterra type integral equations, Tea (1948) and Aprile (1944) especially studied the method of finding the curve of the solution by common integrators in the case of convolution type.

Presently, computers are extensively applied to the numerical solution of integral equations or to that of simultaneous linear algebraic equations. However, this chapter may be useful to study the programming of the solution for integral equations.

4.5. Monte-Carlo method

We assume that a particle collides with others on a straight line at every second. $x(t)$ means the probability of the collision at the point t and $K(t, \xi)$ is the conditional probability of the collision at t after colliding at the point ξ. Then $f(t)$ is the probability that the first collision takes place at the point t.

Since the probability of a collision at the point t is the sum of probability of a first collision at t and the probability of a collision at t after colliding at some other place ξ in the interval (a, b), we have

$$x(t) = f(t) + \int_a^b K(t, \xi)x(\xi)d\xi,$$

where we assume that $f(t)$ and $K(t, \xi)$ are known. Then the equation is an integral equation of Fredholm type.

When the kernel $K(t, \xi)$ has a complicated expression, sometimes it is not easy to solve the equation even numerically.

By the Monte-Carlo method, assuming a random process one can calculate the solution of the integral equation. The Monte-Carlo method is useful for the numerical solution of integral equations when the integral part includes multiple integration.

The integral equation [F_2] can be expressed as

$$x(t) = f(t) + \lambda \int_a^b K(t, \xi)x(\xi)d\xi.$$

We consider that a particle will change its position on a straight line at each second $x(t)$ is the probability density of the particle in the segment (a, b) and $f(t)$ is the initial probability density function. The transient probability or the probability of shifting the position of the particle from ξ to t to $\lambda K(t, \xi)$.

By sampling a particle at the position ξ in the segment (a, b), the new position t can be determined by the transient probability $K(t, \xi)$. If the new position of the particle is outside the interval, then the particle will be neglected; the new position t should be remembered if x remains inside the interval (a, b).

After enough times of sampling, one can find the distribution density function of the particle which is an approximation of the solution $x(t)$ of the integral equation [F_2]. This process will be useful when the kernel $K(t, \xi)$ is complicated and hard to integrate analytically.

Exercises

1. Calculate the following integral equations numerically.

(a) $\int_0^1 (t-\xi)^2 x(\xi)d\xi = t^2 + 2t + 3$

[We may get one solution by the method of Table 4.7];

(b) $\int_0^\pi t\xi x(\xi)d\xi = t - 3$ [ditto];

(c) $\int_0^1 e^{t+\xi} x(\xi)d\xi = e^t(5e+6)$ [ditto];

(d) $x(t) - \int_0^1 K(t, \xi)x(\xi)d\xi = \frac{1}{2}t(1-t)$

$K(t, \xi) = \xi(1-t), (0 \leqq \xi \leqq t),$

$K(t, \xi) = t(1-\xi), (t \leqq \xi \leqq 1),$

$\left[x(t) = \tan\left(\frac{1}{2}\sin t\right) + \cos t - 1 \text{ (Hidaka 1941)}\right];$

(e) $x(t) - \frac{1}{2}\int_0^1 x(\xi)d\xi = e^t - \frac{e}{2} + \frac{1}{2}$, $[x(t) = e^t]$;

(f) $x(t) - \frac{1}{4}\int_0^{\frac{1}{2}\pi} t\xi x(\xi)d\xi = \sin t - \frac{t}{4}$, $[x(t) = \sin t]$;

(g) $\int_0^t (1 + t - \xi)x(\xi)d\xi = t,$ $[x(t) = e^{-t}$ (Hidaka 1941)];

(h) $\int_0^t (t-\xi)^2 x(\xi)d\xi = 2t^{-\frac{3}{2}} J_3(2\sqrt{t})$, $[x(t) = J_0(2\sqrt{t})]$;

(i) $\int_0^t (t\xi + \xi)x(\xi)d\xi = t^2 e^t + t - e^t + 1$

[Has the solution $x(t) = e^t$];

(j) $x(t) + \int_0^t x(\xi)d\xi = 1$ $[x(t) = e^{-t}$ (Hidaka 1941)];

(k) $x(t) + \int_0^t (t-\xi)x(\xi)d\xi = 2\cos t - t$

$[x(t) = 2\cos t - (1+t)\sin t$, (Hidaka 1941)];

(l) $x(t) + \int_0^t (6t - 6\xi - 5)x(\xi)d\xi = -t$, $[x(t) = e^{2t} - e^{3t}]$.

2. When the values of the kernel $K(t, \xi)$ and the disturbance function $f(t)$ are given as in Table 4.20, solve the integral equation

TABLE 4.20 Values of $K(t, \xi)$ and $f(t)$ at reprosentative points

	ξ \ t	0.00000	0.20000	0.40000	0.60000	0.80000	1.00000
$K(t, \xi)$	0.00000	0.00000	0.04000	0.16000	0.36000	0.64000	1.00000
	0.20000	0.04000	0.12000	0.28000	0.52000	0.84000	0.24000
	0.40000	0.16000	0.28000	0.48000	0.76000	1.12000	1.56000
	0.60000	0.36000	0.52000	0.76000	1.08000	1.48000	1.96000
	0.80000	0.64000	0.84000	1.12000	1.48000	1.92000	2.44000
	1.00000	1.00000	1.24000	1.56000	1.96000	2.44000	3.00000
$f(t)$		−102.33	−101.33	−98.33	−93.33	−86.33	−77.33

$$[F_0], [F_1], [F_2]; \ \lambda = 1; \ I(0, 1).$$

(The Newton–Cotes formula is available (Chapter 6), but we may, by the interpolation method, calculate the value at Tschebyscheff's representative points and calculate in a manner similar to Table 4.4.)

3. Determine the eigenvalues of the following integral equations by numerical calculation.

(a) $x(t) = \lambda \int_0^{10} \xi x(\xi) d\xi, \qquad [\lambda = 1/50];$

(b) $x(t) = \lambda \int_0^{1} t\xi x(\xi) d\xi, \qquad [\lambda = 3];$

(c) $K(t, \xi) = \begin{cases} t\xi(t - \xi), \ t > \xi; \\ t\xi(\xi - t), \ t < \xi. \end{cases} \quad I(0, 1) \quad$ [Compare with § 4.3.1].

4. In $K(t, \xi) = K_1(t)K_2(\xi)$, when each of $K_1(t)$, $K_2(\xi)$ is approximated by three terms of Newton's interpolation formula, determine x_0, Δx_0, $\Delta^2 x_0$ obtaining

$$x(t) = x_0 + \frac{t - a}{1!}\frac{\Delta x_0}{h} + \frac{(t - a)(t - a - h)}{2!}\frac{\Delta^2 x_0}{h^2}$$

as the solution.

5. Examine the expression of the error involved in the numerical integral formula and evaluate the errors of the various kinds of solutions treated in § 4.2, and then estimate the precision of the results given by Examples 4.1 to 4.5.

6. Give the reason why the Newton–Cotes and the Maclaurin formulae are used for Volterra type, and the Tschebyscheff and the Gauss formulae are used for Fredholm type in solving an integral equation numerically using numerical integration formulae. We use Maclaurin's formula for the solution of $[V_1]$, yet if we use the Newton–Cotes formula the error grows larger. Give a reason for this.

7. Write a program with which to calculate an integral equation numerically on a computer.
8. Solve the integral equation

$$x(t) - \int_{-1}^{1} K(t-\xi)x(\xi)d\xi = 1.$$

by the Monte-Carlo method.

$$K(u) = \frac{1}{\sqrt{2\pi}} e^{-\frac{u^2}{2}}$$

If the mean number of steps of a particle is $x(t)$, then we have

$$x(t) = 1 \times \left\{ \int_{-\infty}^{-1} K(t-\xi)d\xi + \int_{1}^{\infty} K(t-\xi)d\xi \right\}$$

$$+ \int_{-1}^{1} [1 + x(\xi)]K(t-\xi)d\xi$$

$$= \int_{-\infty}^{\infty} K(t-\xi)d\xi + \int_{-1}^{1} K(t-\xi)x(\xi)d\xi$$

the first term is 1 since $K(t-\xi)$ is the probability density function.

A particle is sampled and the number of steps until the particle steps out from the interval are counted. The random numbers of normal distribution are used to find the position of the particles. The mean value of the steps the approximate solution after enough times of solution.

References

Aprile, G. (1944) Un integrafo per la valutazione delle espression simboliche del calcolo operatorio funzionale. *Comment. Pontificia Acad. Sci.* **8**, 31–44.

Batemann, H. (1921) On the numerical solution of linear integral equations. *Proceedings of the Royal Society* A **100**, 441–449.

Dwyer, P. S. (1951) *Linear Computations*, (John Wiley, New York).

Föttinger, H. (1928) Die Entwickluug der 'Vektorintegratoren' zur machinellen Lösung von Potential und Wirbelproblemen. *Zeitschrift techn. Phys.* **9**, 26–39.

Hidaka, K. (1941) *Theory of Integral Equations* (in Japanese) pp. 268–272.

Tea, P.L. (1948) A mechanical integraph for the numerical solution of integral equations. *Journal of the Franklin Institute* **245**, 403–419.

Wallman, H. (1947) Electronic general transform computer. *Bull. Amer. Math. Soc.* 53/11, No. 421.

5 Applications of integral equations (I)

Initial-value problems of ordinary differential equations can be reduced to Volterra-type integral equations, and boundary-value problems of ordinary differential equations, and partial differential equations can be reduced to Fredholm-type integral equations. Therefore, various problems in mathematical physics, the fundamental equations of which are expressed as differential equations, can also be expressed as integral equations. Although we treat examples of these in §§ 5.1 and 5.2, a more detailed explanation of the association between ordinary differential equations and integral equations will be given in Chapter 10.

Potential theory is widely applied in physics, and its boundary-value problems can be expressed in integral equations. This is a special application of integral equations and, therefore, the last three sections are set aside for its detailed study. That is, the important facts on potential theory are put together in § 5.3. Dirichlet's problem is given in § 5.4, and Neumann's problem in § 5.5. These are the steps that paved the way to the historical study of E.I. Fredholm.

In Chapter 2 we studied the eigenvalues and eigenfunctions of a Fredholm-type integral equation; we now apply these in solving problems, through which we may further understand the physical significance of the eigenvalues and eigen functions.

5.1. Integral equations and ordinary differential equations

The problem of imposing on an ordinary differential a certain initial condition can always lead to the method of solution of Volterra-type integral equations. Since the initial condition is included within the integral equation, the solution of the integral equation agrees with the particular solution of the corresponding differential equation.

Suppose that we solve an ordinary differential equation of the nth order:

$$a_n(t)x^{(n)} + a_{n-1}(t)x^{(n-1)} + \cdots + a_0(t)x = f(t),\ t \geqq 0 \qquad (5.1)$$

under the initial conditions

$$x_0,\ x_0^{(1)},\ \ldots,\ x_0^{(n-1)}. \tag{5.2}$$

Here $a_n(t), a_{n-1}(t), \ldots, a_0(t)$ are continuous functions of t in $t \geqq 0$, and represent

$$x^{(i)} = \frac{d^i}{dt^i}x(t),\ x_0^{(i)} = \left(\frac{d^i x}{dt^i}\right)_{x=0}.$$

Now put

$$x^{(n)} = X(t), \tag{5.3}$$

and integrate successively, then we have

$$x^{(n-1)} = \int_0^t X(\xi)d\xi + x_0^{(n-1)},$$

$$x^{(n-2)} = \int_0^t \int_0^t X(\xi)d\xi\, dt + x_0^{(n-1)}t + x_0^{(n-2)}.$$

As the integral domain is T, if we change the order of integration according to Lemma 1.2, the last equation becomes

$$\begin{aligned} x^{(n-2)} &= \int_0^t X(\xi)d\xi \int_\xi^t dt + x_0^{(n-1)}t + x_0^{(n-2)} \\ &= \int_0^t (t-\xi)X(\xi)d\xi + x_0^{(n-1)}t + x_0^{(n-2)}. \end{aligned}$$

It further follow that

$$\begin{aligned} x^{(n-3)} &= \int_0^t X(\xi)d\xi \int_0^\tau (\tau-\xi)d\tau + \frac{x_0^{(n-1)}}{2!}t^2 + x_0^{(n-2)}t + x_0^{(n-2)} \\ &= \int_0^t \frac{(t-\xi)^2}{2!}X(\xi)d\xi + \frac{x_0^{(n-1)}}{2!}t^2 + x_0^{(n-2)}t + x_0^{(n-3)}. \end{aligned}$$

The same rule applies successively, and we have

$$\begin{aligned} x^{(n-\nu)} &= \int_0^t \frac{(t-\xi)^{\nu-1}}{(\nu-1)!}X(\xi)d\xi + \frac{x_0^{(n-1)}}{(\nu-1)!}t^{\nu-1} + \frac{x_0^{(n-2)}}{(\nu-2)!}t^{\nu-2} \\ &\quad + \cdots + x_0^{(n-\nu+1)}t + x_0^{(n-\nu)}. \end{aligned} \tag{5.4}$$

Put these in (5.1) and we get

$$a_n(t)X + a_{n-1}(t)\int_0^t X(\xi)d\xi + a_{n-1}(t)\int_0^t (t-\xi)X(\xi)d\xi$$

$$+ a_{n-3}(t)\int_0^t \frac{(t-\xi)^2}{2!}X(\xi)d\xi + \cdots$$

$$+ a_0(t)\int_0^t \frac{(t-\xi)^{n-1}}{(n-1)!}X(\xi)d\xi$$

$$= f(t) - a_{n-1}(t)x_0^{(n-1)} - a_{n-2}(t)(x_0^{(n-1)}t + x_0^{(n-2)})$$

$$- a_{n-3}(t)\left(\frac{x_0^{(n-1)}}{2!}t^2 + x_0^{(n-2)}t + x_0^{(n-3)}\right) - \cdots$$

$$- a_0(t)\left\{\frac{x_0^{(n-1)}}{(n-1)!}t^{n-1} + \frac{x_0^{(n-2)}}{(n-2)!}t^{n-1} + \cdots + x_0\right\}.$$

Then put

$$\left.\begin{aligned} &-a_n(t)K(t,\xi) \equiv a_{n-1}(t) + a_{n-2}(t)(t-\xi) \\ &+ a_{n-3}(t)\frac{(t-\xi)^2}{2!} + \cdots + a_0(t)\frac{(t-\xi)^{n-1}}{(n-1)!}, \\ &a_n(t)F(t) \equiv f(t) - a_{n-1}(t)x_0^{(n-1)} - a_{n-2}(t)(x_0^{(n-1)}t \\ &\qquad + x_0^{(n-2)}) - \cdots - a_0(t)\left\{\frac{x_0^{(n-1)}}{(n-1)!}t^{n-1}\right. \\ &\qquad \left. + \frac{x_0^{(n-2)}}{(n-2)!}t^{n-2} + \cdots + x_0\right\}, \end{aligned}\right\} \tag{5.5}$$

and, if $a_n(t) \neq 0$† in $t > 0$, a Volterra-type integral equation of the second kind

$$X(t) - \int_0^t K(t,\xi)X(\xi)d\xi = F(t)$$

is derived. We can solve this equation since the kernel $K(t, \xi)$ is bounded and continuous. Integrate the thus-obtained solution $X(t)$ n times successively and

$$x(t) = \int_0^t \frac{(t-\xi)^{n-1}}{(n-1)!}X(\xi)d\xi + \frac{x_0^{(n-1)}}{(n-1)!}t^{n-1} + \frac{x_0^{(n-2)}}{(n-2)!}t^{n-2}$$
$$+ \ldots + x_0^{(1)}t + x_0 \tag{5.6}$$

is obtained as the solution $x(t)$ of the differential equation (5.1). It is clear that this solution is continuous and satisfies the initial conditions (5.2).

In the differential equation (5.1), (5.6) is taken to be the general

† If there exists a positive value of t, with $a_n(t) = 0$, it leads to a Volterra-type integral equation of the third kind. This will be explained in Chapter 8.

solution containing n integral constants x_0, $x_0^{(1)}$, ..., $x_0^{(n-1)}$, if there are no initial conditions. As for the condition to determine the paticular solution, n linearly independent conditions are sufficient, not only for initial-value problems but also for boundary-value problems.

Example 5.1

From the differential equation

$$x^{(2)} - 5x^{(1)} + 6x = 0 \tag{5.7}$$

with initial conditions

$$x_0 = 0, \qquad x_0^{(1)} = -1 \tag{5.8}$$

we can get the integral equation

$$X(t) + \int_0^t (6t - 5\xi - 5)X(\xi)d\xi = 6t - 5. \tag{5.9}$$

This may of course be solved using the method treated in § 2.4; but here we put

$$X(t) = Ae^{\alpha t} + Be^{\beta t}, \tag{5.10}$$

integrate the regular part, rearrange it, and obtain

$$A\left(1 - \frac{5}{\alpha} + \frac{6}{\alpha^2}\right)e^{\alpha t} + B\left(1 - \frac{5}{\beta} + \frac{6}{\beta^2}\right)e^{\beta t} = 6t\left(1 + \frac{A}{\alpha}\right.$$
$$\left. + \frac{B}{\beta}\right) - 5\left\{1 + \frac{A}{\alpha} + \frac{B}{\beta} - \frac{6}{5}\left(\frac{A}{\alpha^2} + \frac{B}{\beta^2}\right)\right\}.$$

Now put

$$1 - \frac{5}{\alpha} + \frac{6}{\alpha^2} = 0, \quad 1 - \frac{5}{\beta} + \frac{6}{\beta^2} = 0, \quad 1 + \frac{A}{\alpha} + \frac{B}{\beta} = 0,$$
$$1 + \frac{A}{\alpha} + \frac{B}{\beta} - \frac{6}{5}\left(\frac{A}{\alpha^2} + \frac{B}{\beta^2}\right) = 0;$$

then these make

$$\alpha = 2, \quad \beta = 3, \quad A = 4, \quad B = -9,$$

and, therefore, the solution of the integral equation is

$$X(t) = 4e^{2t} - 9e^{3t}. \tag{5.11}$$

The integrals of this are

$$x^{(1)}(t) = 2e^{2t} - 3e^{3t}, \quad x(t) = e^{2t} - e^{3t}. \tag{5.12}$$

This solution agrees with the paticular solution of the differential equation $x(t) = e^{2t} - e^{3t}$, which is decided from $x(t) = Ce^{at} + De^{bt}$ under the initial condition.

Example 5.2

Let us discuss the motion of a particle of unit mass shot perpendicularly upward with initial velocity a, provided there exists a resistance proportional to the speed.

Take t as time and put the x-axis perpendicularly upward; then the equation of motion is

$$x^{(2)} - c\,x^{(1)} = -g, \tag{5.13}$$

with initial conditions

$$x_0 = 0, \qquad x_0^{(1)} = a. \tag{5.14}$$

Let the acceleration be

$$x^{(2)} = X(t); \tag{5.15}$$

then the equation of motion on displacement under the initial conditions becomes an integral equation:

$$X(t) - c\int_0^t X(\xi)d\xi - ca = -g, \tag{5.16}$$

and we obtain a solution:

$$X(t) = (ac - g)e^{ct}. \tag{5.17}$$

From this comes the equation of displacement

$$x(t) = -\frac{ac-g}{c^2} + \frac{g}{c}t + \frac{ac-g}{c^2}e^{ct}. \tag{5.18}$$

Therefore, in the discussion of motion, giving the differential equation of displacement and the initial conditions corresponds to giving the integral equation of acceleration.

5.2. Integral equations and boundary-value problems

The initial-value problem of an ordinary differential equation generally leads to a Volterra-type integral equation, but in the case of a boundary-value problem, it becomes of Fredholm type with boundary conditions. When an original equation is of separated-variable type, the boundary-value problem of a partial differential equation leads to the boundary-value problem of an ordinary differential equation; therefore, here we also have a Fredholm-type integral equation. Another section will be assigned to consider the boundary-value problems of Laplace's differential equations. Generally, we may move from boundary-value problem to an integral equation using Green's function, the detailed explanation of which will be given in Chapter 10.

5.2.1. The Study of Liouville (1837)

In the year 1837, J. Liouville showed that the particular solution of a certain differential equation becomes a solution of an integral equation.

The solution of an inhomogeneous differential equation of the second order

$$x^{(2)} + w_0^2 x = f(t) \tag{5.19}$$

is, as is well known,

$$x(t) = A \sin w_0(t-\alpha) + B \cos w_0(t-\alpha) + \frac{1}{w_0}\int_\alpha^t f(\xi)\sin w_0(t-\xi)d\xi. \tag{5.20}$$

Here w_0 is a constant and $f(t)$ is a continuous function. Now in

$$x^{(2)} + [w_0^2 - \sigma(t)]x = 0, \tag{5.21}$$

if $\sigma(t)$ is continuous, the initial conditions are given at $t = \alpha$ as

$$x_\alpha = 1, \qquad x_\alpha^{(1)} = 0; \tag{5.22}$$

then the constants in the above equation become $A = 0$, $B = 1$ and $x(t)$ becomes

$$x(t) = \cos w_0(t-\alpha) + \frac{1}{w_0}\int_\alpha^t \sigma(\xi)\sin w_0(t-\xi)x(\xi)d\xi, \tag{5.23}$$

which is a [V_2] type integral equation since it contains an unknown function in the integration on the right-hand side.

The purpose of Liouville's study was to find a series solution which rapidly converges for large values of w_0.

By the method of successive substitution, we have

$$x(t) = \cos w_0(t-\alpha) + \frac{1}{w_0}\int_\alpha^t \sigma(\xi)\sin w_0(t-\xi)\cos w_0(\xi-\alpha)d\xi$$

$$+ \frac{1}{{w_0}^2}\int_\alpha^t \sigma(\xi)\sin w_0(t-\xi)d\xi$$

$$\times \int_\alpha^\xi \sigma(\tau)\sin w_0(\tau-\xi)\cos w_0(\tau-\alpha)d\tau$$

$$+ \frac{1}{{w_0}^2}\int_\alpha^t \sigma(\xi)\sin w_0(t-\xi)d\xi \int_\alpha^\xi \sigma(\tau)\sin w_0(\tau-\xi)d\tau$$

$$\times \int_\alpha^\tau \sigma(\mu)\sin w_0(\mu-\tau)x(\mu)d\mu + \cdots. \tag{5.24}$$

In this way we can get the series solution of $1/w_0$, and, since $\sigma(t)$ is continuous, all the integrals in each term converge and when w_0 is large the convergence of (5.24) appears to be good.

5.2.2. *The buckling of a long column*

Select a coordinate axis t, as in Fig. 5.1, for a long column of uniform cross-section of length l under a compression load P in the direction of the axis(section modulus is EI); then for the horizontal displacement $x(t)$ of the central axis of the column, we obtain

$$(x-a)^{(2)} + \lambda(x-a) = 0, \quad \lambda = P/EI. \tag{5.25}$$

The boundary conditions in the case when the lower end is fixed and the upper end is free are given by

$$x_0 = {x_0}^{(1)} = 0, \quad x_l = a. \tag{5.26}$$

Here if we put $(x-a)^{(2)} = x^{(2)} = X(t)$, then we have

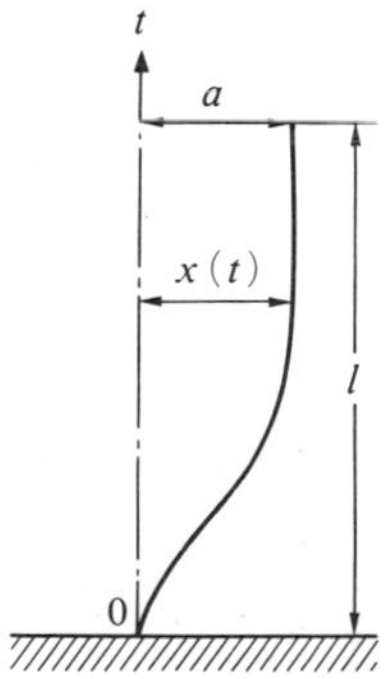

Fig. 5.1. Buckling displacement of a long column

$$x - a = \int_0^t (t - \xi)X(\xi)d\xi + c_1 t + c_2,$$

and from the boundary conditions we get $c_1 = 0$, $c_2 = -a$, and $\int_0^l (t - \xi)X(\xi)d\xi = a$. Therefore the integral equation is

$$X(t) + \lambda \int_0^t (t - \xi)X(\xi)d\xi = \lambda a, \tag{5.27}$$

or

$$X(t) - \lambda\left[\int_0^l (l - \xi)X(\xi)d\xi - \int_0^t (t - \xi)X(\xi)d\xi\right] = 0.$$

Here, putting

$$K(t, \xi) = \begin{cases} l - t, & 0 \leqq \xi < t, \\ l - \xi, & t \leqq \xi \leqq l, \end{cases}$$

we get the integral equation

$$X(t) = \lambda \int_0^l K(t, \xi)X(\xi)d\xi. \tag{5.28}$$

This is a Fredholm-type homogeneous integral equation.

We solve this directly from the differential equation as $X = -\lambda(x - a)$, and put it in (5.27), and then we get

$$x(t) + \lambda \int_0^t (t - \xi)x(\xi)d\xi = \frac{\lambda a}{2}t^2. \tag{5.29}$$

This is a Volterra-type equation. If we make the Laplace transform of it and we obtain

$$X(s) + \frac{\lambda}{s^2}X(s) = \frac{\lambda a}{2}\frac{2!}{s^3} \tag{5.30}$$

from which

$$X(s) = \frac{a\lambda}{s(s^2 + \lambda)}; \tag{5.31}$$

therefore, we have

$$x(t) = a(1 - \cos\sqrt{\lambda}t), \tag{5.32}$$

where a is determined by $x(l) = a$, and consequently when

$\lambda = 1/4(2n - 1)^2\pi^2/l^2$ we get

$$x(t) = \frac{1 - \cos\sqrt{\lambda t}}{1 - \cos\sqrt{\lambda l}}. \tag{5.33}$$

By § 2.2.4 Fredholm's determinant is given by

$$\begin{aligned}
\Delta(\lambda) &= 1 - \lambda\int_0^l K(t_1, t_1)dt_1 \\
&\quad + \frac{\lambda^2}{2!}\int_0^l\int_0^l \begin{vmatrix} K(t_1, t_1) & K(t_1, t_2) \\ K(t_2, t_1) & K(t_2, t_2) \end{vmatrix} dt_1 dt_2 + \ldots \\
&= 1 - \lambda\int_0^l (l - t_1)dt_1 + \frac{\lambda^2}{2!}\int_0^l dt_1 \int_0^{t_1} \begin{vmatrix} (l - t_1) & (l - t_1) \\ (l - t_1) & (l - t_2) \end{vmatrix} dt_2 \\
&\quad + \frac{\lambda^2}{2!}\int_0^l dt_1 \int_{t_1}^l \begin{vmatrix} (l - t_1) & (l - t_2) \\ (l - t_2) & (l - t_2) \end{vmatrix} dt_2 + \cdots \\
&= 1 - \lambda\frac{l^2}{2!} + \frac{\lambda^2 l^4}{4!} - \ldots \\
&= \cos\sqrt{\lambda l};
\end{aligned} \tag{5.34}$$

then, from $\sqrt{\lambda l} = (n - 1/2)\pi$, the eigenvalue is

$$\lambda = \frac{(2n - 1)^2\pi^2}{4l^2}.$$

A similar calculation gives

$$x(t) = \frac{1 - \cos\sqrt{\lambda t}}{1 - \cos\sqrt{\lambda l}}, \tag{5.35}$$

which agrees with (5.33).

The same process applies to cases of other end conditions.

5.2.3. *Free vibration of a string*

This example is apparently a boundary problem of a differential equation, yet by separating variables we may after all lead it to the boundary-value problem of an ordinary differential equation and then we can make it come down to the problem of an integral equation using the method of 5.2.1.

Suppose that there is an infinitesimal vibration of a uniform elastic string of length l. Set the fixed end as the origin, and let the x-axis coincide with the position of rest (Fig. 5.2), and let the transverse displacement of an arbitrary point for the time variable t be denoted by $y(x, t)$; then the equation of motion is well known to be

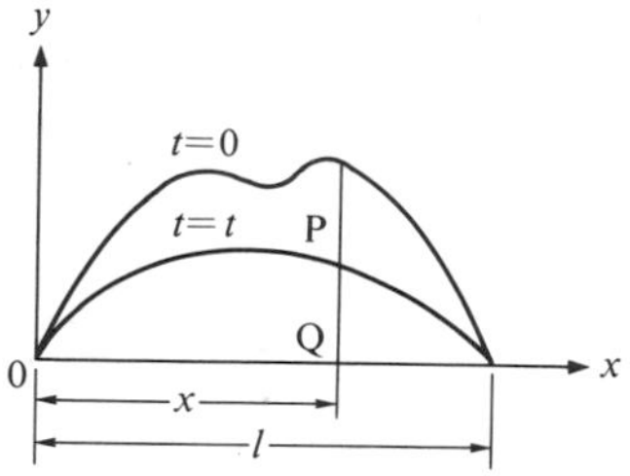

Fig. 5.2. Displacement of a string

$$\frac{\partial^2 y}{\partial t^2} = c^2 \frac{\partial^2 y}{\partial x^2}, \tag{5.36}$$

where the line density is ρ and the tension is T. Then $c^2 = T/\rho$ and this is a constant.

The conditions to make both ends fixed are given by

$$y(0, t) = 0, \qquad y(l, t) = 0, \tag{5.37}$$

and if given an initial displacement $g(x)$ and allowed to move freely, the initial conditions become

$$y(x, 0) = g(x), \qquad y_t(x, 0) = 0. \tag{5.38}$$

Solution of the differential equation by operational calculus.
The problem of solving the differential equation (5.36) under the conditions (5.37) and (5.38) is found in many references as a historically famous exercise of Fourier analysis; therefore, in the following we shall solve the problem using the operational calculus. Transform t into s by Laplace transformation; then (5.36) becomes an ordinary differential equation

$$\frac{d^2 Y}{dx^2} - \frac{s^2}{c^2} Y = -\frac{s^2}{c^2} g(x) \tag{5.39}$$

under (5.38). Let the solution of the homogeneous equation, the left-hand side of which is zero, be

$$Y(x, s) = A(x)\cosh\frac{sx}{c} + B(x)\sinh\frac{sx}{c}, \tag{5.40}$$

and we have only to determine $A(x)$, $B(x)$, so that we may have

$$A'(x)\cosh\frac{sx}{c} + B'(x)\sinh\frac{sx}{c} = 0. \tag{5.41}$$

Putting (5.40) into (5.39) under (5.41), we obtain

$$A'(x)\sinh\frac{sx}{c} + B'(x)\sinh\frac{sx}{c} = -\frac{1}{c}g(x); \tag{5.42}$$

then, from (5.41) and (5.42), we obtain

$$A'(x) = \frac{1}{c}g(x)\sinh\frac{sx}{c}, \quad B'(x) = -\frac{1}{c}g(x)\cosh\frac{sx}{a}.$$

Watching the conditions of both ends (5.37) on (5.40), we get

$$A(0) = 0, \qquad A(l)\cosh\frac{sl}{c} + B(l)\sinh\frac{sl}{a} = 0;$$

consequently we have

$$\begin{aligned} A(x) &= \int_0^x \frac{1}{c}g(\xi)\sinh\frac{s\xi}{c}\,d\xi, \\ B(x) &= -\cosh\frac{sl}{c}\int_0^l \frac{1}{c}g(\xi)\sinh\frac{s\xi}{c}\,d\xi \\ &\quad + \coth\frac{sl}{c}\int_x^l \frac{1}{c}g(\xi)\cosh\frac{s\xi}{a}\,d\xi. \end{aligned}$$

Therefore (5.40) becomes

$$\begin{aligned} Y(x, s) &= \int_0^x \frac{1}{c}g(\xi)\sinh\frac{s\xi}{c}\left(\cosh\frac{sx}{c} - \coth\frac{sl}{c}\sinh\frac{sx}{c}\right)d\xi \\ &\quad + \int_x^l \frac{1}{c}f(\xi)\sinh\frac{sx}{c}\left(\cosh\frac{s\xi}{c} - \coth\frac{sl}{c}\sinh\frac{s\xi}{c}\right)d\xi, \\ &= \int_0^x \frac{1}{c}f(\xi)\sinh\frac{s\xi}{c}\,\frac{\sinh s(l-x)/c}{\sinh sl/c}\,d\xi \\ &\quad + \int_x^l \frac{1}{c}f(\xi)\sinh\frac{sx}{c}\,\frac{\sinh s(l-\xi)/c}{\sinh sl/c}\,d\xi. \end{aligned} \tag{5.43}$$

take the inverse transform of (5.43) and we get a solution, where the poles are,

$$\sinh\frac{sl}{c} = 0, \qquad \frac{sl}{c} = in\pi \qquad (n = 0, \pm 1, \pm 2, \ldots).$$

Therefore,

$$\frac{1}{c}\sinh\frac{s\xi}{c}\,\frac{\sinh s(l-x)/c}{\sinh sl/c}$$

$$\subset \sum_{-\infty}^{\infty} \frac{\frac{in\pi}{l}\sinh\frac{in\pi\xi}{l}\sinh in\pi\frac{l-x}{l}}{in\pi\cosh(in\pi)l/c}\exp\left(in\pi\frac{ct}{l}\right)$$

$$= \frac{2}{l}\sum_{1}^{\infty}\sin\frac{n\pi\xi}{l}\sin\frac{n\pi x}{l}\cos\frac{n\pi ct}{l}.$$

In operational calculas when we write

$$F(S) \subset f(t),$$

$f(t)$ is an original function in original space and $F(S)$ is the image function in image space. The same result comes from the second term of (5.43). Thus the solution is

$$y(x, t) = \sum_{1}^{\infty}\frac{2}{l}\int_0^l g(\xi)\sin\frac{n\pi\xi}{l}d\xi\,\sin\frac{n\pi x}{l}\cos\frac{n\pi ct}{l}. \tag{5.44}$$

If in particular we put $t = 0$, then this becomes a Fourier expansion of $g(\xi)$. We notice that (5.44) satisfies the initial conditions (5.38)

The method of solution by integral equations (1).
Now suppose that the solution is the type of separated variables:

$$y = u(x)v(t); \tag{5.45}$$

putting this into the original equation we obtain

$$\frac{d^2v/dt^2}{v} = c^2\frac{d^2u/dx^2}{u}, \tag{5.46}$$

Setting this fraction equal to $-\lambda c^2$, we obtain the two ordinary differential equations

$$\frac{d^2v}{dt^2} + \lambda c^2 v = 0, \qquad \frac{d^2u}{dx^2} + \lambda u = 0. \tag{5.47}$$

Since $v(t) \not\equiv 0$, by the conditions at both ends we have

$$u(0) = 0, \qquad u(l) = 0. \tag{5.48}$$

Therefore, for $u(x)$ we finally arrive at the solution of an ordinary differential equation:

$$u^{(2)} + \lambda u = 0,$$

under the boundary conditions (5.48). Putting

$$u^{(2)} = U(x),$$

we have

$$u = \int_0^x (x - \xi)U(\xi)d\xi + c_1 x + c_2;$$

and since $u(0) = 0$, $u(l) = 0$, we get

$$c_2 = 0, \qquad c_1 = -\frac{1}{l}\int_0^l (l - \xi)U(\xi)d\xi.$$

Therefore, the differential equation (5.47) leads to the integral equation of $U(x)$:

$$U(x) + \lambda\int_0^x (x - \xi)U(\xi)d\xi - \lambda\frac{x}{l}\int_0^l (l - \xi)U(\xi)d\xi = 0. \quad (5.49)$$

Now, if we put

$$K(x, \xi) = \begin{cases} \xi(l - x), & 0 \leqq \xi < x, \\ x(l - \xi), & x \leqq \xi \leqq l, \end{cases} \quad (5.50)$$

then it becomes [F_0]:

$$U(x) = \lambda\int_0^l K(x, \xi)U(\xi)d\xi. \quad (5.51)$$

We have Theorem 2.5 for the existence of the solution of this integral equation, and as for the eigenvalue and eigenfunction, from

$$\Delta(\lambda) = \frac{1}{\sqrt{\lambda}}\sin\sqrt{\lambda}l, \quad (5.52)$$

$$\Delta\begin{pmatrix} t \\ \xi \end{pmatrix}; \lambda\Big) = A' \sin\frac{n\pi x}{l}, \quad (5.53)$$

there exists a solution not identically zero only when λ is $n^2\pi^2/l^2$ ($n = 1,2,3,\ldots$), and the eigenvalues and eigenfunctions $A' \sin n\pi x/l$ are in one-to-one correspondence. The method now follows the same course as for differential equations. And again (5.44) ends as the solution.

The method of solution by integral equations (2).
Let us try to solve the boundary-value problem by the operational calculus leading to a Volterra-type integral equation. Putting

$$\frac{1}{l}\int_0^l (l - \xi)U(\xi)d\xi = a, \quad (5.54)$$

Then (5.49) becomes

$$U(x) + \lambda\int_0^x (x - \xi)U(\xi)d\xi - \lambda x a = 0.$$

Take the Laplace transform and denote it by $U(x) \supset \bar{U}(s)$; then we have

$$\bar{U}(s) + \lambda \frac{1}{s^2} \bar{U}(s) - \lambda \frac{1}{s^2} a = 0,$$

and so

$$\bar{U}(s) = \frac{\lambda a}{s^2 + \lambda}.$$

Therefore, we have

$$U(x) = \lambda a \sin \sqrt{\lambda} x. \qquad (5.55)$$

To determine a, put this solution into (5.54) and we get

$$a \int_0^l (l - \xi) \sqrt{\lambda} \sin \sqrt{\lambda} \xi d\xi = al.$$

As long as we do not have

$$\int_0^l (l - \xi) \sqrt{\lambda} \sin \sqrt{\lambda} \xi d\xi = l, \qquad (5.56)$$

it must be that $a = 0$. But $U(x) \equiv 0$ when $a = 0$ then $u(x) \equiv 0$ and the solution is trivial. Now from (5.56)

$$l - \frac{\sin \sqrt{\lambda} l}{\sqrt{\lambda}} = l;$$

which implies that $\sin \sqrt{\lambda} l = 0$; that is,

$$\lambda = \frac{n^2 \pi^2}{l^2} \qquad (n = 1, 2, 3, \ldots). \qquad (5.57)$$

Here we have

$$U(x) = A' \sin \frac{n\pi x}{l}, \qquad u(x) = A \sin \frac{n\pi x}{l}$$

from (5.55).

5.2.4. *Forced vibration of a string*

Let us show that the problem of the forced vibration of a string leads to [F_2]. Since the force $\rho H(x, t)$ acting on the mass ρ of a uniform elastic string of length l is in the y-direction in Fig. 5.2, it follows that its equation of motion, boundary condition, and initial condition are

$$\frac{\partial^2 y}{\partial t^2} = c^2 \frac{\partial^2 y}{\partial x^2} + H(x, t), \tag{5.58}$$

$$y(0, t) = 0, \qquad y(l, t) = 0, \tag{5.59}$$

$$y(x, 0) = g(x), \qquad y_t(x, 0) = 0. \tag{5.60}$$

Now, assume that $H(x, t)$ is

$$H(x, t) = C^2 r(x) \cos(\beta t + \gamma) \qquad (C \neq 0), \tag{5.61}$$

and find the solution of the type

$$y(x, t) = u(x) \cos(\beta t + \gamma). \tag{5.62}$$

Putting $C^2\lambda = \beta^2$, we have the ordinary differential equation

$$u^{(2)} + \lambda u + r(x) = 0 \tag{5.63}$$

and the boundary conditions

$$u(0) = 0, \qquad u(l) = 0. \tag{5.64}$$

Now, if we put

$$u^{(2)} = U, \tag{5.65}$$

we may obtain the integral equation

$$U(x) = \lambda \int_0^l K(x, \xi) U(\xi) d\xi - r(x), \tag{5.66}$$

where

$$K(x, \xi) = \begin{cases} \xi(l - x), & 0 \leqq \xi \leqq x; \\ x(l - \xi), & x \leqq \lambda < l, \end{cases} \tag{5.67}$$

as before. From the above differential equation, $U + r = -\lambda u$, with which we transform the above into an integral equation in $u(x)$, and we get

$$u(x) = \lambda \int_0^l K(x, \xi) u(\xi) d\xi + \int_0^l K(x, \xi) r(\xi) d\xi.$$

Here, if we put

$$\int_0^l K(x, \xi) r(\xi) d\xi = f(x), \tag{5.68}$$

Then we can obtain [F_2] as

$$u(x) - \lambda \int_0^l K(x, \xi) u(\xi) d\xi = f(x). \tag{5.69}$$

According to Theorem 2.3, (5.69) has a unique solution:

$$u(x) = f(x) - \lambda \int_0^l \Gamma(t, \xi; \lambda) f(\xi) d\xi, \tag{5.70}$$

$$\Gamma(t, \xi; \lambda) = -\Delta\begin{pmatrix} t \\ \xi \end{pmatrix}; \lambda\Big) \Big/ \Delta(\lambda) \tag{5.71}$$

when $\lambda \neq n^2\pi^2/l^2$. When $\lambda = n^2\pi^2/l^2$, generally there exists no solution. Yet if we regard the eigenfunction of $[\tilde{F}_0]$ to be $\tilde{\varphi}$, then there exist an infinite number of solutions so far as

$$\int_0^l f(x)\tilde{\varphi}(x)dx = 0 \tag{5.72}$$

holds.

Further, since $K(t, \xi) = \tilde{K}(\xi, t)$ from (5.67), we have $[F_0] \equiv [\tilde{F}_0]$, and $\tilde{\varphi}(x)$ agrees with the eigenfunction of $[F_0]$. Therefore, the condition (5.72) is

$$\int_0^l f(x) \sin\frac{n\pi}{l} x \, dx = 0, \tag{5.73}$$

and in that case $[F_2]$, from Theorem 2.8, has the solutions

$$x(t) = f(t) - \frac{n^2\pi^2}{l^2} \int_0^l \Delta\Big(\begin{matrix} t \\ \xi \end{matrix}; n^2\pi^2\Big) f(\xi) d\xi + C \sin\frac{\pi n}{l} x. \tag{5.74}$$

Now (5.68) is an integral equation of the first kind in $r(x)$; if we put this into (5.73) we get

$$\int_0^l \int_0^l K(t, \xi) r(\xi) \sin\frac{n\pi}{l} x \, dx \, d\xi = 0. \tag{5.75}$$

While, since $\sin(n\pi/l)x$ is the eigenfunction of $[F_0]$, $[\tilde{F}_0]$, then we are led to

$$\sin\frac{n\pi}{l} x = \frac{n^2\pi^2}{l^2} \int_0^l K(t, \xi) \sin\frac{n\pi}{l} \xi d\xi,$$

$$\sin\frac{n\pi}{l} \xi = \frac{n^2\pi^2}{l^2} \int_0^l K(t, \xi) \sin\frac{n\pi}{l} x \, dx.$$

Therefore, (5.75) becomes

$$\frac{l^2}{n^2\pi^2} \int_0^l r(\xi) \sin\frac{n\pi}{l} \xi d\xi = 0; \tag{5.76}$$

and finally the necessary and sufficient condition for the boundary-value problems (5.63) and (5.64) to be solved is (5.76).

5.2.5. *The problem of linear heat conduction*

Suppose that there is an infinitely long uniform substance. The cross-sectional area A is constant and the lateral face is fully insulated so that no heat goes in or out, and the heat flows only in the longitudinal direction. This sort of problem is called the problem of heat conduction in one dimension; or linear heat conduction.

Place the origin in an arbitrary section of the substance and put the x-axis in the longitudinal direction; then the temperature u of an arbitrary section is a function $u(x, t)$ of the position x and time t. The equation of heat conduction is obtained by writing down the condition of caloric continuity. That is, in a small part $A \cdot \Delta x$ between the section x and the section $x + \Delta x$, the heat in this part at a certain instant is expressed as

$$s\rho A \Delta x(u + 273),$$

where we neglect infinitesimals of higher order. Here s shows specific heat (cal/g deg), ρ density (g/cm³) and u shows temperature (°C). Under these circumstances, the caloric inflow through the section x in a short time Δt is expressed as

$$-KA\left(\frac{\partial u}{\partial x}\right)_x \Delta t,$$

and the caloric outflow through the section $x + \Delta x$ is

$$-KA\left(\frac{\partial u}{\partial x}\right)_{x+\Delta x} \Delta t,$$

where K is the heat conductivity (cal/cm deg sec) and $\partial u/\partial x$ is the temperature gradient (deg/cm). The caloric continuum is therefore expressed as

$$s\rho A \cdot \Delta x(u + 273) = KA\left\{\left(\frac{\partial u}{\partial x}\right)_{x+\Delta x} - \left(\frac{\partial u}{\partial x}\right)_x\right\}\Delta t.$$

Then put

$$\frac{K}{s\rho^2} = c^2 \text{ (cm}^2\text{/sec)},$$

(where c^2 is the diffusibility) and make $\Delta x \to 0$, $\Delta t \to 0$, and we have

$$\frac{\partial u}{\partial t} = c^2 \frac{\partial^2 u}{\partial x^2}, \tag{5.77}$$

for which we have assumed that derivatives exist. This is the standard equation of heat conduction in one dimension.

Now, suppose that there is a one-dimensional substance of length l, one end of which remains at zero degrees and the other one has a caloric inflow which changes as time passes; then the initial and boundary conditions are expressed as

$$u(x, 0) = 0, \tag{5.78}$$

$$u(0, t) = 0, \tag{5.79}$$

$$\left(\frac{\partial u}{\partial x}\right)_{x=l} = f(t). \tag{5.80}$$

Under the initial condition (5.78), transform t into s and (5.77) becomes

$$sU = c^2 \frac{d^2U}{dx^2}$$

by Laplace transformation, where

$$u(x, t) \supset U(x, s),$$

and its general solution is

$$U = A(s)\exp\left(\frac{\sqrt{s}}{c}x\right) + B(s)\exp\left(-\frac{\sqrt{s}}{c}x\right). \tag{5.81}$$

We have only to take such $A(s)$ and $B(s)$ that satisfy the remaining two conditions (5.79), (5.80). Then, in order to satisfy the condition (5.79), we must have

$$A(s) = -B(s).$$

Therefore, we adopt a new unknown function of s, namely $X(s)$, and put

$$A(s) = -B(s) = \frac{c}{\sqrt{s}}\exp\left(-\frac{\sqrt{s}}{c}l\right)X(s);$$

then (5.81) becomes

$$U(x, s) = \frac{c}{\sqrt{s}}\left\{\exp\left[-\frac{l-x}{c}\sqrt{s}\right] - \exp\left[-\frac{l+x}{c}\sqrt{s}\right]\right\}X(s). \tag{5.82}$$

Then, the problem remaining to be solved is to determine $X(s)$ so that it satisfies the condition (5.80). Differentiate both sides of (5.82) with respect to x and put $x = l$; then we have

$$\left(\frac{dU}{dx}\right)_{x=l} = X(s) + \exp\left[-\frac{2l}{c}\sqrt{s}\right]X(s). \tag{5.83}$$

While by the formula of the operational calculus† we have

$$\exp(-a\sqrt{s}) \subset \frac{a}{2\sqrt{(\pi t^3)}}\exp\left(-\frac{a^2}{4t}\right),$$

then we use the convolution theorem with $X(s) \subset x(t)$ and (5.80) proves to be the integral equation:

$$x(t) + \frac{1}{\sqrt{\pi}}\frac{l}{c}\int_0^t \frac{\exp\left[-\frac{l^2}{c^2(t-\xi)}\right]}{(t-\xi)^{3/2}} x(\xi)d\xi = f(t). \tag{5.84}$$

This integral equation becomes

$$\left[1 + \exp\left(-\frac{2l}{c}\sqrt{s}\right)\right]X(s) = F(s) \tag{5.83a}$$

in the image space of the Laplace transform on condition that $F(s) \subset f(t)$. Therefore, the solution in the image space is

$$X(s) = \frac{F(s)}{1 + \exp\left(-\frac{2l}{c}\sqrt{s}\right)}$$

$$= \left\{1 + \sum_1^{\infty}(-1)^n \exp\left[-\frac{2l}{c}n\sqrt{s}\right]\right\}F(s). \tag{5.85a}$$

In the original space, by the convolution theorem it becomes

$$x(t) = f(t) + \frac{1}{\sqrt{\pi}}\frac{l}{c}\sum_{n=0}^{\infty}(-1)^n \int_0^t \frac{n\exp\left[-\frac{n^2l^2}{c^2(t-\xi)}\right]}{(t-\xi)^{3/2}} f(\xi)d\xi. \tag{5.85}$$

Lastly, putting (5.85a) into (5.81) we find the temperature distribution $u(x, t)$ and obtain

$$U(x, s) = \frac{c}{\sqrt{s}}\left\{\exp\left[-\frac{l-x}{c}\sqrt{s}\right] - \exp\left[-\frac{l+x}{c}\sqrt{s}\right]\right\}F(s)$$

$$+\sum_1^{\infty}(-1)^n\frac{c}{\sqrt{s}}\left\{\exp\left[-\frac{(2n+1)l-x}{c}\sqrt{s}\right]\right.$$

† For example, cit. R.V. Churchill (1944) *Modern Operational Mathematics in Engineering*, McGraw-Hill, N.Y., formula 82.

$$- \exp\left[-\frac{(2n+1)l+x}{c}\sqrt{s}\right]\Bigg\}F(s) \qquad (5.86a)$$

in the image space. By the formula of the operational calculus (Churchill 1941) we have

$$\sqrt{\left(\frac{\pi}{s}\right)}\exp(-a\sqrt{s}) \subset \frac{1}{\sqrt{t}}\exp\left(-\frac{a^2}{4t}\right);$$

so, in the original space (5.86a) becomes

$$u(x,t) = \frac{c}{\sqrt{\pi}}\sum_{n=0}^{\infty}(-1)^n$$

$$\times \int_0^t \frac{\exp\left[-\frac{[(2n+1)l-x]^2}{4c^2(t-\xi)}\right] - \exp\left[-\frac{[(2n+1)l+x]^2}{4c^2(t-\xi)}\right]}{\sqrt{(t-\xi)}} f(\xi)d\xi. \qquad (5.86)$$

It remains to note that since in this method of solution we have omitted the discussion of the change of the order of differentiation and integration or the uniform convergence of the infinite series, it is necessary to examine whether (5.86) is uniformly convergent and is a solution which satisfies the differential equation (5.77).

5.3. Potential

In the two-dimensional problems of hydrodynamics, elastic dynamics, heat conduction, and electromagnetism, the standard equation in most cases takes the form of a Laplace differential equation and is expressed as its boundary-value problem. In general, this has been studied as potential theory (Kellog 1929). In this case there are two ways of giving a boundary condition, which are called boundary-value problems of the first and second kinds. These will be explained precisely in the following two sections, while in this section we shall give the fundamental matters on potential.

5.3.1. Jordan curve

In terms of the parameter t, x and y are expressed as

$$x = \varphi(t), \quad y = \psi(t), \quad a \leqq t \leqq b, \qquad (5.87)$$

and when φ, ψ are continuous functions, the locus of the point P(x, y) makes a curve. In such a case, if

$$\varphi(t_1) = \varphi(t_2), \qquad \psi(t_1) = \psi(t_2) \tag{5.88}$$

do not hold simultaneously for two different values t_1, t_2, then the curve does not contain any multiple point and is called a *Jordan curve.*

Next, a Jordan curve is called a *smooth Jordan curve* when $\varphi(t)$ and $\psi(t)$ have continuous derivatives of the first order, $\varphi'(t)$, $\varphi'(t)$, and these do not become zero simultaneously. A smooth Jordan curve has a tangent at every point and has no singular point. The arc of a curve has a certain length. Therefore, taking the length of the arc s instead of the parameter t, we can express a smooth Jordan curve as

$$x = \xi(s), \quad y = \eta(s), \quad 0 \leq s \leq l; \tag{5.89}$$
$$\xi'^2 + \eta'^2 = 1,$$

where l is the whole length of the curve.

Further, if $\varphi(t)$, $\psi(t)$ or $\xi(s)$, $\eta(s)$ have continuous derivatives of the second order, then the curvature is finite at every point on the curve and it changes continuously on the curve. This sort of curve is called a *Jordan curve of continuous curvature.*

If

$$\varphi(a) = \varphi(b) \quad \text{and} \quad \psi(a) = \psi(b), \tag{5.88a}$$

or

$$x(0) = \xi(l) \quad \text{and} \quad y(0) = \eta(l) \tag{5.89a}$$

on a Jordan curve, then the curve is called a *closed Jordan curve.*

According to Jordan's theorem (Tsuji 1940) '*a closed Jordan curve divides a plane into interior and exterior regions.*' Here the exterior region is the one which contains the point at infinity. Thus the closed Jordan curve forms the boundary of two domains.

5.3.2. *Directional derivatives*

Let a function $f(x, y)$ be defined in the region R as shown in Fig. 5.3. If the limit

$$\lim_{\Delta r \to 0} \frac{f(x_1, y_1) - f(x, y)}{\Delta r} \tag{5.90}$$

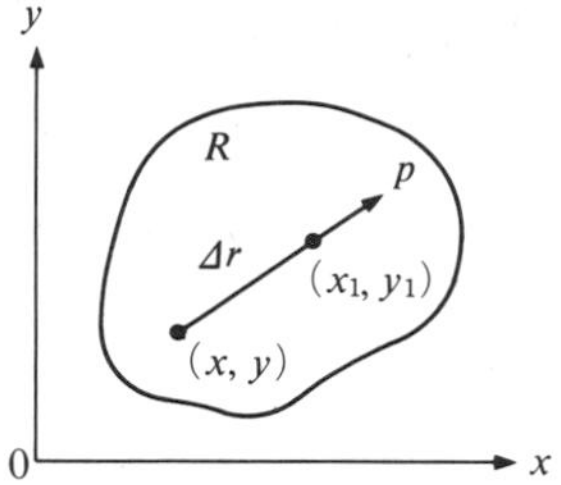

Fig. 5.3. The domain R

exists, we call it a directional derivative in the p direction, and denote it by the symbol $\partial f/\partial p$. Δr is the distance as shown in Fig. 5.3.

When there exist continuous partial derivatives of $f(x, y)$ we have

$$f(x+\Delta x, y+\Delta y)-f(x, y)=f_x(x, y)\Delta x + f_y(x, y)\Delta y+\alpha\Delta x+\beta\Delta y,$$

where α, β converge as $\Delta x \to 0$, $\Delta y \to 0$. Dividing both sides by $\Delta r =((\Delta x)^2+(\Delta y)^2)^{\frac{1}{2}}$ and calculating the limit as $\Delta r \to 0$ we obtain

$$\frac{\partial f(x, y)}{\partial p}=\frac{\partial f}{\partial x}\cos(px)+\frac{\partial f}{\partial y}\cos(py), \tag{5.91}$$

where (px) means the angle between the p-direction and the x-axis, and (py) is defined similarly.

5.3.3. *Green's theorem*

Let C be a closed smooth Jordan curve. Suppose that the curve C intersects lines parallel to the x-axis or the y-axis only finite times. Represent C in

$$x=\xi(s), \quad y=\eta(s), \quad 0 \leqq s \leqq l, \tag{5.89}$$

and write R_e for the exterior region including the boundary C, and R_i for the interior region including the boundary.

LEMMA 5.1 *Green's theorem*
If $P(x, y)$ and $Q(x, y)$ are continuous in R_i to the extent of the partial derivatives of the first order, we have

$$\int_C [P(x, y)dx+Q(x, y)dy]=\int\int_{R_i}\left(\frac{\partial Q}{\partial x}-\frac{\partial P}{\partial y}\right)dx\,dy, \tag{5.92}$$

where the curvilinear integral along C is taken in the positive direction (counter clockwise).

Putting $P = v(\partial u/\partial y)$, $Q = -v(\partial u/\partial x)$, we obtain the following result.

LEMMA 5.2

If $u(x, y)$ is continuous to the extent of the partial derivatives of the second order, and $v(x, y)$ is continuous to the extent of the partial derivatives of the first order, then

$$\int_C v\left(\frac{\partial u}{\partial y}dx - \frac{\partial u}{\partial x}dy\right) = -\iint_{R_i} v\,\nabla^2 u\,dx\,dy$$

$$-\iint_{R_i}\left(\frac{\partial u}{\partial x}\frac{\partial v}{\partial y} + \frac{\partial u}{\partial y}\frac{\partial v}{\partial y}\right)dx\,dy; \qquad (5.93)$$

here, the operational symbol ∇^2 means $\partial^2/\partial x^2 + \partial^2/\partial y^2$.

Now suppose that n is the inner normal at the point (x, y) on the curve C, then we have

$$\frac{\partial u}{\partial n} = \frac{\partial u}{\partial x}\cos(nx) + \frac{\partial u}{\partial y}\cos(ny);$$

but since $\cos(nx) = -dy/ds$, $\cos(ny) = dx/ds$, this becomes

$$\frac{\partial u}{\partial n} = -\frac{\partial u}{\partial x}\frac{dy}{ds} + \frac{\partial u}{\partial y}\frac{dx}{ds},$$

which is written as

$$\int_C v\left(\frac{\partial u}{\partial y}\frac{dx}{ds} - \frac{\partial u}{\partial y}\frac{dy}{ds}\right)ds = \int_C v\frac{\partial u}{\partial n}ds$$

by Green's theorem, and it takes the form given in the following.

LEMMA 5.3

If u is continuous to the extent of the partial derivatives of the second order, and v is continuous to the extent of the partial derivatives of the first order in the region R_i, then

$$\int_C v\frac{\partial u}{\partial n}ds = \iint_{R_i} v\,\nabla^2 u\,dx\,dy$$

$$-\iint_{R_i}\left(\frac{\partial u}{\partial x}\frac{\partial v}{\partial x} + \frac{\partial u}{\partial y}\frac{\partial v}{\partial y}\right)dx\,dy. \qquad (5.94)$$

5.3.4. *Harmonic functions*

When a function $u(x, y)$ is continuous to the extent of the partial derivative of the second order in the region R_i and satisfies

$$\nabla^2 u \equiv \frac{\partial^2 u}{\partial x^2} + \frac{\partial^2 u}{\partial y^2} = 0,\dagger \tag{5.95}$$

then u is called a *harmonic function.*

Put $v = 1$, $\nabla^2 u = 0$ in Green's theorem and we can easily obtain the following lemma on harmonic functions.

LEMMA 5.4

When u is a harmonic function in the region R_i, we have

$$\int_C \frac{\partial u}{\partial n} ds = 0. \tag{5.96}$$

If $v = u$, $\nabla^2 u = 0$, the following lemma is obtained.

LEMMA 5.5

When u is a harmonic function in R_i, we have

$$-\int_C u \frac{\partial u}{\partial n} ds = \int\int_{R_i} \left\{ \left(\frac{\partial u}{\partial x} \right)^2 + \left(\frac{\partial u}{\partial y} \right)^2 \right\} dx\, dy. \tag{5.97}$$

If a harmonic function u is zero everywhere along the boundary C, then by the above equation we have

$$\int\int_{R_i} \left\{ \left(\frac{\partial u}{\partial x} \right)^2 + \left(\frac{\partial u}{\partial y} \right)^2 \right\} dx\, dy = 0,$$

and in R_i we have

$$\frac{\partial u}{\partial x} \equiv 0, \qquad \frac{\partial u}{\partial y} \equiv 0;$$

thus, u is constant in R_i. While it is zero on the boundary it is always zero in the interior; then, finally, we get the following result.

A harmonic function which is zero on the boundary also becomes zero in the interior.

† $\nabla^2 u = 0$ is called Laplace's equation. The operator ∇^2 is called a Laplace operator or a Laplacian. Sometimes ∇^2 is denoted Δ.

In a similar manner we may prove that *a harmonic function for which $\partial u/\partial n = 0$ on the boundary takes a constant value in the region.*

The following lemma can be proved on the exterior region.

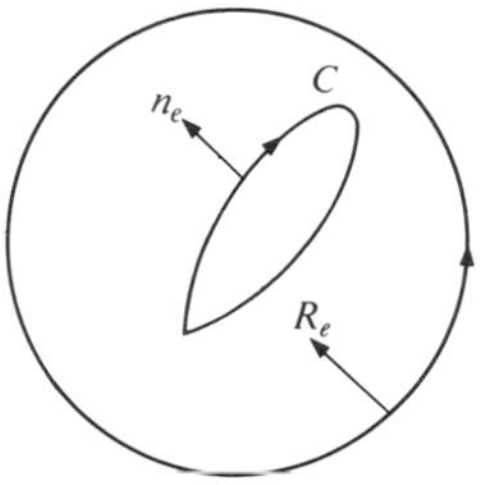

Fig. 5.4 n_e, R_e

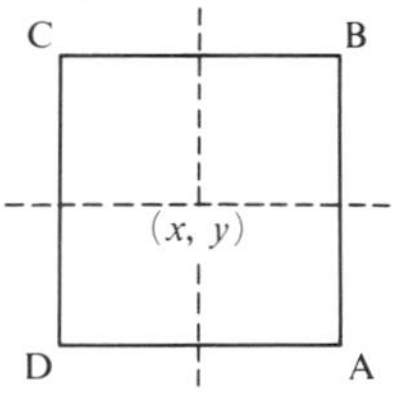

Fig. 5.5 $\delta x \times \delta y$

LEMMA 5.6

Given $u(x, y)$ as a harmonic function in R_e, if $u(r\cos\theta, r\sin\theta) \equiv U(r, \theta)$, with the property that $\lim_{r\to\infty} r\, U\, \partial U/\partial r$ is uniformly convergent to zero on θ, then we have

$$\iint_{R_e} \left[\left(\frac{\partial u}{\partial x}\right)^2 + \left(\frac{\partial u}{\partial y}\right)^2\right] dx\, dy = -\int_C u \frac{\partial u}{\partial n_e} ds. \qquad (5.98)$$

Here, n_e is the outer normal and the curvilinear integral along C is taken in the positive direction in the exterior region (Fig. 5.4).

From this lemma we can easily prove the following two results.

A harmonic function, which is zero on the boundary and in which $\lim_{r\to\infty} r\, U\, \partial U/\partial r = 0$ holds, becomes zero in R_e.

A harmonic function, in which $\partial u/\partial n$ is zero on the boundary, and which itself satisfies the above condition, takes a constant value in R_e.

To prove this lemma, consider a large circle which includes C, and set up Lemma 5.6 on the interior regions of both boundaries and let $r \to \infty$.

Example 5.3 *Velocity potential, stream function*

Potential is often used in the problems of gravity or electrostatics. We shall now give an example of hydrodynamics.

Consider a field of flow of an ideal fluid in a two-dimensional plane, and let u, v be the velocity components of flow parallel to the coordinate axis. Place a little square $(\delta x \times \delta y)$ around the centre (x, y) (Fig. 5.5), and then, denoting density by ρ, the inflow through CD in a unit time is

$$\left(\rho u - \frac{1}{2}\frac{\partial \rho u}{\partial x}\delta x\right)\delta y,$$

and the outflow is

$$\left(\rho u + \frac{1}{2}\frac{\partial \rho u}{\partial x}\delta x\right)\delta y,$$

therefore, the balance becomes

$$-\frac{\partial \rho u}{\partial x}\delta x\,\delta y.$$

Calculate flows along the y-axis in the same manner and we get

$$-\frac{\partial \rho v}{\partial y}\delta x\,\delta y.$$

Under the assumption of incompressible fluid, that there is no change of density, the sum of the two becomes zero; thus

$$\frac{\partial u}{\partial x} + \frac{\partial v}{\partial y} = 0.$$

This is called the condition of continuity.

The integral taken along an arbitrary line AP in the field of stream:

$$\int_A^P (u\,dx + v\,dy),$$

is called a flow. In the present case the flow that goes round the square is the sum of the flows along AB, BC, CD, DA; this is

$$\begin{aligned}&\left\{v + \frac{1}{2}\left(\frac{\partial v}{\partial x}\right)\delta x\right\}\delta y + \left\{-u - \frac{1}{2}\left(\frac{\partial u}{\partial y}\right)\delta y\right\}(-\delta x)\\ &\quad + \left\{-v + \frac{1}{2}\left(\frac{\partial v}{\partial x}\right)\delta x\right\}(-\delta y) + \left\{u - \frac{1}{2}\left(\frac{\partial u}{\partial y}\right)\delta y\right\}\delta x\\ &\quad = \left(\frac{\partial v}{\partial x} - \frac{\partial u}{\partial y}\right)\delta x\,\delta y.\end{aligned}$$

Here, if

$$\frac{\partial v}{\partial x} - \frac{\partial u}{\partial y} = 0,$$

we call the flow irrotational.

Consider a circuit integral of the flow along C in an irrotational flow; using Green's theorem we obtain

$$\int_C (u\,dx + v\,dy) = \int\int_{R_i} \left(\frac{\partial v}{\partial x} - \frac{\partial u}{\partial y}\right) dx\,dy,$$

and this integral always becomes zero. Therefore, in a simply connected region, a flow along the curve from a fixed point A to a point P is independent to the shape of the curve. Therefore, put

$$\varphi = -\int_A^P (u\,dx + v\,dy).$$

Displace P a distance no more than dx, and we have

$$\frac{\partial \varphi}{\partial x} = -u;$$

similarly, we can prove that

$$\frac{\partial \varphi}{\partial y} = -v.$$

By the condition of continuity we get

$$\nabla^2 \varphi = 0$$

We call this φ the velocity potential.

Next, let ds be the line element of the curve connecting AP, and let l, m be the direction cosines of the left-hand side normal; the integral

$$\psi = \int_A^P (l\,u + m\,v) ds$$

is called the flux, which passes along AP from right to left. Displace P a distance no more than dy, and we get

$$\frac{\partial \psi}{\partial y} = -u;$$

displace P again in the same manner, and then

$$\frac{\partial \psi}{\partial x} = v.$$

Therefore, the condition of being irrotational becomes

$$\nabla^2 \psi = 0,$$

and ψ is called the stream function.

Thus we have proved that the velocity potential and the stream function intersect orthogonally:

$$\frac{\partial \varphi}{\partial x}\frac{\partial \psi}{\partial x} + \frac{\partial \varphi}{\partial y}\frac{\partial \psi}{\partial y} = 0.$$

In addition, we find that these correspond to the real and imaginary parts of the analytic function of the complex number $x + iy$ (Lamb 1932, Chapter III).

5.3.5 *Boundary limits of harmonic functions*

We have treated a harmonic function in the region which includes the boundary. Now let us see how Green's theorem is applied when we exclude the boundary. We call the interior region excluding the boundary R_i', and the exterior region with the same property R_e'.

Let C_ε be a curve with the same property as C within R_i' and express it as

$$x = x(s, \varepsilon), \qquad y = y(s, \varepsilon); \tag{5.89b}$$

assume that $\lim_{\varepsilon \to 0} C_\varepsilon = C$ uniformly on s (Fig. 5.6). Then for the harmonic function u in R_i' on s, we assume that

(i) $\lim_{\varepsilon \to 0} u = u_i$,

(ii) $\lim_{\varepsilon \to 0} \partial u/\partial n = \partial u_i/\partial n$

uniformly also. In addition

(iii) $|\partial u/\partial x|\ |\partial u/\partial y|$ is bounded in R_i'.

Call these three conditions (A); then we have the following lemma.

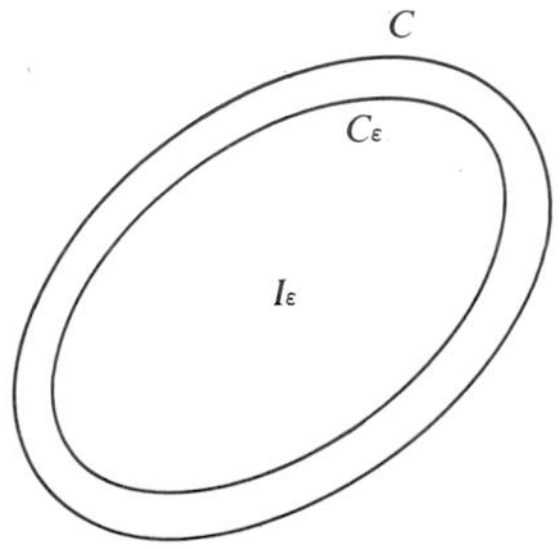

Fig. 5.6. C and C_ε

LEMMA 5.7

If a harmonic function u(x, y) in R_i' satisfies (A), *then we have*

$$\iint_{R_i}\left[\left(\frac{\partial u}{\partial x}\right)^2+\left(\frac{\partial u}{\partial y}\right)^2\right]dx\,dy=-\int_C u_i\frac{\partial u_i}{\partial n}ds. \qquad (5.99)$$

As stated before in page 203, we can prove the following two results.

When $u_i \equiv 0$ *on C*, $u \equiv 0$ *in* R_i'.
If $\partial u_i/\partial n = 0$ *on C, u takes a constant value in* R_i'.

For the exterior region R_e', we can prove the following lemma by introducing three conditions (B) similar to (A) for the interior region R_i'.

LEMMA 5.8

If a harmonic function u(x, y) in R_e' satisfies (B) *then*

$$\iint_{R_e}\left[\left(\frac{\partial u}{\partial x}\right)^2+\left(\frac{\partial u}{\partial y}\right)^2\right]dx\,dy=-\int_C u_e\frac{\partial u_e}{\partial n_e}ds. \qquad (5.100)$$

A similar relationship is found between the boundary value and the value of the function in the region, and we can prove two results.

When $u_e \equiv 0$ *on C*, $u \equiv 0$ *in* R_e'.
When $\partial u_e/\partial n_e \equiv 0$ *on C, u takes a constant value in* R_e'.

5.3.6 *Logarithmic potential of a simple layer*

Let r be the distance between the point Q(ξ, η) on the boundary C and a point P(x, y) not on C, (Fig. 5.7); then

$$r^2 = (x-\xi)^2 + (y-\eta)^2. \qquad (5.101)$$

In this case

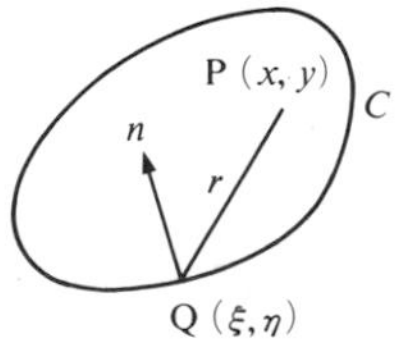

Fig. 5.7. Q on C, P in R_i

$$u(x, y) = \log\frac{1}{r} = -\frac{1}{2}\log\left\{\left[x - \xi(s)\right]^2 + \left[y - \eta(s)\right]^2\right\} \quad (5.102)$$

is a harmonic function and is called a logarithmic potential.

Next, when the continuous line density $\rho(s)$ distributes over the whole length l of the boundary C, the integration

$$u(x, y) = \int_0^l \log\frac{1}{r}\cdot\rho(s)\mathrm{d}s$$

$$= -\frac{1}{2}\int_0^l \log\{[x - \xi(s)]^2 + [y - \eta(s)]^2\}\rho(s)ds \quad (5.103)$$

is also a harmonic function. Here $\rho(0) = \rho(l)$. This $u(x, y)$ is called a logarithmic potential of a simple layer (see Question 12 below).

When P is on the boundary C, represent the coordinates of an arbitrary point on the boundary by $\mathrm{P}_0(x_0, y_0)$ and assume that

$$x_0 = \xi(s_0), \qquad y_0 = \eta(s_0);$$

then by (5.101) $u(x, y)$ becomes

$$u(x_0, y_0) = -\frac{1}{2}\int_0^l \log\{[\xi(s_0) - \xi(s)]^2 + [\eta(s_0) - \eta(s)]^2\}\rho(s)ds; \quad (5.104)$$

If $s = s_0$, that is P_0 concides with Q, the integrand becomes infinite; however, we can prove that the integrated value exists (see Question 13).

Now, place P_i within C and P_e outside of C and we may prove that each of the limits

$$\lim_{\mathrm{P}_i\to\mathrm{P}_0} u(x, y) = u_i(x, y), \qquad \lim_{\mathrm{P}_e\to\mathrm{P}_0} u(x, y) = u_e(x, y), \quad (5.105)$$

exists,

$$u_i(x, y) = u_e(x, y) = u(x_0, y_0); \quad (5.106)$$

and Question 14. That is, a logarithmic potential of a simple layer is continuous on the boundary.

Next, take an inner normal n_i, which passes $\mathrm{P}_0(x_0, y_0)$, and an inner point P_i and an exterior point P_e on n_i, then we have the following lemma on the limit value from the inside and outside of the boundary of $\partial u/\partial n_i$.

LEMMA 5.9

$$\lim_{P_i \to P_0} \frac{\partial u}{\partial n_i} = \frac{\partial u_i}{\partial n_i} = -\pi\rho(s) + \int_0^l \frac{\cos(r_0 n_i)}{r_0}\rho(s)ds, \tag{5.107}$$

$$\lim_{P_e \to P_0} \frac{\partial u}{\partial n_i} = \frac{\partial u_e}{\partial n_i} = \pi\rho(s) + \int_0^l \frac{\cos(r_0 n_i)}{r_0}\rho(s)ds. \tag{5.108}$$

Here, r_0 is the distance between $P_0(x_0, y_0)$ and an arbitrary point $(\xi(s), \eta(s))$ on C.

Since we clearly have

$$\cos(r_0 n_i) = -\frac{\xi(s) - x_0}{r_0}\eta'(s_0) + \frac{\eta(s) - y_0}{r_0}\xi'(s_0),$$

we obtain

$$\frac{\cos(r_0 n_i)}{r_0} = \frac{[\eta(s) - \eta(s_0)]\xi'(s_0) - [\xi(s) - \xi(s_0)]\eta'(s_0)}{[\xi(s) - \xi(s_0)]^2 + [\eta(s) - \eta(s_0)]^2}. \tag{5.109}$$

Example 5.4 *A source of strength m*

The speed of the flow crossing a circle of radius r with the logarithmic potential

$$\varphi = -\frac{m}{2\pi}\log r$$

for its velocity potential is

$$\frac{\partial\varphi}{\partial r} = -\frac{m}{2\pi}\frac{1}{r},$$

and the flux passing across the circle is

$$-\frac{\partial\varphi}{\partial r}\cdot 2\pi r = m \quad \text{(a constant)},$$

where m is the strength of the source.

The stream function, then, is the conjugate function of φ:

$$\psi = \frac{m}{2\pi}\theta,$$

having r, θ for its polar coordinates.

5.3.7. *Logarithmic potential of a double layer*

Let n_i be the normal drawn inside at the point $Q(\xi, \eta)$ on the boundary C; then the directional derivative of the normal direction of $\log(1/r)$:

$$\frac{\partial}{\partial n_i}\left(\log\frac{1}{r}\right) = \frac{d}{dr}\left(\log\frac{1}{r}\right)\frac{\partial r}{\partial n_i} = \frac{\cos(n_i r)}{r}$$

$$= \frac{1}{r}\left[\frac{x-\xi}{r}\cos(n_i\xi) + \frac{y-\eta}{r}\cos(n_i\eta)\right], \quad (5.110)$$

is a harmonic function. This is called the *logarithmic potential of a doublet*. Since $\cos(n\xi) = -\eta'(s)$, $\cos(n\eta) = \xi'(s)$, it is also denoted by

$$\frac{\partial}{\partial n_i}\left(\log\frac{1}{r}\right) = \frac{[y-\eta(s)]\xi'(s) - [x-\xi(s)]\eta'(s)}{r^2}. \quad (5.111)$$

Next, assume that doublets of the continuous density $\rho(s)$ distribute continuously on C, then we call the integration

$$u(x, y) = \int_0^l \frac{\partial}{\partial n_i}\left(\log\frac{1}{r}\right)\rho(s)ds = \int_0^l \frac{\cos(rn_i)}{r}\rho(s)ds, \quad (5.112)$$

the *logarithmic potential of a double layer*. This is derived under the assumption that $\rho(0) = \rho(l)$. This can also be expressed as

$$u(x, y) = \int_0^l \frac{[y-\eta(s)]\xi'(s) - [x-\xi(s)]\eta'(s)}{[x-\xi(s)]^2 + [y-\eta(s)]^2}\rho(s)ds. \quad (5.113)$$

At $P_0(x_0, y_0)$, the value on the boundary becomes

$$u(x_0, y_0) = \int_0^l \frac{[\eta(s_0)-\eta(s)]\xi'(s) - [\xi(s_0)-\xi(s)]\eta'(s)}{[\xi(s_0)-\xi(s)]^2 + [\eta(s_0)-\eta(s)]^2}\rho(s)ds, \quad (5.114)$$

and at $s = s_0$ the integrand takes an indefinite form, and the calculation of the limit is $1/(2R_0)$, provided the radius of curvature at P_0 is R_0. Then, finally, $u(x_0, y_0)$ takes a finite determinate value (see Question 13).

There exist limits on the boundary from the inside and outside, but since

$$\lim_{P_i \to P_0} u(x, y) = u_i(x_0, y_0) = u(x_0, y_0) + \pi\rho(s_0), \quad (5.115)$$

$$\lim_{P_e \to P_0} u(x, y) = u_e(x_0, y_0) = u(x_0, y_0) - \pi\rho(s_0), \quad (5.116)$$

the logarithmic potential of a double layer is discontinuous on the boundary so long as we do not have $\rho(s_0) = 0$.

Example 5.5 *Doublet*

Suppose that there exist a source and a sink each having the strength $\pm m$, and with the distance δs between them; then, the limit, attained by making δs infinitesimally small and m infinitely large keeping $m\delta s = \mu$ constant, is called a double source or a doublet, where μ is the strength and the direction connecting the source and sink is called the axis of the doublet.

In order to find the velocity potential at (x, y) for the doublet of strength μ at (x', y'), let (l, m) be the directional cosine of the axis. And put

$$r^2 = (x - x')^2 + (y - y')^2$$

and let δ be the angle between r and the axis. Neglecting infinitesimally small quantities of higher orders, we get the potential as

$$\varphi = \frac{m}{2\pi}\log\left(r + \frac{1}{2}\delta s \cos \delta\right) - \frac{m}{2\pi}\log\left(r - \frac{1}{2}\delta s \cos \delta\right).$$

Take this as far as the second term of the Taylor expansion and put $\delta s \to 0$, $m\delta s \to \mu$; then we have

$$\varphi = \frac{\mu}{2\pi}\frac{\cos \delta}{r}.$$

And, since $\delta r = -\delta s \cos \delta$, this becomes

$$\varphi = -\frac{\mu}{2\pi}\frac{\partial}{\partial s}(\log r).$$

Now, when doublets of the density $\rho(s)$ are distributed along C and the directions of the axes of doublets agree with those of the inner normals of C, the velocity potential of an arbitrary point is given by

$$\varphi = \frac{1}{2\pi}\int_0^l \frac{\cos(r\pi)}{r}\rho(s)ds.$$

This is an example of the logarithmic potential of a double layer.

5.4. Dirichlet's problem

What we call *Dirichlet's problem* or the first boundary problem is to find the function $u(x, y)$ which satisfies the following two conditions when a function $f(s)$, which has no discontinuous point on C, is given, where $f(0) = f(l)$.

(i) u is a harmonic function in the interior R_i (excluding the boundary) of C.

(ii) When we move the point P(x, y) to an arbitrary point s_0 on the circuit from inside C, u becomes uniformly the function $f(s)$ given in advance. That is,

$$\lim u(x, y) = u_i(x_0, y_0) = f(s_0).$$

Dirichlet's problem can lead to [F_2], and there are two routes from it to the integral equation.

The first method. Denote the solution sought in the problem by

$$u(x, y) = \int_0^l \rho(s) \frac{\partial}{\partial n}\left(\log \frac{1}{r}\right) ds, \tag{5.117}$$

using the logarithmic potential of a double layer; then $u(x, y)$ is a harmonic function in R_i. Take $\rho(s)$ so that it is continuous at $s(0, l)$ and satisfies $\rho(0) = \rho(l)$. Then let $\rho(s)$ satisfy condition (ii); then (5.117) becomes the solution of Dirichlet's problem. Using (5.114), we get

$$u_i(x, y) = u(x_0, y_0) + \pi\rho(s_0);$$

then condition (ii) becomes

$$\int_0^l \frac{[\eta(s_0) - \eta(s)]\xi'(s) - [\xi(s_0) - \xi(s)]\eta'(s)}{[\xi(s_0) - \xi(s)]^2 + [\eta(s_0) - \eta(s)]^2} \rho(s) ds + \pi\rho(s_0) = f(s_0). \tag{5.118}$$

Therefore, putting

$$\frac{1}{\pi} \frac{[\eta(s_0) - \eta(s)]\xi'(s) - [\xi(s_0) - \xi(s)]\eta'(s)}{[\xi(s_0) - \xi(s)]^2 + [\eta(s_0) - \eta(s)]^2} \equiv K(s_0, s), \tag{5.119}$$

$$\frac{1}{\pi} f(s_0) \equiv f(s_0), \tag{5.120}$$

we can obtain the integral equation

$$\rho(s_0) + \int_0^l K(s_0, s)\rho(s) ds = f(s_0). \tag{5.121}$$

This is [F_2] when $\lambda = -1$.

Therefore, so that $u = u(x, y)$ may be the solution of Dirichlet's problem, it is required that $\rho(s)$ should satisfy the integral equation (5.121). While if $\rho(s)$ is the continuous solution of (5.121), then, since $K(0, s) = K(l, s)$, $f(0) = f(l)$, we have $\rho(0) = \rho(l)$ from (5.121), and, thus, (5.121) is also a sufficient condition.

The second method. Using (5.115) and (5.116) we obtain

$$u_i(x_0, y_0) - u_e(x_0, y_0) = 2\pi\rho(s_0), \tag{5.122}$$

$$u_i(x_0, y_0) + u_e(x_0, y_0) = 2u(x_0, y_0) = 2\int_0^l \frac{\cos(r_0 n_i)}{r_0}\rho(s)ds. \tag{5.123}$$

Here let us find a function $u(x, y)$ as the solution of Dirichlet's problem, which is harmonic everywhere except on the boundary C, and satisfies the condition

$$u_i + hu_e = f(s_0) + hg(s_0), \tag{5.124}$$

on C. As h is a parameter, we may take $h = 0$ for the interior problem and $h = \infty$ for the exterior problem.

If we allow the harmonic function $u(x, y)$ to satisfy the condition (5.124), it comes to be the solution of Dirichlet's problem, except on the boundary C. Hence we solve (5.122) and (5.123) in terms of u_i and u_e, then put the result in (5.124) and obtain the integral equation

$$\rho(s_0) - \lambda\int_0^l \frac{\cos(r_0 n_i)}{\pi r_0}\rho(s)ds = F(s_0), \tag{5.125}$$

where

$$F(s_0) = \frac{hg(s_0) + f(s_0)}{\pi(1-h)}, \qquad \lambda = \frac{h+1}{h-1}. \tag{5.126}$$

Equation (5.125) is $[F_2]$ and the kernel agrees with (5.119), and, in the case of the interior problem, $h = 0$ and $\lambda = -1$ from (5.126). Then, finally, (5.126) agrees with (5.121). When $h = \infty$, the case becomes that of the exterior problem and if we put $\lambda = +1$, (5.125) becomes

$$\rho(s_0) - \int_0^l K(s_0, s)\rho(s)ds = -\frac{1}{\pi}g(s_0). \tag{5.127}$$

Solution of the problem. Let us show that the $[F_2]$ type integral equation (5.121) is solved on $\rho(s_0)$. This time if we use a modified kernel, the equation is continuous in the domain $(0 \leqq s_0 \leqq l,\ 0 \leqq s_0 \leqq l)$, and therefore it belongs to $\mathscr{R}$, and $f(s_0)$ is continuous in $I(0, l)$. Therefore, if Fredholm's modified determinant $\Delta_0(\lambda)$ is not zero then (5.121) has a single continuous solution from Theorem 2.3, corollary. Then, if we can prove that

$$\Delta_0(-1) \neq 0,$$

the solution of Dirichlet's problem is given by (5.117), and $\rho(s_0)$ becomes

$$\rho(s_0) = f(s_0) + \int_0^l \Gamma_0(s_0, s; -1)f(s)ds. \tag{5.128}$$

Now, in order to prove that $\Delta_0(-1) \neq 0$, we have only to show that

$$\rho(s_0) = \lambda \int_0^l K_0(s_0, s)\rho(s)ds \qquad [F_0] \tag{5.129}$$

has no continuous solution except the solution identically zero when $\lambda = -1$. When $\lambda = -1$, from (5.129) we have

$$\rho(s_0) + \int_0^l K_0(s_0, s)\rho(s)ds = 0.$$

By (5.115), this means that

$$u_i(x, y) = 0$$

uniformly on s_0. Therefore, by Lemma 5.7 on the boundary-value problem of the potential, we have

$$\rho(s) \equiv 0$$

in the interior. Thus, we could prove that $\Delta_0(-1) \neq 0$.

5.4.1. Poisson integral

In Dirichlet's problem, when the boundary C is a circle of radius 1 the potential is expressed in a famous form called Poisson's integral.

Introduction of the integral equation. If we have

$$\xi(s) = \cos s, \quad \eta(s) = \sin s \quad (0 \leqq s \leqq 2\pi), \tag{5.130}$$

then

$$\begin{aligned} K(s_0, s) &= \frac{1}{\pi} \frac{(\sin s_0 - \sin s)(-\sin s) - (\cos s_0 - \cos s)\cos s}{(\cos s_0 - \cos s)^2 + (\sin s_0 - \sin s)^2} \\ &= \frac{1}{\pi} \frac{1 - \cos(s_0 - s)}{2 - 2\cos(s_0 - s)} = \frac{1}{2\pi} \qquad (s_0 \neq s), \end{aligned}$$

and (5.122) becomes

$$\rho(s_0) + \int_0^{2\pi} \frac{1}{2\pi}\rho(s)ds = \frac{1}{\pi}f(s_0). \tag{5.131}$$

This is the $[F_2]$ type integral equation of

$$\lambda = -\frac{1}{2\pi}, \quad K(s_0, s) = 1.$$

We can solve (5.121) analytically by calculating the iterated kernel. That is,

$$\Delta(\lambda) = 1 - 2\pi \cdot \lambda = 2, \quad \Delta\binom{s_0}{s}, \lambda\Big) = 1,$$

and the solution of the kernel is

$$\Gamma(s_0, s; \lambda) = -\frac{1}{2},$$

then the solution $x(s_0)$ becomes

$$x(s_0) = \frac{f(s_0)}{\pi} - \frac{1}{4\pi}\int_0^{2\pi} \frac{f(s)}{\pi} ds. \tag{5.132}$$

Put $x = r\cos\theta$, $y = r\sin\theta$ and then we have

$$\frac{\partial}{\partial n}\log\frac{1}{r} = \frac{(r\sin\theta - \sin s)(-\sin s) - (r\cos - \theta\cos s)\cos s}{(r\cos\theta - \cos s)^2 + (r\sin\theta - \sin s)^2}$$

$$= \frac{1 - r\cos(\theta - s)}{r^2 + 1 - 2r\cos(\theta - s).} \tag{5.133}$$

Next, denote $u(x, y)$ of (5.117) by $U(r, \theta)$, and use (5.132) and (5.133); then we get

$$U(r, \theta) = \frac{1}{\pi}\int_0^{2\pi}\left\{f(s) - \frac{1}{4\pi}\int_0^{2\pi} f(s)ds\right\}\frac{1 - r\cos(\theta - s)}{r^2 + 1 - 2r\cos(\theta - s)}ds$$

$$= \frac{1}{\pi}\int_0^{2\pi} f(s)\frac{1 - r\cos(\theta - s)}{r^2 + 1 - 2r\cos(\theta - s)}ds$$

$$-\frac{1}{4\pi^2}\int_0^{2\pi} f(s)ds\int_0^{2\pi}\frac{1 - r\cos(\theta - s)}{r^2 + 1 - 2r\cos(\theta - s)}ds,$$

while by the integral formula

$$\int_0^{2\pi}\frac{1 - r\cos(\theta - s)}{r^2 + 1 - 2r\cos(\theta - s)}ds = 2\pi$$

we have

$$U(r, \theta) = \frac{1}{\pi}\int_0^{2\pi} f(s)\left\{\frac{1 - r\cos(\theta - s)}{r^2 + 1 - 2r\cos(\theta - s)} - \frac{1}{2}\right\}ds$$

$$= \frac{1}{2\pi}\int_0^{2\pi} f(s)\frac{1 - r^2}{r^2 + 1 - 2r\cos(\theta - s)}ds.$$

That is, the potential $U(r, \theta)$ which takes the boundary value $f(s)$ on the circuit, is denoted by

$$U(r, \theta) = \frac{1}{2\pi}\int_0^{2\pi} \frac{1 - r^2}{r^2 + 1 - 2r\cos(\theta - s)} f(s)ds. \tag{5.134}$$

This is called Poisson's integral.

5.4.2. *The case of an elliptic boundary*

Poisson's integral is also used when the boundary is an ellipse in Dirichlet's problem. Let a be the long axis and b be the short one; then the equation of an ellipse is

$$x = a\cos s, \quad y = b\sin s \quad (a \geqq b, 0 \leqq s \leqq 2\pi). \tag{5.135}$$

In this case the kernel of an integral equation corresponding to (5.121) is

$$K(t, s) = \frac{1}{\pi}\frac{b(\sin s_0 - \sin s)(-a\sin s) - a(\cos s_0 - \cos s)(b\cos s)}{a^2(\cos s_0 - \cos s)^2 + b^2(\sin s_0 - \sin s)^2}$$

$$= \frac{1}{\pi}\frac{\dfrac{ab}{a^2 - b^2}}{\dfrac{a^2 + b^2}{a^2 - b^2} - \cos(s_0 + s)}.$$

Since this kernel is not simple, we have to perform numerical calculations in order to obtain the solution. Divide the ellipse into two by the long axis and calculate only the upper part (this sort of method is allowed when the ellipse is symmetrical either side of the long axis) and then the simultaneous equations are

$$\rho_i + \frac{1}{\pi}\cdot\frac{ab}{a^2 - b^2}\sum_{i=1}^{n}\frac{w_i^n\rho_i \cdot \pi}{\dfrac{a^2 + b^2}{a^2 - b^2} - \cos(s_{0j} + s_j)}$$

$$= \frac{1}{\pi}f_j \qquad (j = 1, 2, \ldots, 2n), \tag{5.136}$$

where w_i^n is the weight of the ith point of n representative points.

5.4.3. *The eigenvalue*

$\lambda = +1$. Consider the integral in the complex domain

$$\int_C \frac{d\zeta}{\zeta - z}; \tag{5.137}$$

if we rewrite this with $\zeta = \xi + i\eta$, $z = x + iy$, we obtain

$$\int_C \frac{(\xi - x)d\xi + (\eta - y)d\eta}{(x-\xi)^2 + (y-\eta)^2} + i\int_C \frac{(y-\eta)d\xi - (x-\xi)d\eta}{(x-\xi)^2+(y-\eta)^2};$$

then, comparing the imaginary part with (5.119), we obtain

$$i\pi\int_C K(s_0, s)ds.$$

By Cauchy's theory of integration, (5.137) becomes zero when z is outside C, and $2\pi i$ when z is inside C; therefore,

$$\pi\int_C K(s_0, s)ds = \begin{cases} 2\pi, & \text{interior,} \\ 0, & \text{exterior.} \end{cases}$$

Next, let us find the boundary value of the potential at $\rho(s) \equiv 1$ along the curve C. By (5.112), the logarithmic potential of a double layer is

$$u(x, y) = \int_0^l \frac{\cos(rn_i)}{r} ds.$$

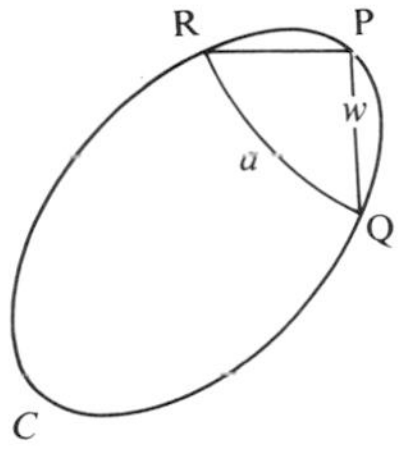

Fig. 5.8 Elliptic domain

Consider a small circle a with centre $P(x_0, y_0)$. Eliminate the arc QPR from C and substitute the arc a for the boundary C: C' then P becomes an outer point and the potential is zero, while

$$\int_{C'} \frac{\cos(r, n_i)}{r} ds + \int_a \frac{\cos(r, n_i)}{r} ds = 0.$$

The second integration clearly becomes $-w$, where the angle QPR is w. At the limit as the little circle a becomes smaller, we have $C' \to C$, $w \to \pi$, and the first integration coincides with the boundary value of the potential at P. Then

$$u(x_0, y_0) = \pi, \tag{5.138}$$

or

$$\pi\int_0^l K(s_0, s)ds = \pi.$$

Thus, we could prove that [F_0]:

$$\rho(s_0) = \lambda\int_0^l K(s_0, s)\rho(s)ds \tag{5.139}$$

has the eigenvalue $\lambda = +1$, and has the eigensolution $\rho(s) \equiv 1$ not identically zero.

In a similar manner,

$$\rho(s) = c \text{ (an arbitrary constant)} \tag{5.140}$$

becomes the eigensolution for $\lambda = +1$ of (5.139). Thus, it is shown that there are an infinite number of eigenfunctions for $\lambda = +1$, and since there is no solution except (5.140), the index of $\lambda = +1$ becomes 1.

Example 5.6 *Stream function of a uniform flow around a cylinder*

Given the source distribution of the line density $\rho(s)$ on a closed curve C, the stream function ψ_P of an arbitrary point P is

$$\psi_P = \frac{1}{2\pi}\int_0^l \rho(s)\theta\, ds = \frac{1}{2\pi}\int \theta\, d\rho.$$

Integrate this, and we get

$$\psi_P = \frac{1}{2\pi}[\theta\rho] - \frac{1}{2\pi}\int \rho d\theta = -\frac{1}{2\pi}\int \rho d\theta.$$

Let θ_0 be the direction of the tangent at the point Q on the boundary; then if we move P_i toward Q, it is easily proved that if we approach it from the inside, we get

$$\lim_{P_i \to Q} \psi_P = -\frac{1}{2\pi}\int_{\theta_0}^{\theta_0+\pi} \rho d\theta - \frac{1}{2}\rho_Q, \tag{5.141}$$

and from outside, we get

$$\lim_{P_e \to Q} \psi_P = -\frac{1}{2\pi}\int_{\theta_0}^{\theta_0+\pi} \rho d\theta + \frac{1}{2}\rho_Q \tag{5.142}$$

(see Question 15).

If ψ_P takes the given value $f(\theta_0)$ on each circuit of (5.141), and (5.142)

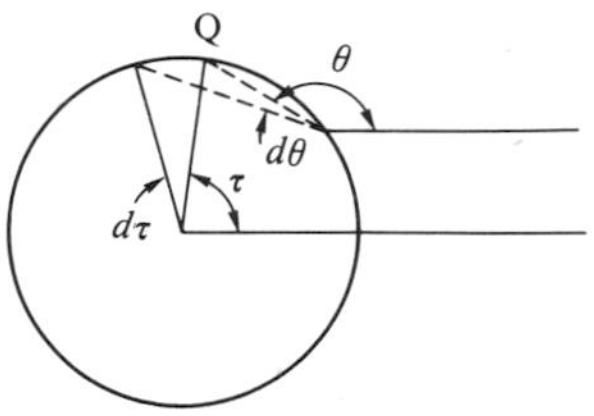

Fig. 5.9. $d\theta = 1/2\, d\tau$

becomes an integral equation, then using the iterated kernel we can express it in

$$\frac{1}{2}\rho(\theta_0) = -f(\theta_0) + \frac{1}{\pi}\int_{\theta_0}^{\theta_0+\pi} f(\theta)d\theta - \frac{1}{\pi^2}\int_{\theta_0}^{\theta_0+\pi}\int_{\theta_0}^{\theta_0+\pi} f(\theta)d\theta d\theta_1 = \cdots, \tag{5.143}$$

$$\frac{1}{2}\rho(\theta_0) = f(\theta_0) + \frac{1}{\pi}\int_{\theta_0}^{\theta_0+\pi} f(\theta)d\theta + \frac{1}{\pi^2}\int_{\theta_0}^{\theta_0+\pi}\int_{\theta_0}^{\theta_0+\pi} f(\theta)d\theta d\theta_1 + \cdots. \tag{5.144}$$

Now let us find the source distribution along the circuit of the cylinder of radius a, moving with constant velocity V in an ideal fluid which is still at infinity. The stream function becomes

$$\rho(\tau) = Va \sin \tau \tag{5.145}$$

on the circuit, and from Fig. 5.9 we clearly have $d\theta = (1/2)d\tau$; then on the right-hand side of (5.143) we get

$$\int_{\theta_0}^{\theta_0+\pi} Va \sin \tau d\theta = \frac{1}{2}\int_0^{2\pi} Va \sin \tau\, d\tau = 0,$$

and so on. From (5.144) we have

$$\rho(\tau) = 2Va \sin \tau.$$

When P is given as the outside point (τ_1, θ_1), the stream function yields

$$\psi_P = -\frac{Va}{\pi}\int \sin \tau\, d\theta = \frac{Va^2}{\tau_1}\sin \theta_1$$

by a simple calculation. This is a well-known equation, and it becomes (5.145) when the point is on the circuit.

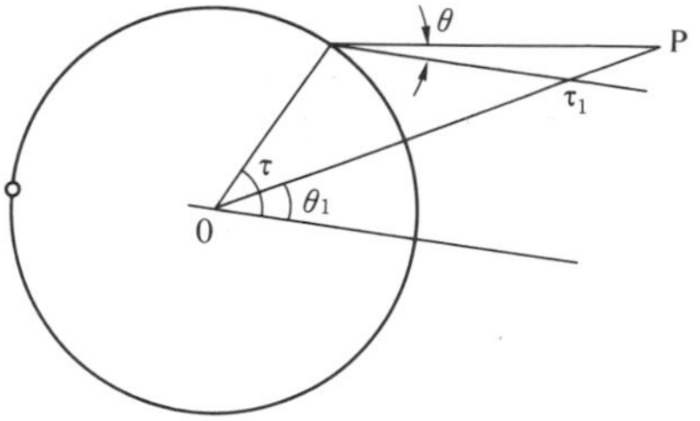

Fig. 5.10. P (τ_1, θ_1)

5.5. Neumann's problem

What we call Neumann's problem, or the second boundary problem for Dirichlet's boundary problem, is to find a function $u(x, y)$ which satisfies conditions (i) and (ii) below when given the curve

$$C\colon\ x = \xi(s)\quad y = \eta(s),\ 0 \leqq s \leqq l;\ \xi(0) = \xi(l),\ \eta(0) = \eta(l),$$

and a function which has no discontinuous point on C:

$$f(s),\quad f(0) = f(l).$$

(i) u is a harmonic function in the interior R_i (excluding the boundary) of C.

(ii) When we approach the point P(x, y) toward an arbitrary point s_0 on C from inside C, $\partial u_i/\partial n_i$ uniformly becomes the function $f(s)$ given in advance. Namely

$$\lim \frac{\partial u_i}{\partial n_i} = f(s_0).$$

This leads to the integral equation [F_2] also. For this, there are two methods.

The first method. Denote the solution required by the question using the potential of a simple layer by

$$u(x, y) = \int_0^l \log \frac{1}{r} \cdot \rho(s) ds; \tag{5.147}$$

this is a harmonic function in R_i. Take $\rho(s)$ so that it is continuous at $s(0, l)$ and $\rho(0) = \rho(l)$. Then select $\rho(s)$ so that it satisfies condition (ii), and (5.147) becomes the solution to Neumann's problem.

By Lemma 5.9, condition (ii) becomes

$$\int_0^l \frac{[\eta(s)-\eta(s_0)]\xi'(s_0)-[\xi(s)-\xi(s_0)]\eta'(s_0)}{[\xi(s)-\xi(s_0)]^2+[\eta(s)-\eta(s_0)]^2}\rho(s)ds = f(s_0)+\pi\rho(s_0). \tag{5.148}$$

Then putting

$$\frac{1}{\pi}\frac{[\eta(s)-\eta(s_0)]\,\xi'(s_0)-[\xi(s)-\xi(s_0)]\eta'(s_0)}{[\xi(s)-\xi(s_0)]^2+[\eta(s)-\eta(s_0)]^2} \equiv K(s_0, s), \tag{5.149}$$

$$\frac{1}{\pi}f(s_0) = f(s_0), \tag{5.150}$$

we can obtain the integral equation

$$\rho(s_0) - \int_0^l K(s_0, s)\rho(s)ds = f(s_0). \tag{5.151}$$

This is $[F_2]$ for $\lambda = 1$.

Therefore, in order for $u = u(x, y)$ to be the solution of Neumann's problem, it is necessary for the density $\rho(s)$ to satisfy the integral equation (5.151). On the other hand, if $\rho(s)$ is the continuous solution of (5.151), we have $K(0, s) = K(l, s)$, $f(0) = f(l)$; and, from (5.151), $\rho(0) = \rho(l)$ is derived and thus (5.151) is sufficient.

The second method. From Lemma 5.9 we have

$$\frac{\partial u_i}{\partial n_i} = -\pi\rho(s_0) + \int_0^l \frac{\cos(r_0 n_i)}{r_0}\rho(s)ds, \tag{5.152}$$

$$\frac{\partial u_e}{\partial n_i} = +\pi\rho(s_0) + \int_0^l \frac{\cos(r_0 n_i)}{r_0}\rho(s)ds. \tag{5.153}$$

Then, for the solution of Neumann's problem let us find a function $u(x, y)$ which is harmonic everywhere except on the boundary C, and satisfies

$$\frac{\partial u_e}{\partial n_i} + h\frac{\partial u_i}{\partial n_i} = g(s_0) + hf(s_0) \tag{5.154}$$

on C. Take h as a parameter, and let $h = \infty$ in the case of the interior problem of the boundary, and $h = 0$ in the case of the exterior problem.

If we make the harmonic function $u(x, y)$ so that it satisfies the condition (5.154), this becomes the solution of Neumann's problem, except in the case of the boundary C. We solve (5.152), (5.153) on $\partial u_i/\partial n_i$, $\partial u_e/\partial n_i$, and put the result in (5.154), to obtain the integral equation

$$\rho(s_0) - \lambda \int_0^l \frac{\cos(r_0\, n_i)}{\pi r_0} \rho(s) ds = F(s_0). \tag{5.155}$$

In this we put

$$F(s_0) = \frac{g(s_0) + hf(s_0)}{\pi(1-h)}, \qquad \lambda = \frac{h+1}{h-1}; \tag{5.156}$$

(5.155) is [F_2], and the kernel agrees with (5.149), and in the interior problem we get $\lambda = 1$ from (5.156) with $h = \infty$. Then, finally, (5.156) agrees with (5.151). When $h = 0$ in the exterior problem, we get $\lambda = -1$. This time, (5.155) becomes

$$\rho(s_0) - \int_0^l K(s_0, s)\rho(s) ds = \frac{1}{\pi} g(s_0). \tag{5.157}$$

Solution of the problem. Comparing (5.119) with (5.149)′, it appears that

$$K(s, s_0) = \tilde{K}(s_0, s). \tag{5.158}$$

Therefore, the integral equation (5.151) is the associate integral equation of the integral equation of Dirichlet's problem (5.151). Hence, the eigenvalues should agree.

As we have already shown that $\lambda = +1$ is the eigenvalue for the index 1 of $K(s_0, s)$, $\lambda = 1$ is also the eigenvalue for the associate kernel $\tilde{K}(s_0, s)$, and the index is 1. Therefore, from Lemma 3.3, the solution exists only when the condition

$$\int_0^l f(s)\varphi(s_0) = 0 \tag{5.159}$$

holds for the eigenfunction $\varphi(s_0)$ of $\tilde{K}(s_0, s)$. But, since $\varphi(s_0) \equiv 1$ as we showed in § 5.4.3, (5.159) becomes

$$\int_0^l f(s) ds = 0, \tag{5.160}$$

and, by (5.150), it is necessary to have the condition

$$\int_0^l f(s) ds = 0$$

for the given function $f(s)$. In this case $u(x, y)$, given by (5.147), becomes the solution of Neumann's problem. It is clear that $u(x, y)$ satisfies conditions (i) and (ii).

Now, if we let

$$u_1(x, y), \quad u_2(x, y)$$

be the solution of Newmann's problem, the function $v(x, y)$, which is defined as

$$v(x, y) = u_1(x, y) - u_2(x, y),$$

is harmonic in the region I' and becomes $\partial v_i/\partial n_i = 0$ on C; consequently we have

$$v = c$$

in I' by the note in Lemma 5.6 in § 5.3.4. Thus, Neumann's problem has an infinite number of solutions

$$u(x, y) + c.$$

5.5.1. The stream around an arbitrary body

The velocity potential $\psi(x, y)$ around an arbitrary body (for instance a profile), corresponds to Neumann's exterior problem when there is a double-source distribution of the density $\rho(s)$ along the boundary C of this body. That is,

(i) $\psi(x, y)$ is a harmonic function of the exterior region of C, and
(ii) on C

$$\lim \frac{\partial \psi}{\partial n_i} = 0$$

uniformly. n_i means an inward normal.

In this case, the integral equation which $\rho(s)$ must satisfy is

$$\rho(s_0) + \int_0^l \tilde{K}(s_0, s)\rho(s)ds = 0, \tag{5.161}$$

which is $[F_2]$ with $\lambda = -1$.

Prager (1928) proposed the following result. Construct a stream function ψ:

$$\psi = \psi_0 + \psi_1$$

as the sum of the stream function ψ_0 of a uniform stream when there is no body, and the stream function ψ_1 which should be placed upon ψ_0 so that a body may exist. Then, ψ_1 must satisfy the following conditions:

(i) ψ_1 is a harmonic function in the exterior region of the body;
(ii) $\psi_1 = -\psi_0$ on the boundary of the body;

(iii) $\psi_0 = 0$ at the infinitely far point ($r \to \infty$) when the rotation around the body is zero; $\psi_1 = (c/2\pi)\log(1/r)$ at the infinitely far point when the rotation is c.

If the limit on the boundary

$$\lim \frac{\partial \psi_i}{\partial n_1} = v_1(s_0)$$

is known, the case becomes Neumann's exterior problem, and we have

$$\rho(s_0) + \int_0^l \frac{\cos(r_0, n_i)}{\pi r_0}\rho(s)ds = \frac{1}{\pi}v_1(s_0). \tag{5.162}$$

In order to consider the physical significance of $\rho(s)$, denote the left-hand sides of (5.152), (5.153) by $\partial\psi_{1i}/\partial n_i$, $\partial\psi_{1e}/\partial n_i$, and on the boundary we get

$$\frac{\partial \psi_{1e}}{\partial n_i} - \frac{\partial \psi_{1i}}{\partial n_i} = 2\pi\rho(s_0). \tag{5.163}$$

As for the interior, we have

$$\frac{\partial \psi}{\partial n_i} = 0, \quad \frac{\partial \psi_{1i}}{\partial n_i} + \frac{\partial \psi_{1i}}{\partial n_i} = 0;$$

and, since the normal derivative of ψ_2 can be considered continuous, we get

$$\frac{\partial \psi_{1e}}{\partial n_i} = \frac{\partial \psi_{2i}}{\partial n_i} = -\frac{\partial \psi_{1i}}{\partial n_i}.$$

Hence, (17) becomes

$$\frac{\partial \psi_{1e}}{\partial n_i} + \frac{\partial \psi_{2e}}{\partial n_i} = 2\pi\rho(s_0). \tag{5.164}$$

Then, the left-hand side of this equation shows the tangent velocity on the boundary, and to find $\rho(s_0)$ from (5.162) corresponds to finding the velocity distribution $v(s_0)$ around the body. If we take $v_0(s_0)$ as the velocity distribution on the boundary by only the uniform stream ψ_0, we clearly have

$$v(s_0) = 2\pi\rho(s_0) = v_0(s_0) + v_1(s_0). \tag{5.165}$$

Hence, (5.162) becomes the integral equation for $v(s_0)$:

$$v(s_0) - \int_0^l \frac{\cos(r_0, n_i)}{\pi r_0}v(s)ds = 2v_0(s_0). \tag{5.166}$$

Prager performed the numerical calculations on a circle and an ellipse. He solved (5.166) in the case of the ellipse $x = 2\cos\alpha, h = \sin\alpha$, with a uniform stream of the velocity component $(u, 0)$ by Tscheby-scheff's numerical integration, and found the velocity distribution and pressure distribution on the elliptic circumference.

When there is only rotation around a body without uniform flow, we have $v_0(s_0) \equiv 0$; and then (5.166) becomes a homogeneous integral equation. Moriya (1936) pointed out that it is necessary to give the rotation around the body:

$$\int v(s)ds = C,$$

in order to establish the solution $v(s_0)$ by Joukowski's condition, since $\lambda = 1$ corresponds to the eigenvalue in (5.166).

Exercises

1. Find the integral equations which satisfy the following ordinary differential equations and the initial conditions.

(a) $x^{(2)} + x^{(1)} + x = 1, \quad [x_0 = x_0{}^{(1)} = 0]$;

(b) $x^{(2)} - 5x^{(1)} + 6x = 0, \quad [x_0 = 0, x_0{}^{(1)} = -1]$;

(c) $x^{(2)} + n^2x^{(1)} + ax = f(t), \quad [x_0 = 0, x_0{}^{(1)} = V]$;

(d) $x^{(1)} - x = 0, \quad [x_0 = x_0{}^{(1)} = 1]$;

(e) $x^{(3)} - 3x^{(2)} - 6x^{(1)} + 5x = 0$. [See the relationship between the initial condition and the form of the integral equation].

2. What are the differential equations equivalent to the following integral equations and their initial and boundary conditions?

(a) $x(t) + \int_0^t (t - \xi)x(\xi)d\xi = 2\cos t - t$,

(b) $x(t) - \int_0^t (t - \xi)x(\xi)d\xi = \cos t - t - 2$,

(c) $x(t) - \int_0^t (6t - 6\xi + 5)x(\xi)d\xi = 29 + 6t$,

(d) $x(t) - \int_0^t (t - \xi)x(\xi)d\xi = t$,

(e) $x(t) - \lambda \int_0^1 K(t, \xi)x(\xi)d\xi = t + \xi + 1.$

$$K(t, \xi) = t - \xi,\ t > \xi;\ \xi - t,\ t \leqq \xi.$$

3. Express the problem of the yielding of a long bar under the various support conditions at both ends using Fredholm's integral equation.
4. Given a uniform beam, both ends ($x = 0$, $x = 2c$) of which are fixed, make an integral equation for deflection with a uniformly distributed load (load w_0 for a unit length), and solve it.
5. What will the result be in the previous question if both ends are simply supported and the distributed load is expressed as $f(x)$, in general?
6. Given a spring of strength k with a weight of mass m, make the integral equation of the motion formed when this spring is gently released with the initial deformation x_0 in the medium in which resistance is proportional to velocity.
7. Establish the integral equation of the case with the force $F_0 \sin wt$ in the previous question and solve it.
8. Prove that, if $u(x)$ and its first and second derivatives are continuous, the solution $u(x)$ of the boundary problems (5.47), (5.48) is continuous and satisfies [F_0] and the kernel is given by (5.50). [Lovitt]
9. Prove that if $u(x)$ is continuous and satisfies (5.50) and [F_0], $u(x)$ is continuous as far as the derivative of the second order and satisfies the boundary-value problems (5.47), (5.48). [Lovitt]
10. Prove that [F_0] and (5.50) have a solution identically non-zero:

$$u(x) = A \sin\frac{n\pi}{l}x,$$

only when $\lambda = n^2\pi^2/l^2$ (for n an integer). [Lovitt]
11. Prove, in the same way, the equality of the boundary-value problem of a differential equation and an integral equation. If $u(x)$ is continuous as far as the derivative of the second order and serves as the solution of the boundary-value problems (5.63), (5.64), it serves as the continuous solution of the integral equations (5.66), (5.65), and inversely, if $u(x)$ is the continuous solution of the integral equation it is the solution of the boundary-value problems (5.63), (5.64) and is continuous as far as the derivative of the second order.
12. Prove that the logarithmic potential of a simple layer and that of a double layer are harmonic functions.
13. Find the values of the logarithmic potential for a simple layer and for a double layer on the boundary C.

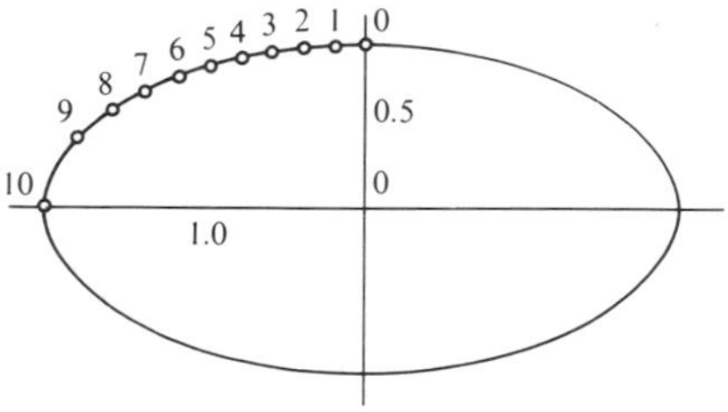

Fig. 5.11. Section of an elliptic cylinder

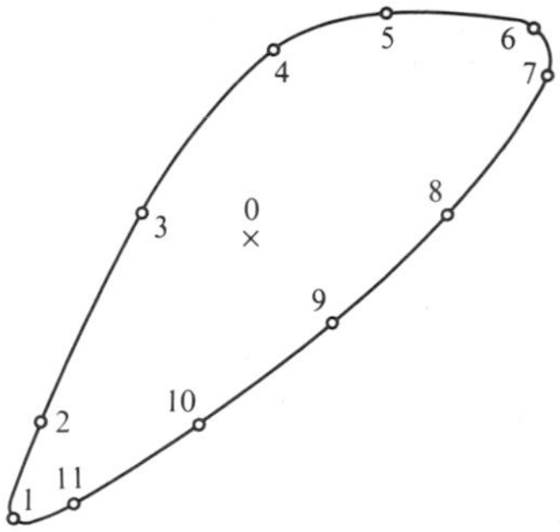

Fig. 5.12. Propeller shaft

14. Prove that the logarithmic potential of a simple layer is continuous on the boundary.
15. Prove (5.141), (5.142) with regard to Example 5.6.
16. For the stress function of the torsion $\psi(x, y)$, we have

$$\tau_x = -\frac{\partial \psi}{\partial x}, \qquad \tau_y = \frac{\partial \psi}{\partial y},$$

and on the perimeter $\psi =$ constant, in the interior $\Delta\psi = -2G\theta$, where G is a constant, θ is the torsion angle. Now if we take

$$\psi = G\theta\left[\varphi - \frac{1}{2}(x^2 + y^2)\right],$$

we get on φ,

$$\tau_x = G\theta(x - \varphi_x), \quad \tau_y = G\theta(\varphi_y - y),$$

and on the perimeter φ becomes $\varphi - \frac{1}{2}(x^2 + y^2) =$ constant, and in the interior we have $\Delta\varphi = 0$. Calculate the torsion stress of the elliptic cylinder shown in the Fig. 5.11. [Bairstow L., Berry A.]
17. Calculate the torsion stress of the propeller shaft shown in Fig. 5.12.

Miscellaneous problems (I)

1. Prove the Goursat–Heywood theorem, that the reciprocal function of the set of kernels is the set of the reciprocal functions of each kernel. The product of Fredholm's determinant of a kernel is the Fredholm determinant of the product of kernels.

2. Solve the Fredholm integral equation of the first kind:

$$\int_0^1 K(t, \xi)x(\xi)d\xi = 67t^3 - 40t^2 + 52t + 17,$$

$$K(t, \xi) = \begin{cases} t\xi(t-\xi)^2 + 1, & t > \xi; \\ t\xi(t-\xi)^2, & t < \xi \end{cases}$$

by numerical calculation, and compare the result with the exact solution $x(t) = 60t^2 + 60t + 24$. (See § 2.5 Example 2.7.)

3. Prove that Tschebyscheff's formula does not lead to a solution by the numerical method in Chapter 4 (Example 4.1). Then consider what sort of caution is necessary in leading to simultaneous equations by Tschebyscheff's formula in the case of symmetrical kernels of [F_1], in general.

In order to examine the properties possessed by the solution in Table 4.7 solved by Gauss's formula, let $x(t) = at^3 + bt^2 + ct + d$ be the solution of [F_1], select two of the coefficients from it so that we have $x(t) = 73.620\,687, -182.925\,814$ when $t = 0.500\,000 \pm 0.169\,991$, and decide the remaining two by the method of unknown coefficients as in § 3.3. Then compare the exact solution with the numerical solution.

4. When we denote the solution of the difference–integral equation

$$x(t-1) - \int_0^t K(t-\xi)x(\xi)d\xi = f(t),$$

as

$$x(t) = f(t+1) - \int_0^t G(t-\xi)f(\xi)d\xi,$$

prove that the following equation can be deduced:

$$K(t+1) + G(t-1) = \int_0^t K(t-\xi)G(\xi)d\xi = \int_0^t G(t-\xi)K(\xi)d\xi$$

[According to the operational calculus, a difference–integral equation becomes $(e^{-2} - K)X = F$. If we denote it by $X = (e^2 - G)F$ we get $(e^2 - G)(e^{-2} - K) = 1$.]

5. Solve the following difference–integral equations.

(a) $x(t-1) - \int_0^t (1 + t - \xi)x(\xi)d\xi = t + 1,$

(b) $x(t-1) - \int_0^t (t - \xi)x(\xi)d\xi = t,$

$$\text{(c)}\quad x(t-1) - \int_0^t [c + d(t-\xi)]x(\xi)d\xi = a + bt.$$

References

Churchill, R.V. (1941) *Fourier Series and Boundary Value Problems* (McGraw, New York) Formula 83.

Kellogg, O.D. (1929) *Foundations of Potential Theory* (Dover, NewYork).

Lamb, H. (1932) *Hydrodynamics* (Cambridge University Press).

Liouville, J. (1837) *Mathematical Journal* **2**, 24.

Moriya, T. (1936) Method to calculate pressure distribution of an arbitrary profile. *Journal of the Japan Aeronautical Society*, **4**, pp.365–366 (in Japanese). *The Aviation Society Magazine*. **4**, 365–366.

Prager, W. (1928) Druckverteilungen an Körpern in ebene Potentialströmung. *Physikalische Zeitschrift* **29**, 865–869.

Tsuji, M. (1940) *Set Theory*, Kyoritsushuppan, p. 20 (in Japanese).

Part II

Special theories of integral equations: summary of special theories

In Part I 'Fundamental theories' we explained the fundamental theories of integral equations and their applications. In order to clarify the mathematical characteristics, we restricted the types and kinds of equations and in some cases the kernels themselves, and on that basis developed precise theories and fundamental treatments. Equations applied in practice throughout the sciences, however, are not always of these types. In addition, characteristics of theories are sometimes more clearly shown in specific cases. Thus, it appears to be necessary, considering both the application of integral equations and the theoretical study, to examine more general as well as more specific cases. Therefore, the aim of the lectures in Part II 'Special Theories' is to show the theories and applications of the methods of solution (a) in the more general cases, in which these restrictions on the kernels and the types of equation are removed, and, (b) in the more specific cases, in which the characteristics of problems are more sharply defined, necessitating tighter restrictions. Many of the equations we shall study in this part will be treated in a similar fashion to those in the fundamental theories, with the addition of a few amendments and simple extensions; therefore we shall explain them briefly without considering them in the forms of theorems or definitions when the theory is sufficiently clear. In our discussion of the fundamental theories, we treated the methods of solution in three chapters. In this part we have divided the chapters according to the kinds and types of equation and in each chapter we shall study the existence theorem, the special method of solution, the numerical method of calculation, and so on, for each of these integral equations.

Chapter 6 contains the theories of symmetric kernels, in which we shall show that the mathematical properties of the eigensolution for Fredholm-type equations become clearer under the assumption that a kernel is symmetric. In Chapter 7 we treat what is called the singular integral equation, which stands out from the types treated in the fundamental theories, integral equations that include multiple integrals, simultaneous integral equations, difference-integral equations, and so on. Chapter 8 deals with a singular kernel, which we often see in practical applications, and the method of solution is seen as important. Brief explanations on nonlinear integral equations, integro-differential equations and so on will be given in Chapter 9. In Chapter 10 we again treat the applications of integral equations discussed in Part II.

Chapters 8 and 9 treat one of the central problems in the study of integral equations today. Because the theories relating to these have still not been systematically established and lie beyond the level of this book, we shall touch only on the fringe of these questions while considering the primary applications.

6 Symmetric kernels

If we consider the kernel by assuming a special characteristic, that is, symmetry, in addition to the conditions we have previously given to it in the discussion of the fundamental theories, the essential characteristics of the kernel become clearer. This is the subject we shall mainly develop in this chapter. In the the sphere of mathematics, we usually aim to achieve generalizations by removing as many restrictions as possible; however, we can sometimes get a clearer aspect of the essential characteristics of a problem by treating special cases under some given conditions. The theories of the symmetric kernel that we shall chiefly discuss in the present chapter were studied by D. Hilbert, E. Schmidt, and their students; their studies have given us many remarkable results. As we may reduce other types of integral equations, their study serves to considerably advance the general theoretical study of integral equations, and we can find symmetric kernels in many applications. Thus, the kernel is regarded as an important factor.

6.1. Symmetric kernels

Where the kernel $K(t, \xi)$ of an integral equation belongs to $\mathscr{R}$ and we always have

$$K(t, \xi) = K(\xi, t), \qquad K = \tilde{K}$$

everywhere in the domain, we call it a *symmetric kernel*. (From now on we denote the form obtained by exchanging t and ξ in $K(t, \xi)$ by $\tilde{K}(\xi, t)$. Hence by symmetry $K = \tilde{K}$.)

Therefore, Fredholm's theories are all effective. But under the assumption of symmetry we may still obtain many mathematical properties.

When we extend the linear transform

$$\sum_{1}^{n}{}_{j}\, a_{ij} x_j = y_j \qquad (i, j = 1, 2, \ldots, n),$$

which is used to standardize the n-dimensional quadratic form

$\sum_{ij} a_{ij} x_i x_j$, $a_{ij} = a_{ji}$ into ∞-dimensional functional space, the quadratic form becomes

$$\int_a^b \int_a^b K(t, \xi) x(t) x(\xi) dt d\xi,$$

and the linear transform becomes a Fredholm-type integral equation with a symmetric kernel. In this way, Hilbert studied the quadratic form of the functional by the principle of passage from discontinuity to continuity and introduced the important result on symmetric kernels. We are going to explain his achievement minutely, in due course.

Now let us prove that all sorts of integral equations can be related to the cases of symmetric kernels. In

$$(E - \lambda K) \circ x = f, \qquad [F_2]$$

multiply by $(E - \lambda \tilde{K})$ from the left and we get

$$(E - \lambda \tilde{K}) \circ (E - \lambda K) \circ x = (E - \lambda \tilde{K}) \circ f; \qquad (1a)$$

it then follows that

$$[E - \lambda(K + \tilde{K} - \lambda \tilde{K} \circ K)] \circ x = (E - \lambda \tilde{K}) \circ f. \qquad (1)$$

Now (1) is a Fredholm-type equation with kernel $K + \tilde{K} - \lambda \tilde{K} \circ K$, which is clearly a symmetric kernel since it is symmetric in K, $\tilde{K}$.

Further, let us show that the solutions of (1) and $[F_2]$ coincide. From (1), viz. (1a) denote

$$(E - \lambda \tilde{K}) \circ \{(E - \lambda K) \circ x - f\} = 0;$$

then it is a homogeneous integral equation with an unknown function $(E - \lambda K) \circ x - f$. Therefore, let $\Delta(\lambda) = 0$ be the characteristic function, and for the value of λ of $\Delta(\lambda) \neq 0$ we have

$$(E - \lambda K) \circ x - f = 0$$

and this agrees with $[F_2]$. When we take eigenvalues, if φ is the eigenfunction corresponding to one of the eigenvalues, we get $(E - \lambda K) \circ x - f = \varphi$. Yet the expression $\Delta(\lambda)$ changes neither in the case of $K(t, \xi)$ nor in that of $\tilde{K}(\xi, t)$. Therefore, φ may be regarded as the eigenfunction of $[\tilde{F}_0]$, and, so that there exists the solution in $[F_2]$, f and φ intersect orthogonally by Theorem 2.8. That is, $f \circ \varphi = 0$.

If we write $(E - \lambda K) \circ x = f + \varphi$, the eigenfunction φ must intersect orthogonally with $f + \varphi$ also; then we get $(f + \varphi) \circ \varphi = 0$ and from this, $\mathring{\varphi}^2 = 0$. Hence, it must always be that

$$\varphi = 0.$$

Therefore there is no eigenvalue, and the solution of (1) agrees with that of $[F_2]$. In other words, we only have to solve equation (1) with a symmetric kernel instead of solving $[F_2]$.

In the case of Fredholm-type integral equations, for

$$K \circ x = f \qquad [F_1]$$

multiply by $\tilde{K}$ from the left and we get

$$\tilde{K} \circ K \circ x = \tilde{K} \circ f.$$

Since the right-hand side is known, if we take $\tilde{K} \circ K$ as a kernel this expression is symmetric in t, ξ (see Question 1).

Symmetric kernels possess the following mathematical properties. First, *when a kernel is symmetric it is commutative for the convolution of the second kind.* For example, since we have

$$\int_a^b K(t, \xi)f(\xi)d\xi = \int_a^b K(\xi, t)f(\xi)d\xi = \int_a^b f(\xi)K(\xi, t)d\xi$$

for an arbitrary function $f(t)$, we always have

$$K \circ f = f \circ K.$$

Therefore, in the case of a symmetric kernel, an associated function is the same as the original one.

All the iterated kernels are symmetric. For example, if for $n \leqq m$ $\mathring{K}^n$ are all symmetric, that is $\mathring{K}^m(\xi, t) = \mathring{K}^m(t, \xi)$, we have

$$\mathring{K}^{m+1}(\xi, t) = \mathring{K}^m(\xi, t) \circ K(\xi, t) = \mathring{K}^m(t, \xi) \circ K(t, \xi) = \mathring{K}^{m+1}(t, \xi).$$

Hence, by induction all the iterated kernels are symmetric.

Next, since the kernel of the solution is

$$\Gamma(t, \xi; \lambda) = -\sum_0^\infty \lambda^n \mathring{K}^{n+1},$$

the kernel of the solution is symmetric.

Lastly, we give here the following theorem.

THEOREM 6.1

If $K(t, \xi)$ is a symmetric kernel and not identically zero at the continuous points, every iterated kernel is not identically zero.

Proof. If $\mathring{K}^n$ is the first iterated kernel which becomes identically zero, we have

$$\mathring{K}^{n+1} = \mathring{K}^{n} \circ K,$$

therefore $\mathring{K}^{n+1} \equiv 0$. Let $2m$ be whichever of n, $n+1$ is even; then

$$\mathring{K}^{2m} = \int_a^b [\mathring{K}^m(t, \xi)]^2 d\xi \; (= \mathring{K}^m \circ \mathring{K}^m) \equiv 0.$$

While the iterated kernel is continuous from Lemma 1.1, we must have

$$\mathring{K}^m \equiv 0.$$

This turns out to be contradictory. Hence, the assumption that $\mathring{K}^n \equiv 0$ is false.

THEOREM 6.1-a

If $K(t, \xi)$ is a symmetric kernel and is not identically zero at the continuous points, we have

$$k_{2n} \equiv \int_a^b \mathring{K}^{2n}(t, t)dt > 0.$$

6.1.1. Numerical method of solution

In the case of a symmetric kernel, if we make use of Tschebyscheff's formula for a Fredholm-type integral equation, the coefficient determinant of the simultaneous equations then introduced has the symmetric property. Consequently, we can generally write the simultaneous equations as

$$\begin{aligned}
a_{11}x_1 + a_{12}x_2 + \cdots + a_{1n}x_n &= f_1, \\
a_{21}x_1 + a_{22}x_2 + \cdots + a_{2n}x_n &= f_2, \\
\cdots \\
a_{n1}x_1 + a_{n2}x_2 + \cdots + a_{nn}x_n &= f_n,
\end{aligned}$$

$a_{ij} = a_{ji} \qquad (i, j = 1, 2, \ldots, n).$

In this case, we have the Gauss–Doolittle method of calculation using the symmetry with the orthogonal line element as the pivot. For example, the calculation in Chapter 4 (Example 4.1) is handled as in the following. That is, we have Table 6.1 (A) corresponding to Table 4.5 (A) and Table 6.1 (B) as the example of numerical calculation to Table 4.5 (B). Here we have

$$b_{ij} = a_{ij}/a_{ii}, \qquad b_{ij.1} = g_{ij.1}/g_{ii.1},$$
$$b_{ij.12} = g_{ij.12}/g_{ii.12}, \qquad b_{ij.123} = g_{ij.123}/g_{ii.123};$$

TABLE 6.1 (A) Example of the method of solution of integral equations (symmetric)

x_1	x_2	x_3	x_4		Sum of the row	Proof
a_{11}	a_{12}	a_{13}	a_{14}	a_{15}	a_{1T}	a'_{15}
*	a_{22}	a_{23}	a_{24}	a_{25}	a_{2T}	a'_{25}
*	*	a_{33}	a_{34}	a_{35}	a_{3T}	a'_{35}
*	*	*	a_{44}	a_{45}	a_{4T}	a'_{45}
1	b_{12}	b_{13}	b_{14}	b_{15}	b_{1T}	b'_{1T}
	$g_{22.1}$	$g_{23.1}$	$g_{24.1}$	$g_{25.1}$	$g_{2T.1}$	$g'_{2T.1}$
	*	$g_{33.1}$	$g_{34.1}$	$g_{35.1}$	$g_{3T.1}$	$g'_{3T.1}$
	*	*	$g_{44.1}$	$g_{45.1}$	$g_{4T.1}$	$g'_{4T.1}$
	1	$b_{23.1}$	$b_{24.1}$	$b_{25.1}$	$b_{2T.1}$	$b'_{2T.1}$
		$g_{33.12}$	$g_{34.12}$	$g_{35.12}$	$g_{3T.12}$	$g'_{3T.12}$
		*	$g_{44.12}$	$g_{45.12}$	$g_{4T.12}$	$g'_{4T.12}$
		1	$b_{34.12}$	$b_{35.12}$	$b_{3T.12}$	$b'_{3T.12}$
			$g_{44.123}$	$g_{45.123}$	$g_{4T.123}$	$g'_{4T.123}$
			1	$b_{45.123}$	$b_{4T.123}$	$b'_{4T.123}$
$b_{15.234}$	$b_{25.134}$	$b_{35.124}$	$b_{45.123}$			

$$g_{ij.1} = a_{ij} - a_{i1}b_{1j},$$

$$g_{ij.12} = g_{ij.1} - g_{i2}b_{2j.1} = a_{ij} - g_{i1}b_{1j} - g_{i2.1}b_{2j.1},$$

$$g_{ij.123} = g_{ij.12} - g_{i3.12}b_{3j.12} = a_{ij} - a_{i1}b_{1j} - g_{i2.1}b_{2j.1} - g_{i3.12}b_{3j.12};$$

where

$$x_4 = g_{45.123}/g_{44.123} = b_{45.123}.$$

Every determinant that appears in the process of calculation is symmetric, and we need not calculate the part indicated by the asterisks.

6.2. Existence of eigenvalues

The following theorems concern the eigenvalue and the eigenfunction of the integral equation having a symmetric kernel.

THEOREM 6.2. *Orthogonality of eigenfunctions*

Two eigenfunctions for different eigenvalues intersect orthogonally.

Proof. Let λ_1, λ_2 be the eigenvalues and $\varphi_1(t)$, $\varphi_2(t)$ be their related eigenfunctions; then we get

$$\varphi_1 = \lambda_1 K \circ \varphi_1, \qquad \varphi_2 = \lambda_2 K \circ \varphi_2.$$

TABLE 6.1. (B) Solution of simultaneous equations (symmetric)

x_1	x_2	x_3	x_4	a_{i5}	Sum of the row	Verification
1.010542	1.041706	1.060967	1.092131	12.821584	17.026930	12.821590
*	1.165002	1.241202	1.364498	15.249632	19.020334	15.249630
*	*	1.352594	1.532829	16.750368	19.635791	16.750384
*	*	*	1.805196	19.178656	20.983852	19.178424
1.000000	1.030839	1.049899	1.080738	12.687829	16.849305	
	0.091170828666	0.147515912306	0.238686740972	2.032644403726	2.510017885670	
	*	0.238685807667	0.386201646354	3.289000129357	3.913887583378	
	*	*	0.624888527322	5.321884626401	5.946773153723	
	1.000000	1.618017	2.618017	22.294899	27.530932	
		0.000002553795	0.000002480182	0.000147763736	0.000152945229	
		*	0.000002581709	0.000387843167	0.000390424876	
		1.000000	0.971175	57.860453	59.873452	
			0.000000173018	0.000244338713	0.000241927818	
					→ 4511731	
			1.000000	141.221557		
169.201386	− 219.132334	− 79.290393	141.221557			

Multiply the first expression by $\lambda_2\varphi_2$ and the second by $\lambda_1\varphi_1$ from the left, and we have

$$\lambda_2\varphi_2 \cdot \varphi_1 = \lambda_1\lambda_2\varphi_2 \circ K \circ \varphi_1, \qquad \lambda_1\varphi_1 \cdot \varphi_2 = \lambda_1\lambda_2\varphi_1 \circ K \circ \varphi_2.$$

While the right-hand sides of these equations are equal because of the commutability of a symmetric kernel, it should be that

$$(\lambda_2 - \lambda_1)\varphi_1 \cdot \varphi_2 = 0.$$

By the supposition $\lambda_2 \neq \lambda_1$, we obtain

$$\varphi_1 \cdot \varphi_2 = 0.$$

(This is essentially equal to Theorem 2.7.)

Example 6.1 *Orthogonality of eigenfunctions*

The symmetric kernel

$$K(t, \xi) = \begin{cases} (1 - \xi)t, & 0 \leqq \xi \leqq t, \\ (1 - t)\xi, & t \leqq \xi \leqq 1 \end{cases}$$

on the interval (0.1) has the eigenvalues

$$\lambda_n = n^2\pi^2, \qquad n = 1, 2, \ldots,$$

and the eigenfunctions are

$$\varphi_n = \sin n\pi t.$$

Since we have

$$\int_0^1 \sin n\pi t \sin m\pi t \, dt = 0 \qquad (n \neq m),$$

this becomes

$$\varphi_n \cdot \varphi_m = 0.$$

THEOREM 6.3 *Existence of the eigenvalue*

There exists at least one eigenvalue for a symmetric kernel not identically zero at every continuous point.

Proof. This theorem was first proved by D. Hilbert, but we give here the proof by Kneser (1924, pp. 70–74).

If $K(t, \xi)$ has no eigenvalue, the characteristic equation $\Delta(\lambda) = 0$ has neither a real root nor a complex one; then $\Delta\binom{t}{\xi}; \lambda\Big) \Big/ \Delta(\lambda)$ can be expanded into a uniformly convergent series of λ, $\sum_{n=0}^{\infty} c_n\lambda^n$, as we proved in Theorem 2.3. From (3) in § 2.2.

$$\lambda\Gamma(t, \xi; \lambda) = \lambda K(t, \xi) + \lambda^2 \mathring{K}^2(t, \xi) + \cdots \tag{6.1}$$

is uniformly convergent in S.

If (6.1) is uniformly convergent, it naturally follows that

$$\lambda^2 \mathring{K}^2(t, t) + \lambda^3 \mathring{K}^3(t, t) + \cdots \tag{6.2}$$

is uniformly convergent for every λ in I. Each term can be integrated individually since it is continuous. If we express

$$\int_a^b \mathring{K}^n(t, t)dt \equiv k_n \tag{6.3}$$

and integrate (6.2) from a to b term by term, then

$$k_2\lambda^2 + k_3\lambda^3 + \cdots \tag{6.4}$$

is uniformly convergent for every value of λ. Since this is a power series in λ it converges uniformly. Therefore, its partial series

$$k_2\lambda^2 + k_4\lambda^4 + \cdots \tag{6.5}$$

is convergent for every value of λ.

Since K is symmetric, we have

$$k_{2n} = \int_a^b \int_a^b [\mathring{K}^n(t, \xi)]^2 d\xi dt. \tag{6.6}$$

Now, if we expand the left-hand side of the inequality

$$\int_a^b \int_a^b [pK^{n+1}(t, \xi) + qK^{n-1}(t, \xi)]^2 d\xi dt \geqq 0 \tag{6.7}$$

for arbitrary real numbers p, q, we get

$$p^2 k_{2n+2} + 2pqk_{2n} + q^2 k_{2n-2} \geqq 0.$$

Therefore, we ought to have

$$k_{2n+2}k_{2n-2} - k^2{}_{2n} \geqq 0.$$

While we have $k_{2n} > 0$, (see 11.1) we get

$$\frac{k_{2n+2}}{k_{2n}} \geqq \frac{k_{2n}}{k_{2n-2}}. \tag{6.8}$$

The ratio of neighbouring terms in the power series (6.5) is $\lambda^2 k_{2n+2}/k_{2n}$, and there is none smaller than $\lambda^2 k_4/k_2$; then if $\lambda = (k_2/k_4)^{\frac{1}{2}}$, the terms of (6.5) do not converge to zero by increasing the number of terms. This contradicts the hypothesis that (6.5) is convergent regardless of λ. Hence, the assumption that (6.1) is uniformly convergent is false.

Given a circle C of radius $(k_2/k_4)^{\frac{1}{2}}$ in the λ-plane, and that

$$\sum_1^\infty k_{2n}|\lambda|^{2n}$$

diverges for any value of λ outside C. Then $\Delta\left(\begin{matrix}t\\ \xi\end{matrix};\ \lambda\right)\Big/\Delta(\lambda)$ is divergent outside this circle, and at least one eigenvalue is either in the interior of C or on the circumference.

THEOREM 6.4. *Reality of an eigenvalue*

If the kernel $K(t, \xi)$ is real and symmetric, its eigenvalue is real.

Proof. Suppose that there exists a complex eigenvalue $\lambda = \mu + i\nu$. Then the eigenfunction $\varphi(t)$ exists, and when $\varphi \equiv 0$ we have

$$\varphi = (\mu + i\nu)K \circ \varphi. \tag{6.9}$$

(i) *φ is a complex function.* If φ is a real function, divide (6.9) into real and imaginary parts, and we get

$$\varphi = \mu K \circ \varphi, \tag{6.10}$$

$$0 = \nu K \circ \varphi. \tag{6.11}$$

Since $\nu \neq 0$ we get $K \circ \varphi = 0$ from (6.11) and $\varphi \equiv 0$ from (6.10); these results contradict the proposition.

(ii) *Let $\varphi = \varphi_1 + i\varphi_2$ and $\varphi_1(t)$, $\varphi_2(t)$ be real functions.* Then we get

$$\varphi_1 + i\varphi_2 = (\mu + i\nu)K \circ [\varphi_1 + i\varphi_2]. \tag{6.12}$$

Dividing this into real and imaginary parts, we have

$$\varphi_1 = \mu K \circ \varphi_1 - \nu K \circ \varphi_2, \tag{6.13}$$

$$\varphi_2 = \mu K \circ \varphi_2 + \nu K \circ \varphi_1. \tag{6.14}$$

Multiply both sides of (6.14) by $-i$ and add the result to (6.13); this gives

$$\varphi_1 - i\varphi_2 = (\mu - i\nu)K \circ [\varphi_1 - i\varphi_2],$$

and $\bar{\varphi} = \varphi_1 - i\varphi_2$ becomes the eigenfunction for $\bar{\lambda} = \mu - i\nu$. Then from Theorem 12 we get

$$\varphi \circ \bar{\varphi} = 0,$$

viz.

$$[\varphi_1 + i\varphi_2] \circ [\varphi_1 - i\varphi_2] = 0, \qquad \mathring{\varphi}_1{}^2 + \mathring{\varphi}_2{}^2 = 0.$$

From this expression comes $\varphi_1 \equiv \varphi_2 \equiv 0$, and $\varphi \equiv 0$ contradicts the assumption. Hence, there cannot exist a complex eigenvalue.

As for the real property of the eigenvalue, it corresponds to the fact that if in the n-dimensional simultaneous equations § 2.2 (f)

$$k_{ij} = k_{ji}$$

holds, then all the roots of λ are real.

By Fredholm's theory, when the index of an eigenvalue $\lambda = \lambda_0$ is p and the multiplicity is m, we have $p \leqq m$; while in the case of a symmetric kernel we can have simply $p = m$. Hence, with regard to the symmetric kernel, the degree at the zero point in a Fredholm determinant agrees with the number of linearly independent eigenfunctions.

THEOREM 6.5 *Index of an eigenvalue and multiplicity*

With respect to the symmetric kernel, the index and multiplicity of an eigenvalue agree.

Proof. See (Volterra and Pérès 1936, p. 287).

6.3. Complete orthogonal eigenfunction systems

We now construct the complete orthogonal system of an eigenfunction. By Theorem 2.5, when λ_0 is the eigenvalue of an index ν, there exist ν linearly independent eigenfunctions:

$$\varphi_i(t) = \Delta\begin{pmatrix} t'_1 & t'_2 \ldots t'_{i-1} & t & t'_{i+1} \ldots t'_\nu \\ \xi'_1 & \xi'_2 \ldots \xi'_{i-1} & \xi'_i & \xi'_{i+1} \ldots \xi'_\nu \end{pmatrix}; \lambda_0\Big) \qquad (i = 1, 2, \ldots, \nu).$$

From (2.27) it is evident that when $K(t, \xi)$ is a real function, $\varphi_i(t)$ are all real functions. Besides, an arbitrary eigenfunction ψ_j is expressed in the linear combination of $\varphi_i(t)$:

$$\psi_j(t) = \sum_{i=1}^{\nu} c_{ij}\varphi_i(t) \qquad (i, j = 1, 2, \ldots, \nu).$$

Then let us form a normalized orthogonal function system in the manner used in § 3.1.

First, put

$$\psi_1(t) = c\varphi_1(t).$$

If we take

$$c = 1/||\varphi_1||,$$

we have $\dot{\psi}_1^2 = 1$. Then take

$$\psi_2 = \frac{\alpha_1}{c}\varphi_1 + \alpha_2\varphi_2 = \alpha_1\psi_1 + \alpha_2\varphi_2,$$

and determine the values of α_1, α_2 so that they satisfy the condition for normalized orthogonality. Here we have

$$\psi_1 \cdot \psi_2 = \alpha_1\dot{\psi}_1^2 + \alpha_2\psi_1 \cdot \varphi_2,$$

where $\dot{\psi}_1^2 = 1$; then we have $\alpha_1 = -\alpha_2\psi_1 \cdot \varphi_2$ since we have $\psi_1 \cdot \psi_2 = 0$ by orthogonality, it follows that

$$\psi_2 = \alpha_2[\varphi_2 - (\psi_1 \cdot \varphi_2)\psi_1].$$

Now we have only to decide α_2 so that we can get $\dot{\psi}_2^2 = 1$. Here it follows that

$$\varphi_2 - (\psi_1 \cdot \varphi_2)\psi_1 \not\equiv 0.$$

(If $\varphi_2 - (\psi_1 \cdot \varphi_2)\psi_1 \equiv 0$, then $\varphi_2 - c'\varphi_1 = 0$ follows and φ_1, φ_2 contradict the assumption of linear independence.)

Similarly, take

$$\psi_3 = \beta_1\psi_1 + \beta_2\psi_2 + \beta_3\varphi_3,$$

and by the orthogonality condition $\psi_1 \cdot \psi_3 = 0$ we obtain

$$\beta_1 = -\beta_3\psi_1 \cdot \varphi_3,$$

and from $\psi_2 \circ \psi_3 = 0$,

$$\beta_2 = -\beta_3\psi_2 \cdot \varphi_3;$$

then we obtain

$$\psi_3 = \beta_3[\varphi_3 - (\psi_1 \cdot \varphi_3)\psi_1 - (\psi_2 \cdot \varphi_3)\psi_2].$$

TABLE 6.2 Eigenvalue and eigenfunction system

Eigenvalue	Index	Normalized orthogonal function system		
λ_1	ν_1	$\psi_1^1, \psi_2^1,$	$\cdots,$	$\psi_{\nu_1}^1$
λ_2	ν_2	$\psi_1^2, \psi_2^2,$	$\cdots,$	$\psi_{\nu_2}^2$
$\cdot$	$\cdot$	$\cdot \quad \cdot$	$\cdots$	$\cdot$
$\cdot$	$\cdot$	$\cdot \quad \cdot$	$\cdots$	$\cdot$
λ_n	ν_n	$\psi_1^n, \psi_2^n,$	$\cdots,$	$\psi_{\nu_n}^n$
$\cdot$	$\cdot$	$\cdot \quad \cdot$	$\cdots$	$\cdot$

Now β_3 is determined by the condition of normalization $\psi_3^2 = 1$. In this case too we can prove that

$$\varphi_3 - (\psi_1 \cdot \varphi_3)\psi_1 - (\psi_2 \cdot \varphi_3)\psi_2 \not\equiv 0.$$

By following a similar method, we can obtain ν normalized orthogonal function systems. The ψ_i are all real functions, linearly independent (see § 3.2).

Thus, to every root of the characteristic function

$$\Delta(\lambda) = 0,$$

there corresponds the same number of normalized orthogonal function systems with the same index (multiplicity). Even if there are infinitely many eigenvalues, they are countable, and then the whole number of normal orthogonal function systems is also enumerable. Therefore we can arrange them in a new series

$$\psi_1, \psi_2, \ldots, \psi_n, \cdots$$

and the corresponding eigenvalues into

$$\lambda_1, \lambda_2, \ldots, \lambda_n, \ldots.$$

This series exhausts all the orthogonal functions. (These eigenvalues are, of course, not always different from each other.)

Take two arbitrary functions ψ_i, ψ_j in this function system, and if they both correspond to the same eigenvalue, they intersect orthogonally as we have explained above, while if each of them corresponds to a different eigenvalue, they still intersect orthogonally by Theorem 6.2. Therefore, we have

$$\psi_i \cdot \psi_j = \begin{cases} 1, & i = j; \\ 0, & i \neq j. \end{cases}$$

Since a function system like this is complete, it is called a *complete normalized orthogonal system of eigenfunctions* (*characteristic functions*).

These theories are summed up in the following theorem.

THEOREM 6.6. *The existence of the complete normalized orthogonal function system*

If the kernel $K(t, \xi)$ belongs to $\mathscr{R}$ in S and is continuous in t, and if $K(t, t)$ is integrable in I, and is a real function and symmetric, there exists a complete normalized orthogonal system of eigenfunctions $\{\varphi_i\}$ that consists of eigenfunctions of real functions, and every eigenfunction is expressed as a linear combination of the φ_i.

The index of the eigenvalue $\lambda_n = n^2\pi^2$ of the symmetric kernel on the interval (0, 1)

$$K(t, \xi) = \begin{cases} (1-\xi)t, & 0 \leqq \xi \leqq t, \\ (1-t)\xi, & t \leqq \xi \leqq 1, \end{cases}$$

which we treated in Example 6.1 is 1 and the norm of the eigenfunction

$$\varphi_n = \sin n\pi t$$

is

$$\sqrt{\dot{\varphi}_n{}^2} = \left(\int_0^1 \sin^2 n\pi t \, dt\right)^{\frac{1}{2}} = 1/\sqrt{2};$$

then the normalized orthogonal system becomes

$$\psi_n = \sqrt{2} \sin n\pi t \qquad (n = 1, 2, \ldots).$$

6.4. Orthogonal system expansion of a kernel

In the previous section, we derived the normalized orthogonal system of eigenfunctions $\{\psi_i\}$ and the eigenvalue series $\{\lambda_i\}$:

$$\psi_1, \psi_2, \psi_3, \ldots, \psi_i, \ldots;$$
$$\lambda_1, \lambda_2, \lambda_3, \ldots, \lambda_i, \ldots.$$

According to the explanation in § 3.1, *if a kernel is expansible into the form $K(t,\xi) = \sum C_i\psi_i$, and the series on the right-hand side is uniformly convergent, we can express it as*

$$K(t, \xi) = \sum \frac{\psi_i(t)\psi_i(\xi)}{\lambda_i}. \tag{6.15}$$

Though we have $\psi_i = \lambda_i K \circ \psi_i$, yet the kernel becomes as in (6.15), since we have $K \circ \psi_i = \sum C_i \psi_i \circ \psi_j = C_i$ and $C_i = \psi_i/\lambda_i$. Kneser called (6.15) a *bilinear formula.*

We can obtain an orthogonal system expansion for a kernel without assuming the expansibility of the kernel. The theorem concerning this is as follows.

THEOREM 6.7. *Expansion theorem of the kernel*

Where the kernel $K(t,\xi)$ belongs to $\mathscr{R}$ in S and is continuous on t, and $K(t, t)$ is an integrable, symmetric real function, then with respect to the complete normalized orthogonal system of eigenfunctions $\{\psi_i\}$, if

$$\frac{1}{\lambda_i} \sum \psi_i(t)\psi_i(\xi)$$

is uniformly convergent in S, we have

$$K(t, \xi) = \sum_i \frac{\psi_i(t)\psi_i(\xi)}{\lambda_i} \tag{6.15}$$

at continuous points.

Proof. When $\{\varphi_i\}$ consists of finite or enumerable infinite functions, if we put

$$K(t, \xi) = \sum \frac{\psi_i(t)\psi_i(\xi)}{\lambda_i} + H(t, \xi), \tag{6.16}$$

$H(t, \xi)$ is a symmetric real kernel. The whole function is continuous, since $\sum$ in the right side is uniformly convergent and ψ_i is a continuous function. Then there is at least one eigenvalue ρ; let φ be the eigenfunction, so that

$$\varphi = \rho H \circ \varphi. \tag{6.17}$$

Since from (2) we have

$$H = K - \sum \frac{\psi_i}{\lambda_i} \times \psi_i,$$

it follows that

$$\varphi = \sigma K \circ \varphi - \sigma \sum \frac{\psi_i}{\lambda_i} \times \psi_i \cdot \varphi. \tag{6.18}$$

On multiplying both sides by ψ_j from the right, we get

$$\varphi \cdot \psi_j = \sigma K \circ \varphi \cdot \psi_j - \rho \sum \frac{\psi_i}{\lambda_i} \cdot \psi_j \times \varphi \cdot \psi_i$$

$$= \frac{\rho}{\lambda_j} \psi_j \cdot \varphi - \frac{\rho}{\lambda_j} \psi_j \cdot \varphi$$

$$= 0 \qquad (j = 1, 2, \ldots, m).$$

Therefore, from (4) we have

$$\varphi = \rho K \circ \varphi.$$

Here we find that φ is the eigenfunction of K also. Since $\{\varphi_i\}$ form a complete system it is expressed as

$$\varphi = \sum C_i \psi_i,$$

from which

$$\varphi \cdot \psi_i = C_i.$$

If we put $\varphi \cdot \psi_i = 0$ in the above, so that $C_i = 0$, then $\varphi \equiv 0$. This is a contradiction. Therefore it must be that $H(t, \xi) \equiv 0$.

In the eigenfunction expansion, $\psi_i(t)\psi_i(\xi)/\lambda_i$ is called a *principal element* of the kernel for λ_i.

6.5. Hilbert's theorem

If an arbitrary function $f(t)$ is expressed by the normal orthogonal system of eigenfunctions $\{\psi_i\}$ of the symmetric kernel $K(t, \xi)$ in

$$f(t) = \sum c_i \psi_i(t), \tag{6.19}$$

and the right-hand side is uniformly convergent, then the coefficient c_i is

$$c_i = \int_a^b f(t)\psi_i(t)dt \equiv f \cdot \psi_i. \tag{6.20}$$

This poses the question of which types of function $f(t)$ are expansible into expressions of this form. D. Hilbert gave the following theorem in 1912.

THEOREM 6.8. *Hilbert's theorem*

If a function $f(t)$ can be expressed by the symmetric kernel $K(t, \xi)$, which *has a normalized orthogonal system of eigenfunctions* $\{\psi_i\}$ *and a continuous function* $g(t)$ *in* I *such that*

$$f(t) = \int_a^b K(t, \xi)g(\xi)d\xi \equiv = K \circ g, \tag{6.21}$$

then $f(t)$ *is expressed as*

$$f(t) = \sum c_i\psi_i \tag{6.19}$$

in I, *and the series on the right-hand side is absolutely uniformly convergent in* I, *and* c_i *is*

$$c_i = f \cdot \psi_i. \tag{6.20}$$

Proof. (i) *Absolute uniform convergence. Let*

$$g(t) = \sum e_i\psi_i(t), \qquad e_i = g \cdot \psi_i. \tag{6.22}$$

While making use of the symmetry of K, we get from (6.20), (6.21) that

$$c_i = f \cdot \psi_i = K \circ g \cdot \psi_i = g \cdot K \circ \psi_i = g \cdot \frac{\psi_i}{\lambda_i} = e_i/\lambda_i. \tag{6.23}$$

Then Schwartz's inequality† leads to

$$\sum_m^n{}_i |c_i\psi_i| = \sum\left|e_i\frac{\psi_i}{\lambda_i}\right| \leqq \sum|e_i|\left|\frac{\psi_i}{\lambda_i}\right| < \left(\sum_m^n e_i^2 \sum_m^n \psi_i^2/\lambda_i^2\right)^{\frac{1}{2}}. \tag{6.24}$$

According to Bessel's inequality,† for any value of m, n we have

$$\sum\varphi_i^2/\lambda_i^2 \leqq \int_a^b K^2(t, \tau)d\tau; \qquad \sum_m^n e_i^2 < \varepsilon, \qquad m, n > N, \tag{6.25}$$

for some sufficiently large number N, and since the right-hand side of the first expression is bounded, $\sum_m^n |c_i\psi_i|$ becomes very small; hence finally (6.19) is absolutely uniformly convergent.

(ii) *Sum.* Suppose that

$$S(t) = \sum c_i\psi_i(t) = \sum f \cdot \psi_i \times \psi_i, \tag{6.26}$$

and this is continuous since ψ_j is continuous and the right-hand side

---

† When φ, ψ are continuous in I, Schwartz's inequality gives $(\varphi \cdot \psi)^2 \leqq \mathring{\varphi}^2 \times \mathring{\psi}^2$.

† When f is continuous in I, and $\{\psi\}$ is a normalized orthogonal system, Bessel's inequality gives $\sum_1^m [f \cdot \psi_i]^2 \leqq \mathring{f}^2$.

is uniformly convergent. Besides, we have already supposed that $g(t)$ is continuous, so $f(t)$ is continuous from Theorem 1.1*. Then

$$R(t) \equiv f(t) - S(t) = f - \sum f \cdot \psi_i \times \psi_i \tag{6.26a}$$

is also continuous in I. On multiplying both sides by R from the right, we have

$$\mathring{R}^2 = f \cdot R - \sum f \cdot \psi_i \times \psi_i \cdot R. \tag{6.27}$$

Again multiplying both sides of (6.26a) by ψ_j from the right we find

$$\begin{aligned} R \cdot \psi_j &= f \cdot \psi_j - \textstyle\sum_i f \cdot \psi_i \times \psi_i \cdot \psi_j, \\ &= f \cdot \psi_j - f \cdot \psi_j = 0 \end{aligned} \tag{6.28}$$

(from orthogonal normality, $\sum_i$ leaves the term of $i = j$ only). Then, from (6.21) we have

$$f \cdot R = K \circ g \cdot R = g \cdot K \circ R,$$

while (6.28) gives us $K \circ R = 0$, which makes

$$f \cdot R = 0. \tag{6.29}$$

From (6.28) and (6.29), (6.27) becomes $\mathring{R}^2 \equiv 0$. Then we have $R \equiv 0$. Hence we obtain

$$f(t) = S(t) = \sum c_i \psi_i(t).$$

6.5.1. *Expansion of a symmetric kernel*

In making an assumption in Hilbert's theorem, if we assume the possibility of expansion as in (6.22), instead of assuming the continuity of $g(t)$, and assume also the continuity of $f(t)$, we can still prove the theory.

The symmetric kernel $K(t, \xi)$ we are considering now is expansible in I to a Fourier expansion with parameter ξ and the iterated kernels are all continuous. Then, for example, suppose that ξ is a parameter in

$$\mathring{K}^2 = \int_a^b K(t, \tau)K(\tau, \xi)d\tau;$$

then this expression satisfies Hilbert's theorem, and, therefore, can be expanded into a Fourier series of eigenfunctions which is represented by

$$\mathring{K}^2 = \frac{\psi_1(t)\psi_1(\xi)}{\lambda_1^2} + \frac{\psi_2(t)\psi_2(\xi)}{\lambda_2^2} + \cdots + \frac{\psi_i(t)\psi_i(\xi)}{\lambda_i^2} + \cdots,$$

and when ξ is constant, this is uniformly convergent in t. If t is constant, it is natural that this expression is uniformly convergent in ξ; thus it is uniformly convergent in t, ξ in the domain S.

On generalizing the expression, we have

$$\mathring{K}^n = \frac{\psi_1(t)\psi_1(\xi)}{\lambda_1^n} + \frac{\psi_2(t)\psi_2(\xi)}{\lambda_2^n} + \cdots + \frac{\psi_i(t)\psi_i(\xi)}{\lambda_i^n} + \cdots.$$

Further, we have

$$\frac{\lambda}{\lambda_i - \lambda} = 1 - \left(1 - \frac{\lambda}{\lambda_i}\right)^{-1} = -\left(\frac{\lambda}{\lambda_i}\right) - \left(\frac{\lambda}{\lambda_i}\right)^2 - \left(\frac{\lambda}{\lambda_i}\right)^3 - \cdots,$$

and the kernel of the solution is

$$\Gamma(t, \xi; \lambda) = -\{K + \lambda\mathring{K}^2 + \cdots + \lambda^{n-1}\mathring{K}^n + \cdots\}.$$

Hence

$$\Gamma(t, \xi; \lambda) = -K(t, \xi) - \sum_i \frac{\lambda\psi_i(t)\psi_i(\xi)}{\lambda_i(\lambda_i - \lambda)}.$$

6.5.2. Closed kernels

Definition. Closed kernel

If a symmetric kernel $K(t, \xi)$ has no function $h(t) \not\equiv 0$ such that

$$\int_a^b K(t, \xi)h(\xi)d\xi = 0, \qquad \text{i.e. } K \circ h = 0,$$

it is called a *closed kernel.*

Closed kernels have the following mathematical properties.

(1) *The eigenfunction system of closed kernels is complete.* Suppose that $\{\psi_i\}$ are incomplete; then there exists a function h which orthogonally intersects every one of these. That is, we have

$$\psi_i \cdot h = 0.$$

But the calculation of $K \circ h$ shows us that

$$K \circ h = \frac{\psi_i}{\lambda_i} \times \psi_i \cdot h,$$

and then the right-hand side becomes zero, and we obtain a function of the form $K \circ h = 0$; this is contradictory to the assumption of the closed kernel. (The converse is also true (see Question 4).

(2) *A closed kernel has always infinitely many eigenvalues.* If the number of eigenvalues is finite the eigenfunctions are also finite, therefore we can make a function $h(t)$ which is orthogonal to them all. Then we have $K \circ h = 0$, which contradicts the ussumption.

The symmetric kernel used in Example 6.1 is closed.

6.6. Disymmetric kernels

If the kernel $K(t, \xi)$ is such that

$$K(t, \xi) = A(t)B(\xi)k(t, \xi),$$

with $A(t)B(t) > 0$ in $I(a, b)$ and $k(t, \xi) = k(\xi, t)$ in S, then we call the kernel a *disymmetric kernel*. Here, if we let Φ_i, Ψ_i be the eigenfunctions for the eigenvalue λ_i, on the function systems $\{\Phi_i\}$, $\{\Psi_i\}$, we have

$$\Phi_i \cdot \Psi_j = \begin{cases} 1, & i = j; \\ 0, & i \neq j. \end{cases}$$

Such a function system is called a *biorthogonal system*, and is dealt with in the following theorem.

THEOREM 6.9. *Biorthogonal system*

Suppose that $K(t, \xi)$ is a disymmetric kernel whith is not always zero at every continuous point; then the eigenvalues are all real and for each there exist two eigenfunctions which comprise a biorthogonal function system.

Proof. (i) *Connect K with the symmetric kernel k_1.* By a transformation we get

$$K(t, \xi) = \left(\frac{A(t)B(\xi)}{A(\xi)B(t)}\right)^{\frac{1}{2}} k(t, \xi)(A(t)B(t)A(\xi)B(\xi))^{\frac{1}{2}}.$$

If we put

$$\left(\frac{A(t)}{B(t)}\right)^{\frac{1}{2}} \equiv m(t), \qquad k(t, \xi)(A(t)B(t)A(\xi)B(\xi))^{\frac{1}{2}} \equiv k_1(t, \xi),$$

then $k_1(t, \xi)$ is a symmetric kernel, and we have

$$K(t, \xi) = \frac{m(t)}{m(\xi)} k_1(t, \xi).$$

(ii) *The eigenvalue of K agrees with that of* k_1. Since the iterated kernel of $m(t)/m(\xi)k_1(t, \xi)$ is $m(t)/m(\xi)\dot{k}_1{}^h$, if we make $\gamma_1(t, \xi; \lambda)$ the kernel of the solution of $k_1(t, \xi)$, then the kernel of the solution is $m(t)/m(\xi)\gamma_1(t, \xi; \lambda)$. Hence, the eigenvalue of K agrees with that of k_1. From Theorem 6.4, all the eigenvalues of K are real since k_1 it a symmetric kernel.

(iii) *An eigenfunction makes a biorthogonal function system.* Express the eigenfunction of k_1 for the eigenvalue λ_i as

$$(A(t)B(t))^{\frac{1}{2}}\varphi_i(t),$$

and we may assume that this makes a normalized orthogonal function system; then we have

$$\int_a^b A(t)B(t)\varphi_i(t)\varphi_j(t)dt = \begin{cases} 0, & i \neq j; \\ 1, & i = j. \end{cases} \tag{6.30}$$

And since we have

$$(A(t)B(t))^{\frac{1}{2}}\varphi_i(t) = \lambda_i \int_a^b k_1(t, \xi)(A(\xi)B(\xi))^{\frac{1}{2}}\varphi_i(\xi)d\xi, \tag{6.31}$$

multiply both sides by $(A(t)/B(t))^{\frac{1}{2}}$ and we get

$$A(t)\varphi_i(t) = \lambda_i \int_a^b \left(\frac{A(t)B(\xi)}{A(\xi)B(t)}\right)^{\frac{1}{2}} k_1(t, \xi)A(\xi)\varphi_i(\xi)d\xi,$$

i.e.

$$A(t)\varphi_i(t) = \lambda_i \int_a^b K(t, \xi)A(\xi)\varphi_i(\xi)d\xi. \tag{6.32}$$

Therefore, suppose that

$$A(t)\varphi_i(t) \equiv \Phi_i(t), \tag{6.33}$$

then $\Phi_i(t)$ becomes the eigenfunction of K for λ_i. Similarly,

$$B(t)\varphi_i(t) \equiv \Psi_i(t) \tag{6.33a}$$

is at eigenfunction for λ_i. And since for λ_i we can make only one eigenfunction of k_1, it follows that Φ_i and Ψ_i correspond with regard to K. Then from (6.30)

$$\int_a^b \Phi_i(t)\Psi_j(t)dt = \begin{cases} 0, & i \neq j, \\ 1, & i = j, \end{cases} \tag{6.34}$$

from which we find that we have a biorthogonal function system. The main elements of the kernel for the eigenvalue λ_i are

$$\frac{1}{\lambda_i}\varphi_i(t)\varphi_i(\xi) \quad \text{for } K(t, \xi),$$

$$\frac{1}{\lambda_i}(A(t)B(t)A(\xi)B(\xi))^{\frac{1}{2}}\varphi_i(t)\varphi_i(\xi) \quad \text{for } k_1(t, \xi),$$

$$\frac{1}{\lambda_i}A(t)B(\xi)\varphi_i(t)\varphi_i(\xi) \quad \text{for } K(t, \xi).$$

Example 6.2 *Eigenvalues and eigenfunctions*

$$K(t, \xi) = (t\xi^2)^{\frac{1}{2}}(t + \xi), \qquad I(0, 1).$$

This is a disymmetric kernel and $A(t) = \sqrt{t}$, $B(\xi) = \sqrt{\xi^3}$, and

$$k_1(t, \xi) = t\xi(t + \xi).$$

The characteristic equation of k_1 is

$$\lambda^2 + 120\lambda - 240 = 0,$$

and the eigenvalues are

$$\lambda_1 \simeq 2, \quad \lambda_2 \simeq -122.$$

Normalized eigenfunctions are

$$\psi_1 = \frac{1}{108.2}(120t^2 + 96t) = (1.110t + 0.887)t \quad \text{for } \lambda_1,$$

$$\psi_2 = \frac{1}{852}(7560t^2 - 5856t) = (8.873t - 6.873)t \quad \text{for } \lambda_2.$$

Hence, the eigenvalues of $K(t, \xi)$ are 2, $-$ 122 and the eigenfunctions are

$$\Phi_1(t) = (1.110t + 0.887)\sqrt{t}, \quad \Psi_1(t) = (1.110t + 0.887)\sqrt{t^3}$$
$$\text{for } \lambda_1 = 2,$$
$$\Phi_2(t) = (8.873t - 6.873)\sqrt{t}, \quad \Psi_2(t) = (8.873t - 6.87fl)\sqrt{t^3}$$
$$\text{for } \lambda_2 = -122.$$

These make a normalized biorthogonal system.

6.7. Skew symmetric kernels

With regard to the kernel $K(t, \xi)$, when

$$K(t, \xi) = -K(\xi, t) \tag{6.35}$$

we call it a skew symmetric kernel. The iterated kernels $\mathring{K}^{(2)}$, $\mathring{K}^{(2n)}$ are symmetric and $\mathring{K}^{(3)}$, $\mathring{K}^{(2n+1)}$ are skew symmetric. Besides, for an arbitrary function $f(t)$, we have

$$K \circ f = -f \circ \tilde{K}, \qquad \tilde{K} \circ f = -f \circ K. \tag{6.36}$$

The following theorem concerns skew symmetric kernels.

THEOREM 6.10 *Imaging eigenvalue*

Suppose that $K(t, \xi)$ is a skew symmetric kernel that is not always zero at every continuous point; then the eigenvalue is a purely imaginary number with only one corresponding eigenfunction, and its conjugate function is the eigenfunction of the associated integral equation.

Proof. (i) *The eigenvalue is a purely imaginary number.* Let $\varphi(t)$ be the eigenfunction and we have

$$\varphi = \lambda K \circ \varphi. \tag{6.37}$$

From (6.36) we obtain

$$\varphi = -\lambda \varphi \circ \tilde{K}; \tag{6.38}$$

let $\bar{\varphi}$ be the conjugate function of φ and let $\bar{\lambda}$ be the complex conjugate of λ. Then

$$\bar{\varphi} = -\bar{\lambda} \varphi \circ \tilde{K}. \tag{6.39}$$

Multiply (6.37) by $\bar{\lambda}\varphi$ from the left and multiply (6.39) by $\lambda\varphi$ from the right, then sum the two results, to obtain

$$(\lambda + \bar{\lambda})\bar{\varphi} \cdot \varphi = 0. \tag{6.40}$$

If $\bar{\varphi} \cdot \varphi = 0$, this surely leads to $\varphi \equiv 0$, which is inconvenient; then we have

$$\lambda + \bar{\lambda} = 0 \quad \text{or} \quad \lambda = -\bar{\lambda}.$$

This shows that λ is a purely imaginary number.

(ii) *There exists only one eigenfunction.* Suppose that there are two eigenfunctions φ_1, φ_2 and the difference between the two is $r(t)$; then r is also an eigenfunction, which is such that

$$r = \lambda K \circ r \quad \text{and} \quad r = -\lambda r \circ K. \tag{6.41}$$

Multiply this by r from the left and we have

$$r \cdot r = \lambda r \cdot K \circ r. \tag{6.42}$$

On the right $\lambda r \circ K$ becomes $-r$ from (6.41); it then follows that

$$r \cdot r = -r \cdot r,$$

and this leads to $r \cdot r = 0$. Hence,

$$r \equiv 0,$$

and thus the uniqueness of the eigenfunction has been proved.

(iii) $\bar{\varphi}$ becomes the solution of $x = \lambda x \circ \tilde{K}$ —this is immediate from (6.39) since $-\bar{\lambda} - \lambda$. Hence $\bar{\varphi}$ is an associated eigenfunction.

Example 6.3 *Imaginary eigenvalue*

$$K(t, \xi) = t - \xi, \qquad I(0, 1).$$

The eigenvalues are purely imaginary numbers at $\lambda = \pm 2\sqrt{3}i$. The eigenfunction for $\lambda = 2\sqrt{3}i$ is $\varphi_1(t) = (\sqrt{3} + 3i)t + 2 = (\sqrt{3}t + 2) + 3it$ and the eigenfunction for $\lambda = -2\sqrt{3}i$ is $\varphi_2(t) = (-\sqrt{3} + 3i)t + 2 = (-\sqrt{3}t + 2) + 3it$.

Then, for example, $\bar{\varphi}_1 = (\sqrt{3}t + 2) - 3it$ serves as the eigenfunction of $\tilde{K}(\xi, t) = \xi - t$.

6.8. Schmidt's eigenfunction system—in the case of non-symmetric kernels

By applying the theory on symmetric kernels, Schmidt (1906) gave an explicit method for the case of non-symmetric kernels. First, in the simultaneous integral equations

$$\left.\begin{aligned} u(t) &= \lambda \int_a^b K(t, \xi)v(\xi)d\xi, \\ v(t) &= \lambda \int_a^b u(\xi)K(\xi, t)d\xi, \end{aligned}\right\} \tag{6.43}$$

determine the eigenvalue λ such that u, v are not always zero and determine the eigenfunctions. The system (6.43) easily leads to an integral equation with symmetric kernels:

$$u(t) = \lambda^2 \int_a^b K_{2\cdot}(t, \xi)u(\xi)d\xi, \qquad K_{2\cdot}(t, \xi) = \int_a^b K(t, \tau)K(\xi, \tau)d\tau,$$

$$v(t) = \lambda^2 \int_a^b K_{\cdot 2}(t, \xi)v(\xi)d\xi, \qquad K_{\cdot 2}(t, \xi) = \int_a^b K(\tau, t)K(\tau, \xi)d\tau.$$

In addition (6.43) may lead to the case of the symmetric kernel in the following way (see § 7.3 below). To simplify the discussion, put $a = 0$, $b = 1$, and let the integral interval be $t(0, 2)$. Then we have

$$w(t) = \lambda^2 \int_0^2 H(t, \xi)w(\xi)d\xi, \tag{6.44}$$

provided

$$w(t) = \begin{cases} u(t), & 0 < t < 1, \\ v(t-1), & 1 \leqq t < 2, \end{cases}$$

$$H(t, \xi) = \begin{cases} 0, & \\ 0, & \\ K(t, \xi - 1), & 0 < t < 1, \quad 1 < \xi < 2, \\ K(\xi, t-1), & 1 < t < 2, \quad 0 < \xi < 1 \end{cases} \tag{6.45}$$

(see Fig. 6.1). Since $H(t, \xi)$ is a symmetric kernel it has eigenfunctions. Let λ_k be one of the eigenvalues, w_k its eigenfunction, and choose their values so that

$$\int_0^2 w_k(t)w_i(t)dt = 0, \qquad i \neq k, \tag{6.46}$$

and

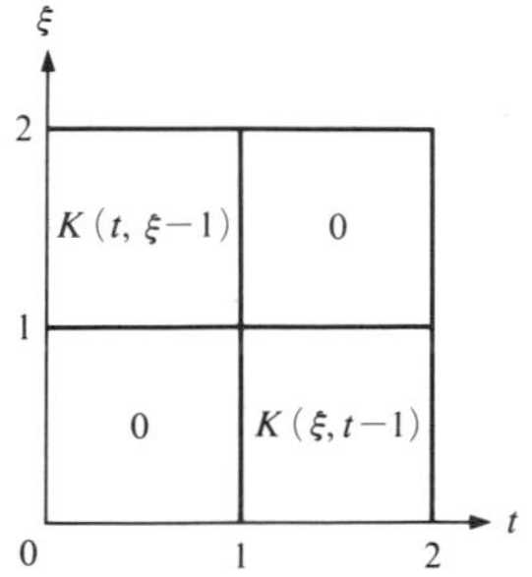

Fig. 6.1. The kernel *H* (*t*, *ξ*)

$$\int_0^2 \{w_k(t)\}^2 dt = 2. \tag{6.47}$$

We call w_k a *Schmidt's eigenfunction* for the eigenvalue λ_k.

If we suppose λ_k to be an eigenvalue, then $-\lambda_k$ is also an eigenvalue. That is, suppose that the eigenfunctions for λ_k are u_k, v_k, then those for $-\lambda_k$ are u_k, $-v_k$. But the above pairs do not yield any new eigenfunctions and therefore it is enough for us to consider the positive eigenvalue only.

To the finite or infinite eigenvalue system series

$$\lambda_1, \lambda_2, \ldots, \lambda_k, \ldots \tag{6.48}$$

we have the corresponding eigenfunction series:

$$\begin{aligned} &u_1(t), u_2(t), \ldots, u_k(t), \ldots \\ &v_1(t), v_2(t), \ldots, v_k(t), \ldots \end{aligned} \tag{6.49}$$

every one of which is normalized and orthogonal. From (6.46), (6.47) we derive

$$u_i \cdot u_k + v_i \cdot v_k = \begin{cases} 0, & i \neq k, \\ 2, & i = k, \end{cases}$$

but from (6.43) we have $u_k = \lambda_k K \circ v_k$, $v_i = \lambda_i u_i \circ K$. From the second expression put

$$v_i \circ v_k = \lambda_i u_i \circ (K \circ v_k),$$

and using the first expression we get $\lambda_k v_i \cdot v_k = \lambda_i u_i \cdot u_k$; then we obtain

$$u_i \cdot u_k = \begin{cases} 0, \; i \neq k, \\ 1, \; i = k, \end{cases} \qquad v_i \cdot v_k = \begin{cases} 0, \; i \neq k, \\ 1, \; i = k. \end{cases}$$

We call $\{u_i\}$, $\{v_i\}$ *Schmidt's eigenfunction systems*. We can apply the properties of the symmetric kernel to the non-symmetric kernel $K(t, \xi)$ since the integral equation (6.43) may be written in the form (6.44) by means of the symmetric kernel $H(t, \xi)$.

For example, we make use of Schmidt's theorem.

THEOREM 6.11. *Schmidt's theorem* (1906)

If the function $f(t)$ is continuous in I and is expressed in the form

$$f(t) = \int_a^b K(t, \xi) h(\xi) d\xi \quad \textit{or} \quad \int_a^b h(\xi) K(\xi, t) d\xi,$$

then $f(t)$ can be expanded into the series in $u_n(t)$ or $v_n(t)$ respectively.

Further, the study of the expansion of the kernel H allows us quite easily to obtain the following result for the general kernel $K(t, \xi)$. We shall omit the proof. E. Schmidt proofed the theorem in 1906 (see paper 42).

THEOREM 6.12. *Expansion of the kernel*

If the series

$$\sum_i \frac{u_i(t)v_i(\xi)}{\lambda_i}$$

is uniformly convergent, the expansion of the kernel becomes

$$K(t, \xi) = \sum_i \frac{u_i(t)v_i(\xi)}{\lambda_i}.$$

In this case, two orthogonal systems $u_i(t)$, $v_i(t)$ are completely independent, and we may select them independently. In particular we can take one as a complete system and leave the other free.

6.9. Fredholm-type integral equations of the first kind

We can study Fredholm type integral equations of the first kind.

$$\int_a^b K(t, \xi)x(\xi)d\xi = f(t), \qquad [F_1]$$

making use of Schmidt's eigenfunctions. This is given by the following theorem.

THEOREM 6.13

With regard to the integral equation of the first kind, $[F_1]$, *when the kernel is expressed by the complete normalized system series of eigenfunctions* $\{u_i(t)\}$, $\{v_i(t)\}$ *for the eigenvalue system series* $\{\lambda_i\}$ *in*

$$K(t, \xi) = \sum \frac{u_i(t)v_i(\xi)}{\lambda_i}, \qquad (6.50)$$

the necessary and sufficient condition for $[F_1]$ *to have a solution is that*

$$\sum \lambda_i^2 c_i^2$$

is convergent. Here c_i is the Fourier coefficient of $f(t)$ on $\{u_i(t)\}$. The solution is

$$x(t) = \sum \lambda_i c_i v_i(t) \tag{6.51}$$

and this is unique.

Proof. (i) *Necessity.* Suppose that $[F_1]$ has a solution which may be expressed as

$$x(t) = \sum d_i v_i(t),$$

provided d_i is the Fourier coefficient on $\{v_i(t)\}$. So that this solution may hold, it is necessary and sufficient that $\sum d_i^2$ is convergent. Since $f(t) = \sum c_i u_i(t)$, $[F_2]$ becomes

$$\int_a^b \sum \frac{u_i(t)v_i(\xi)}{\lambda_i} \sum d_i v_i(\xi) d\xi = \sum c_i u_i(t);$$

then comparing the coefficients, we get

$$c_i = d_i/\lambda_i.$$

Therefore, since $\sum d_i^2$ is convergent, $\sum \lambda_i^2 c_i^2$ is also convergent.

(ii) *Sufficiency.* If $\sum \lambda_i^2 c_i^2$ is convergent there necessarily exists a function with $\lambda_i c_i$ as its Fourier coefficient (Whittaker and Watson 1953, p.163). Let us denote this function by $h(t)$. Then with regard to

$$f(t) - \int_a^b K(t, \xi)h(\xi)d\xi,$$

all its Fourier coefficients on $\{u_i(t)\}$ are zero. Therefore, when $\{u_i(t)\}$ is a complete system, the above expression is zero at almost every point. This shows that $h(t)$ is the solution of $[F_1]$.

(iii) *Uniqueness.* An arbitrary function $h(t)$ which satisfies

$$\int_a^b K(t, \xi)h(\xi)d\xi = 0$$

intersects orthogonally with all of $\{v_i(t)\}$. Then if $\{v_i(t)\}$ is a complete system, it must be that $h(t) \equiv 0$. Hence, uniqueness of the solution follows.

6.9.1. *An incomplete system*

In the case where $\sum \lambda_i^2 c_i^2$ is convergent, yet $\{u_i(t)\}$ is not a complete system, take a function $h(t)$ of which $\lambda_i c_i$ is the Fourier coefficient on $\{v_i(t)\}$, then we may put

$$\int_a^b K(t, \xi)h(\xi)d\xi = f(t) + f_1(t). \tag{6.52}$$

The function $f_1(t)$ on the right-hand side is orthogonally intersecting with all of $\{u_i\}$. Changing $f_1(t)$ does not affect c_i or $h(t)$.

Moreover, when $\{v_i(t)\}$ is not a complete system, (6.51) with the arbitrary function $k(t)$ intersecting orthogonally with all these put together gives the solution of $[F_1]$. Hence, the solution is not unique.

6.9.2. The case when $\sum_i \lambda_i^2 c_i^2$ diverges

In this case we cannot solve the integral equation $[F_1]$; yet if we take the sum of the form

$$h_n(t) = \sum_1^n {}_i \lambda_i c_i v_i(t),$$

then when

$$f_n = \sum_1^n {}_i c_i u_i(t),$$

we have

$$f(t) - \int_a^b K(t, \xi)h_n(\xi)d\xi = f - f_n.$$

Therefore, we may take n sufficiently large so that

$$f(t) - \int_a^b K(t, \xi)h_n(\xi)d\xi$$

becomes sufficiently small.

6.9.3. The case of the symmetric kernel

We have concentrated on general kernels according to Schmidt's theorem in this section. The following corollary, for the case of symmetric kernels, comes directly from Hilbert's theorem.

THEOREM 6.13.1

In order that a Fredholm-type integral equation of the first kind with the symmetric kernel

$$\int_a^b K(t, \xi)x(\xi)d\xi = f(t) \qquad [F_1]$$

has a solution, the known function, for the complete system of the kernel $\psi_n(t)$, *should take the form*

$$f(t) = \sum_1^\infty c_n\psi_n(t).$$

In addition, if

$$\sum_1^\infty c_n\lambda_n\psi_n(t)$$

is uniformly convergent in I, or at least if termwise integration is possible, then $[F_1]$ *always has a solution. this is*

$$x(t) = \sum_1^\infty c_n\lambda_n\psi_n(t). \qquad (6.53)$$

(To see this, put (6.53) in the left-hand side of $[F_1]$.)

Example 6.4 $[F_1]$

In

$$\int_0^1 K(t, \xi)x(\xi)d\xi = \frac{1}{2}\sin \pi t + \frac{1}{3}\sin 2\pi t, \qquad [F_1]$$

put

$$K(t, \xi) = \sum_1^5 \frac{\sin n\pi t \cos n\pi t}{n^2\pi^2}.$$

The solution is

$$x(t) = \frac{\pi}{2}\cos \pi t + \frac{2\pi}{3}\cos 2\pi t.$$

6.10. The solution of Fredholm-type integral equations of the second kind (symmetric kernel)

This section describes Schmidt's study on Fredholm-type integral equations of the second kind with the symmetric kernel

$$x(t) - \lambda\int_a^b K(t, \xi)x(\xi)d\xi = f(t), \qquad x = f + \lambda K \circ x. \qquad [F_2]$$

6.10.1. The case when λ is not an eigenvalue

If we have

$$x(t) - f(t) \equiv g(t), \qquad x - f \equiv g, \tag{6.54}$$

[F_2] becomes

$$g = \lambda K \circ (f + g). \tag{6.55}$$

Then from Hilbert's theorem, g is expansible in the form of the complete system $\{\psi_i\}$ of the kernel, and is read as

$$g = \sum_{i=1}^{\infty} (\psi_i \cdot g)\psi_i. \tag{6.56}$$

On multiplying both sides of (6.55) by ψ_i from the right, we obtain

$$g \cdot \psi_i = \lambda K \circ (f + g) \cdot \psi_i = \lambda (f + g) \cdot K \circ \psi_i.$$

Yet, since $\psi_i = \lambda_i K \circ \psi_i$, the above expression becomes

$$g \cdot \psi_i = \frac{\lambda}{\lambda_i}(f + g) \cdot \psi_i;$$

that is,

$$g \cdot \psi_i = \frac{\lambda}{\lambda_i - \lambda} f \cdot \psi_i. \tag{6.57}$$

then from (6.54), (6.56) we have

$$x = f + \lambda \sum \frac{\psi_i}{\lambda_i - \lambda}(f \cdot \psi_i), \tag{6.58}$$

viz.

$$x(t) = f(t) - \lambda \sum_{1}^{\infty} \left[\frac{1}{\lambda - \lambda_i} \int_a^b f(\xi)\psi_i(\xi)d\xi \right] \psi_i(t).$$

Putting

$$\Gamma(t, \xi; \lambda) \equiv \sum_{1}^{\infty} \frac{\psi_i(t)\psi_i(\xi)}{\lambda - \lambda_i}, \tag{6.59}$$

we may write the solution as

$$x(t) = f(t) - \int_a^b \Gamma(t, \xi; \lambda) f(\xi) d\xi \tag{6.60}$$

(see Question 8).

6.10.2. *The case when λ is an eigenvalue*

When $\lambda = \lambda_1$ is the eigenvalue of the index q, from (6.57) we have

$$(\lambda_i - \lambda) g \cdot \psi_i = \lambda f \cdot \psi_i.$$

Therefore,

$$f \cdot \psi_i = 0 \qquad (i = 1, 2, \ldots, q)$$

is necessary. As for $i > q$, we have

$$g \cdot \psi_i = \frac{\lambda}{\lambda_i - \lambda} f \cdot \psi_i \qquad (i = q+1, q+2, \ldots), \qquad (6.57a)$$

because $\lambda \neq \lambda_i$. For (6.58) wc have

$$x = f + c_1\psi_1 + c_2\psi_2 + \cdots + c_q\psi_q - \sum_{q+1}^{\infty} \frac{\lambda}{\lambda - \lambda_i}(f \cdot \psi_i)\psi_i; \qquad (6.58a)$$

that is,

$$x(t) = f(t) + c_1\psi_1(t) + c_2\psi_2(t) + \cdots + c_q\psi_q(t)$$
$$- \lambda \sum_{q+1}^{\infty} \left[\frac{1}{\lambda - \lambda_i} \int_a^b f(\xi)\psi_i(\xi) d\xi \right] \psi_i(t).$$

Here $c_1, c_2, \ldots, c_q$ are used as arbitrary constants.

Example 6.5 $[F_2]$

$$[F_2]: \; x(t) - \lambda \int_0^1 K(t, \xi) x(\xi) d\xi = e^t.$$

$$K(t, \xi) = \begin{cases} (1 - \xi)t, & 0 \leqq \xi \leqq t, \\ (1 - t)\xi, & t \leqq \xi \leqq 1. \end{cases}$$

By the method used in Example 6.1, we get the eigenvalue and eigenfunction as

$$\lambda_n = n^2\pi^2, \qquad \psi_i = \sqrt{2} \sin n\pi t \qquad (n = 1, 2, \ldots).$$

This leads to

$$\Gamma(t, \xi; \lambda) \equiv 2 \sum_1^{\infty} \frac{\sin n\pi t \sin n\pi \xi}{\lambda - n^2\pi^2},$$

and the solution is

$$x(t) = e^t - 2\lambda\sum_1^\infty\left[\frac{1}{\lambda - n^2\pi^2}\int_0^1 e^\xi \sin n\pi\xi\, d\xi\right]\sin n\pi t,$$

$$= e^t - 2\lambda\sum_1^\infty \frac{n\pi(1 -(-1)^n e)}{(1 + n^2\pi^2)(\lambda - n^2\pi^2)}\sin n\pi t.$$

Example 6.6 [F_2]

$$x(t) - n^2\pi^2\int_0^1 K(t, \xi)x(\xi)d\xi - \sin m\pi t,\ n \neq m$$

$$K(t, \xi) = \begin{cases}(1 - \xi)t, & 0 \leqq \xi \leqq t,\\ (1 - t)\xi, & t \leqq \xi \leqq 1.\end{cases}$$

Here λ agrees with the eigenvalue $n^2\pi^2$. Thus its index is 1. And the condition

$$f \cdot \psi = 0, \qquad \text{viz.}\ \int_0^1 \sqrt{2}\sin n\pi t \sin m\pi t\, dt = 0$$

is satisfied. Hence, the solution is

$$x(t) = \sin m\pi t + C\sin n\pi t$$
$$- \sqrt{2}n^2\pi^2{\sum_1^\infty}_\nu\Big/\left[\frac{1}{n^2\pi^2 - \upsilon^2\pi^2}\int_0^1 \sin \nu\pi\xi \sin m\pi\xi\, d\xi\right]\sin m\pi t.$$

Here $\sum_\nu$ means the sum ever $\nu = 1, 2, \ldots, n - 1, n + 1, \ldots$ with $\nu = n$ omitted. Then we have

$$\int_0^1 \sin \nu\pi\xi \sin m\pi\xi\, d\xi = \begin{cases}0, & \nu \neq m;\\ 1, & \nu = m.\end{cases}$$

Finally, this expression becomes

$$x(t) = \sin m\pi t + C\sin n\pi t - \frac{n^2}{n^2 - m^2}\sin m\pi t,$$

which serves as an arbitrary constant.

6.11. Approximation of the minimum eigenvalue

Sometimes, it is important to determine the minimum eigenvalue for the symmetric kernel.

The process is follows:

(i) Calculate the iterated kernel $\mathring{K}^m$

(ii) Calculate the trace of mth degree

$$T_m \equiv \int_a^b \mathring{K}_m(t, t)dt$$

(iii) Then we have

$$|\lambda_1| \simeq (T_{2m}/T_{2m+2})^{\frac{1}{2}}$$

or

$$|\lambda_1| \simeq (T_{2m})^{-1/(2m)}$$

Example 6.7

$$K(t, \xi) = \begin{cases} \xi, & 0 \leq \xi \leq t \\ t, & t \leq \xi \leq 1 \end{cases}$$

Then we have for $0 \leq \xi \leq t$

$$\mathring{K}^2(t, \xi) = \int_0^t K(t, \tau)K(\tau, \xi)d\tau$$

$$= -\frac{1}{6}\xi^2 - \frac{1}{2}\xi t^2 + \xi t$$

and for $t \leq \xi \leq 1$

$$\mathring{K}^2(\xi, t) = -\frac{1}{6}t^2 - \frac{1}{2}t\xi^2 + t\xi$$

Then

$$T_2 = \int_0^1 \mathring{K}^2(t, t)dt$$

$$= \int_0^1 (-\frac{2}{3}t^3 + t^2)dt = \frac{1}{6}$$

Similarly we have

$$\mathring{K}^4(t, \xi) = -\frac{1}{5040}t^7 - \left(\frac{\xi^2}{240} - \frac{\xi}{120}\right)t^5$$

$$-\left(\frac{\xi^4}{144} - \frac{\xi^3}{36} + \frac{\xi}{18}\right)t^3 - \left(\frac{\xi^6}{720} - \frac{\xi^5}{120} + \frac{\xi^3}{18} - \frac{2\xi}{15}\right)t$$

$$(t \leq \xi \leq 1)$$

and

$$T_4 = \frac{17}{630}$$

Hence we have

$$(T_2/T_4)^{\frac{1}{2}} = 2.483\,9$$

or

$$(T_4)^{-\frac{1}{4}} = 2.467\,8.$$

The exact value for λ_1 is

$$\lambda_1 = \left(\frac{1}{2}\pi\right)^2 = 2.467\,4$$

Exercises

1. Derive the integral equations with symmetric kernels from $[F_1]$, $[F_0]$, then prove that the solutions of those integral equations agree with those of of the original equations. Solve in the manner descrived in § 6.1.

2. Derive the integral equation with the symmetric kernel from each of the following integral equations, then show that the solutions agree:

(a) $x(t) = \lambda \int_0^1 t\xi^2 x(\xi) d\xi,$

(b) $x(t) - \lambda \int_0^1 t\xi^2 x(\xi) d\xi = t + 3,$

(c) $\int_0^1 t\xi^2 x(\xi) d\xi = 2t + 3.$

3. Make a normalized orthogonal function system out of the eigenfunction system of the symmetric kernel $\{\varphi_i(t)\}$ and find the complete system expansion of the kernel.

(a) $K(t, \xi) = 1,$ $I(0, 1),$

(b) $K(t, \xi) = \sin t \sin \xi,$ $I(0, 2\pi),$

(c) $K(t, \xi) = t + \xi,$ $I(0, 1),$

(d) $K(t, \xi) = e^{t+\xi},$ $I(0, \log 2),$

(e) $K(t, \xi) = t\xi + t^2\xi^2,$ $I(0, 1).$

4. When the eigenfunction system of a symmetric kernel forms a complete system, the symmetric kernel is closed.
5. When $f(t)$ satisfies the conditions of Hilbert's theorem, prove that

$$\lambda_i f \cdot \psi_i = g \cdot \psi_i$$

[Lovitt].
6. When $f(t)$ satisfies the conditions of Hilbert's theorem, prove that

$$K \cdot g = \sum \frac{(g \cdot \psi_i)}{\lambda_i} \cdot \psi_i$$

and that the right-hand side is uniformly convergent [Lovitt].
7. When g, f are continuous in I, prove that with regard to the symmetric kernel $K(t, \xi)$,

$$\int_a^b \int_a^b K(t, \xi) g(t) h(\xi) dt\, d\xi = \sum \frac{(g \cdot \psi_i)(f \cdot \psi_i)}{\lambda_i}$$

holds [Lovitt].
8. Show that (6.59) is absolutely uniformly convergent, and (6.60) satisfies $[F_2]$.
9. Solve the following integral equations:

(a) $x(t) + 2\int_0^1 x(\xi) d\xi = f(t)$,

(b) $x(t) - \lambda \int_0^1 (t^2 + t\xi + \xi^2) x(\xi) d\xi = 5t^2 + 4t^2 + 3t + 2$,

(c) $x(t) = \lambda \int_0^2 (1 + t\xi) x(\xi) d\xi$,

(d) $x(t) - \lambda \int_0^1 x(\xi) d\xi = t$, $(\lambda \neq = 1)$,

(e) $x(t) - \lambda \int_0^1 x(\xi) d\xi = \frac{1}{2} - t$,

(f) $x(t) - \lambda \int_0^1 (t + \xi) x(\xi) d\xi = t$,

(g) $x(t) - (6 - 4\sqrt{3}) \int_0^1 (t + \xi) x(\xi) d\xi = 1 - \sqrt{3} t$,

(h) $x(t) - \int_0^1 (t^2 + t\xi + \xi) x(\xi) d\xi = 25t^2 - 102\frac{1}{3}$

$[x(t) = 2t^2 - t - 7/3]$.

10. Calculate numerically the integral equations:

$$x(t) = 2\int_0^1 (t+\xi)x(\xi)d\xi,$$

$$x(t) = 2\int_0^1 (t+\xi)x(\xi)d\xi = t+2.$$

11. Find the Schmidt eigenfunction systems of the following kernels.

(a) $K(t,\xi) = t\xi^2, \quad I(0,1),$

(b) $K(t,\xi) = 2t+3\xi, \quad I(0,1),$

(c) $K(t,\xi) = \sin t \cos\xi, \quad I(0,\pi).$

12. Prove that we can solve simultaneous equations such as (4.4) in the way shown in Table 6.1A.

13.

$$x(t) = \lambda\int_{-\pi}^{\pi}\left\{\frac{1}{4}\pi^{-1}(t-\xi)^2 - \frac{1}{2}|t-\xi|\right\}x(\xi)d\xi.$$

14.

$$x(t) = \frac{\lambda}{2\pi}\int_{-\pi}^{\pi}\frac{1-h^2}{1-2h\cos(t-\xi)+h^2}x(\xi)d\xi, \quad |h|<1.$$

15. In $[F_2]$ with the symmetric kernel, if $\Delta(\lambda) = 0$, prove that it has no solution unless $f(t)$ intersects orthogonally with all eigenfunctions.

References

Horn, J. *Einführung in die Theorie der Partiellen Differential-Gleichungen* (Berlin). (See: Volterra (Book—Ref. 25))

Kneser, A. (1924) *Die Integralgleichungen und ihre Anwendungen in der mathematischen Physik* (Braunschweig).

Schmidt, E. (1906) Entwicklung willkürlicher Funktionen nach systemen Vorgeschriebener. *Mathematische Annalen* **63**, 433–476.

Volterra, V. and Pérès, J. (1936) *Théorie générale des fonctionnelles* (Paris).

Whittaker, E.T. and Watson, G.N.: (1953), *A Course of Modern Analysis* (Cambridge).

7 Singular integral equations

This chapter introduces the theories of integral equations in which the types of the integral of its regular part are extended in various ways. Since the integral in an infinite integral domain is called a singular integral, we can generally call all the integral equations covered in this chapter singular integral equations. In addition, we shall also treat here the cases including multiple integrals.

The case including a singular integral where the integrand becomes infinite within its integral domain, which is the case when the bounded condition of the kernel is dropped, will be explained in the next chapter as the singular kernel.

Of all the types of integral equations, those of the second kind are the easiest to deal with. Volterra's method of solving iterated kernels and Fredholm's theorem are easily extended to these cases. Many forms other than those of the second kind can be reduced to this case. Sometimes, however, quite new methods are necessary. We shall thus make some comments on the simultaneous integral equations.

7.1. Fredholm-type singular integral equation

In a Fredholm-type integral equation of the second kind, when $t = b$, $\xi = b$, the boundary of definition S of the kernel $K(t, \xi)$ is infinitely large; that is, when $b \to \infty$, the equation becomes

$$x(t) - \lambda \int_a^\infty K(t, \xi)x(\xi)d\xi = f(t). \tag{7.1}$$

The regular part of (7.1) is assumed to be an improper Riemann integral, which is considered as

$$\lim_{b \to \infty} \int_a^b K(t, \xi)x(\xi)d\xi,$$

and its value is assumed to exist. This kind of integral equation is called a *singular Fredholm-type integral equation of the second kind.*

The same procedure is applied to the other kinds of integral equations, such as that of the first kind and the homogeneous one.

When S expands over the whole plane, put $a \to -\infty$, $b \to \infty$ and the above expression becomes

$$x(t) - \lambda \int_{-\infty}^{\infty} K(t, \xi)x(\xi)d\xi = f(t). \tag{7.2}$$

We may, in such cases, extend the convolution of the second kind or the iterated kernel by taking the form

$$K \circ G = \int_{a}^{\infty} K(t, \tau)G(\tau, \xi)d\xi \left(\equiv \lim_{b\to\infty} \int_{a}^{b} K(t, \tau)G(\tau, \xi)d\xi\right)$$

or

$$\mathring{K}^{n} = \int_{a}^{\infty} \mathring{K}^{n-1}(t, \tau)K(\tau, \xi)d\xi \left(\equiv \lim_{b\to\infty} \int_{a}^{b} \mathring{K}^{n-1}(t, \tau)K(\tau, \xi)d\tau\right).$$

Yet in the methods used in Chapter 2, for example, in Volterra's method of solution, we may not find the condition for the series

$$K(t, \xi) + \lambda \int_{0}^{\infty} K(t, \xi_1)K(\xi_1, \xi)d\xi_1 + \lambda^2 \int_{0}^{\infty}\int_{0}^{\infty} K(t, \xi_1)K(\xi_1, \xi_2)K(\xi_2, \xi)d\xi_1 d\xi_2 + \cdots \tag{7.3}$$

to be absolutely uniformly convergent so easily as in the case where the integral domain is bounded. In Fredholm's method of solution, too, we cannot prove that a series corresponding to $\Delta(\lambda)$ or $\Delta\begin{pmatrix} t \\ \xi \end{pmatrix}; \lambda)$ is absolutely uniformly convergent. Therefore, we may not be able to apply all the theories we have previously established to the singular integral equations, and this makes a great difference for the eigenvalue and the eigenfunction. But, if (7.3) is absolutely uniformly convergent, then let it be $\Gamma(t, \xi; \lambda)$. And if the integral involved in

$$x(t) = f(t) - \lambda \int_{a}^{\infty} \Gamma(t, \xi; \lambda)f(\xi)d\xi \tag{7.4}$$

is convergent, this becomes one of the solutions.

Example 7.1

In a singular integral equation of the second kind:

$$x(t) - \lambda\int_1^\infty \frac{1}{t\xi^2}x(\xi)d\xi = \frac{1}{t}$$

when $|\lambda| < 2$, we have

$$\begin{aligned}\Gamma(t, \xi; \lambda) &= \frac{-1}{t\xi^2}\left(1 + \lambda\int_1^\infty \tau^{-3}d\tau + \frac{\lambda^2}{2}\int_1^\infty \tau^{-3}d\tau + \cdots\right)\\ &= \frac{-1}{t\xi^2}\left(1 + \frac{\lambda}{2} + \left(\frac{\lambda}{2}\right)^2 + \cdots\right)\\ &= \frac{-2}{2-\lambda}\frac{1}{t\xi^2}.\end{aligned}$$

Therefore, we obtain

$$x(t) = \frac{1}{t}\left(1 - \frac{2\lambda}{2-\lambda}\int_1^\infty \xi^{-3}d\xi\right) = \frac{2}{2-\lambda}\frac{1}{t},$$

which satisfies the original equation. We have $\lambda = 2$ for the eigenvalue, and then there are an infinite number of eigenfunctions which are expressed as $\varphi(t) = C/t$, for C an arbitrary constant.

The general theory of the singular integral equation is not yet complete, but we shall explain some special cases in the following.

(1) *The convolution-type case.* In the Fredholm-type integral equation, when the integration interval is a, ∞ and the kernel is a function of $(t - \xi)$ only, then we can solve it making use of the Fourier transform. Here we can assume that $f(t) = 0$ in $t < a$, and then and it is defined in $(-\infty, \infty)$.

First, in the case of the first kind, multiply both sides of

$$\int_a^\infty K(t - \xi)x(\xi)d\xi = f(t) \tag{7.5}$$

by $\cos\pi\mu(t - z)$ and take the integral from a to ∞, then we have

$$\int_a^\infty \cos\pi\mu(t - z)f(t)dt = \int_a^\infty \cos\pi\mu(t - z)dt\int_a^\infty K(t - \xi)x(\xi)d\xi. \tag{7.6}$$

Put $t - \xi = \tau$ on the right hand side and change the variable from t to τ and we have

$$\int_a^\infty \cos\pi\mu(t - z)f(t)dt = \int_0^\infty K(\tau)\cos\pi\mu(\tau + \xi - z)d\tau\int_a^\infty x(\xi)d\xi.$$

Put

$$\left.\begin{aligned} K_c(\mu) &= \int_0^\infty K(\tau)\cos\pi\mu\tau \; d\tau, \\ K_s(\mu) &= \int_0^\infty K(\tau)\sin\pi\mu\tau \; d\tau \end{aligned}\right\}, \tag{7.7}$$

where $K_c(\mu)$, $K_s(\mu)$ are the Fourier cosine transform and the Fourier sine transform respectively. Then

$$\int_a^\infty \cos\pi\mu(t-z)f(t)dt = K_c(\mu)\int_a^\infty \cos\pi\mu(\xi - z)x(\xi)d\xi$$
$$- K_s(\mu)\int_a^\infty \sin\pi\,\mu(\xi - z)x(\xi)d\xi. \tag{7.8}$$

Similarly, we get

$$\int_a^\infty \sin\pi\mu(t-z)f(t)dt = K_s(\mu)\int_a^\infty \cos\pi\mu(\xi - z)x(\xi)d\xi$$
$$+ K_c(\mu)\int_a^\infty \sin\pi\mu(\xi - z)x(\xi)d\xi; \tag{7.9}$$

and thus we have

$$\int_a^\infty \cos\pi\mu(\xi - z)x(\xi)d\xi$$
$$= \int_a^\infty \frac{K_c(\mu)\cos\pi\mu(t-z) + K_s(\mu)\sin\pi\mu(t-z)}{[K_s(\mu)]^2 + [K_c(\mu)]^2} f(t)dt. \tag{7.10}$$

Replace z with t and the variable t on the right-hand side with ξ, then the above expression becomes

$$\int_a^\infty \cos\pi\mu(t-\xi)x(\xi)d\xi$$
$$= \int \frac{K_c(\mu)\cos\pi\mu(\xi - t) + K_s(\mu)\sin\pi\mu(\xi - t)}{[K_s(\mu)]^2 + [K_c(\mu)]^2} f(\xi)d\xi.$$

Now, according to Fourier's theorem on the integral, if $f(t)$ satisfies Dirichlet's condition in $t > a$, and

$$\int_a^\infty |f|\, dt$$

is bounded, then we have

$$\int_0^\infty d\mu \int_a^\infty \cos\pi\mu(t-\xi)f(\xi)d\xi = \begin{cases} f(t), & t > a, \\ 0, & t < a; \end{cases} \tag{7.11}$$

therefore, when $t > a$ we get

$$x(t) = \int_0^\infty d\mu \int_a^\infty \frac{K_c(\mu)\cos\pi\mu(\xi - t) + K_s(\mu)\sin\pi\mu(\xi - t)}{[K_s(\mu)]^2 + [K_c(\mu)]^2} f(\xi)d\xi. \quad (7.12)$$

The integral on the right-hand side becomes zero in $t < a$.

In the case of $I(a, b)$ instead of (7.5), calculate

$$\int_a^b K(t - \xi)x(\xi)d\xi = f(t) \quad [F_1]$$

in a similar way, and use

$$\left.\begin{aligned} K_c(\eta) &= \int_{-\infty}^\infty K(\tau)\cos \pi\eta\tau \, d\tau, \\ K_s(\eta) &= \int_{-\infty}^\infty K(\tau)\sin \pi\eta\tau \, d\tau \end{aligned}\right\} \quad (7.13)$$

instead of (7.7), and we have

$$x(t) = \int_{-\infty}^\infty d\eta \int_a^b \frac{K_c(\eta)\cos \pi\mu(\xi - t) + K_s(\eta)\sin \pi\mu(\xi - t)}{[K_s(\eta)]^2 + K_c(\eta)]^2} f(\xi)d\xi \quad (7.14)$$

in $a < t < b$. The integral on the right becomes zero in $t < a, b < t$.

In the case of the integral equation of the second kind, calculate

$$x(t) - \lambda \int_a^\infty K(t - \xi)x(\xi)d\xi = f(t), \quad (7.15)$$

and the solution in $t > a$ becomes

$$x(t) = \int_0^\infty d\mu \int_a^\infty \frac{K_c(\mu)\cos \pi\mu(\xi - t) + K_s(\mu)\sin \pi\mu(\xi - t)}{[K_s(\mu)]^2 + [K_c(\mu)]^2 - \lambda[K_s^2 + K_c^2]} f(\xi)d\xi. \quad (7.16)$$

When the integral interval is from $-\infty$ to $+\infty$ the convolution theorem of the Fourier integral can be applied. For example, in the case of the first kind, such as

$$\int_{-\infty}^\infty K(t - \xi)x(\xi)d\xi = f(t), \quad (7.17)$$

let $X(u)$, $K(u)$, $F(u)$ be the Fourier transforms of $x(t)$, $K(t)$, $f(t)$, and as we did in § 3.5 put

$$1/K(u) = L(u),$$

then we obtain

$$x(t) = \int_{-\infty}^{\infty} L(t-\xi)f(\xi)d\xi.$$

In the case of the second kind such as

$$x(t) - \lambda \int_{-\infty}^{\infty} K(t-\xi)x(\xi)d\xi = f(t), \tag{7.18}$$

put

$$-\frac{\mathrm{K}(u)}{1-\lambda \mathrm{K}(u)} = \Gamma(u,\lambda)$$

and then

$$x(t) = f(t) - \lambda \int_{-\infty}^{\infty} \Gamma(t-\xi;\lambda)f(\xi)d\xi.$$

There is also a reciprocal relation:

$$\begin{aligned} K(t) + \Gamma(t,\lambda) &= \lambda \int_{-\infty}^{\infty} K(t-\xi)\Gamma(\xi,\lambda)d\xi \\ &= \lambda \int_{-\infty}^{\infty} \Gamma(t-\xi,\lambda)K(\xi)d\xi. \end{aligned} \tag{7.19}$$

If $K(t)$ is identically zero on the negative side of t, the kernel $K(t-\xi)$ is zero in $t-\xi<0$; that is, it is zero outside the triangular domain T. Therefore, it is enough to take the integration of the regular part from 0 to t, and, consequently, this agrees with the convolution-type integral equation of Volterra type:

$$x(t) - \int_0^t K(t-\xi)x(\xi)d\xi = f(t).$$

Hence, in this case, it turns out that the integral equation (7.5) is of Volterra type (see § 10.3).

(2) *Eigenvalues and eigenfunctions.* Now let us study the eigenvalue and the eigenfunction of a homogeneous singular integral equation when the kernel takes a special form.

(a) In the Laplace kernel $K(t,\xi) = e^{-t\xi}$, the regular part of

$$x(t) - \lambda \int_0^{\infty} e^{-t\xi}x(\xi)d\xi = 0, \qquad t>0 \tag{7.20}$$

takes the form of the Laplace transform. Accordingly we call this sort of integral equation a *homogeneous integral equation with Laplace's kernel.* If we consider the integral formula (Churchill 1941)

$$\int_0^\infty e^{-t\xi}\xi^\nu\, d\xi = \Gamma(\nu+1)t^{-\nu-1}, \quad \nu > -1$$

and the property of the gamma function

$$\Gamma(1-\alpha)\Gamma(\alpha) = \frac{\pi}{\sin \pi\alpha},$$

we can immediately find

$$(\Gamma(\alpha))^{\frac{1}{2}}t^{-\alpha} \pm (\Gamma(1-\alpha))^{\frac{1}{2}}t^{\alpha-1}$$
$$= \pm\left(\frac{\sin \pi\alpha}{\pi}\right)^{\frac{1}{2}}\int_0^\infty e^{-t\xi}\{(\Gamma(\alpha))^{\frac{1}{2}}\xi^{-\alpha} \pm (\Gamma(1-\alpha))^{\frac{1}{2}}\xi^{\alpha-1}\}d\xi,$$
$$0 < \alpha < 1. \tag{7.21}$$

That is, the eigenvalues are $\pm\{(\sin \pi\alpha)/\pi\}^{\frac{1}{2}}$ and the eigenfunctions are

$$(\Gamma(\alpha))^{\frac{1}{2}}t^{-\alpha} \pm (\Gamma(1-\alpha))^{\frac{1}{2}}t^{\alpha-1};$$

then the eigenfunction is able to take continuous values in the interval

$$-\frac{1}{\sqrt{\pi}} \leqq \lambda \leqq \frac{1}{\sqrt{\pi}},$$

with 0 omitted. This kind of continuous eigenvalue is called a *continuous spectrum of an integral equation.*

(b) In Fourier's sine kernel $K(t, \xi) = \sin t\xi$,

$$x(t) - \lambda\int_0^\infty \sin t\xi\; x(\xi)d\xi = 0$$

is a homogeneous integral equation. From the integral formula

$$\int_0^\infty \sin t\xi \cdot e^{-a\xi}d\xi = \frac{t}{a^2+t^2}, \qquad \int_0^\infty \sin t\xi \cdot \frac{\xi}{a^2+\xi^2}d\xi = \frac{1}{2}\pi e^{-at},$$

we get $\lambda = \sqrt{(\pi/2)}$ for the eigenvalue, and

$$\psi(t) = \frac{t}{a^2+t^2} + \frac{\pi}{2}e^{-at}$$

for the eigenfunction. Here a serves as an arbitrary constant, and therefore we have an infinite number of eigenfunctions corresponding to one eigenvalue.

(c) In Fourier's cosine kernel $K(t, \xi) = \cos t\xi$, the eigenvalue of the homogeneous integral equation is $\lambda = \sqrt{(\pi/2)}$, and its eigenfunctions are

$$\frac{t}{a^2+t^2} + \sqrt{\frac{\pi}{2}}\, e^{\,at},$$

which are infinite in number, since a is an arbitrary constant.

(d) Now take the Gauss kernel

$$K(t, \xi) = \frac{1}{\sigma\sqrt{(2\pi)}} \exp\left\{-\frac{(\xi - t)^2}{2\sigma^2}\right\}.$$

The transform

$$\frac{1}{\sigma\sqrt{(2\pi)}}\int_{-\infty}^{\infty} \exp\left\{-\frac{(t-\xi)^2}{2\sigma^2}\right\} f(\xi)d\xi = f_\sigma(t)$$

is called Gauss's transform with parameter σ. According to the formula on Hermite's polynomial,

$$\frac{1}{\sqrt{(2\pi)}}\int_{-\infty}^{\infty} \exp\{-(t-\xi)^2/2\} H_n(\xi)d\xi = t^n,$$

we know that the Gauss transform of $H_n(t)$ at $\sigma = 1$ is t^n. Further, according to the formulae

$$\frac{1}{\sqrt{(2\pi)}}\int_{-\infty}^{\infty} \exp\{-(t-\xi)^2/2\}\cos\sqrt{2}\xi d\xi = \frac{1}{e}\cos\sqrt{2}t,$$

$$\frac{1}{\sqrt{(2\pi)}}\int_{-\infty}^{\infty} \exp\{-(t-\xi)^2/2\}\sin\sqrt{2}\xi d\xi = \frac{1}{e}\sin\sqrt{2}t$$

or, more generally, according to

$$\frac{1}{\sigma\sqrt{(2\pi)}}\int_{-\infty}^{\infty} \exp\left\{-\frac{(t-\xi)^2}{2\sigma^2}\right\}\cos \alpha\xi d\xi = \exp\left(-\frac{\alpha^2\sigma}{2}\right)\cos \alpha t,$$

$$\frac{1}{\sigma\sqrt{(2\pi)}}\int_{-\infty}^{\infty} \exp\left\{-\frac{(t-\xi)^2}{2\sigma^2}\right\}\sin \alpha\xi d\xi = \exp\left(-\frac{\alpha^2\sigma}{2}\right)\sin \alpha t,$$

the Gauss transform of $\sum(a_n\cos nt + b_n\sin nt)$ with the parameter α becomes

$$\sum\left[\exp\left(-\frac{n^2\sigma}{a}\right)\cdot(a_n\cos nt + b_n\sin nt)\right].$$

Hence, expanding the given function in the Fourier series

$$f(t) = \sum(a_n\cos nt + b_n\sin nt)$$

for the homogeneous integral equation

$$\frac{1}{\sigma\sqrt{(2\pi)}}\int_{-\infty}^{\infty} \exp\left(-\frac{(t-\xi^2)}{2\sigma^2}\right)x(\xi)d\xi = f(t),$$

the solution becomes

$$x(t) = \Sigma\left[\exp\left(-\frac{n^2\sigma}{2}\right)(a_n \cos nt + b_n \sin nt)\right].$$

7.1.1. *Numerical method of solution*

Take an integral equation of integral interval $(0, \infty)$, for example. Divide the domain $(t > 0, \xi > 0)$ into four parts as shown in Fig. 7.1 and let K_{11}, K_{12}, K_{21}, K_{22} be the values of the kernel in these parts; then the singular integral equation is

$$x(t) - \lambda\left(\int_0^1 K_{11}(t, \xi)x(\xi)d\xi + \int_1^{\infty} K_{21}(t, \xi)x(\xi)d\xi + \int_0^1 K_{12}(t, \xi)x(\xi)d\xi + \int_1^{\infty} K_{22}(t, \xi)x(\xi)d\xi\right) = f(t).$$

For the interval $(1, \infty)$ put

$$\frac{1}{\xi} = \zeta$$

and the result will be

$$\int_1^{\infty} K_{21}(t, \xi)x(\xi)d\xi = \int_0^1 K_{21}\left((t, \frac{1}{\zeta}\right)x\left(\frac{1}{\zeta}\right)\frac{d\zeta}{\zeta^2}.$$

Then select the lattice points as shown in Fig. 7.1 and introduce the simultaneous equations by the method of representative points.

Example 7.2

In the singular integral equation of the second kind,

$$x(t) - \int_1^{\infty} e^{-2(t+\xi)}x(\xi)d\xi = \frac{1}{t+1},$$

put

$$1/t = \tau, \quad 1/\xi = \zeta, \; x(1/\tau) = X(\tau), \quad f(1/\tau) = F(\tau),$$

and we get

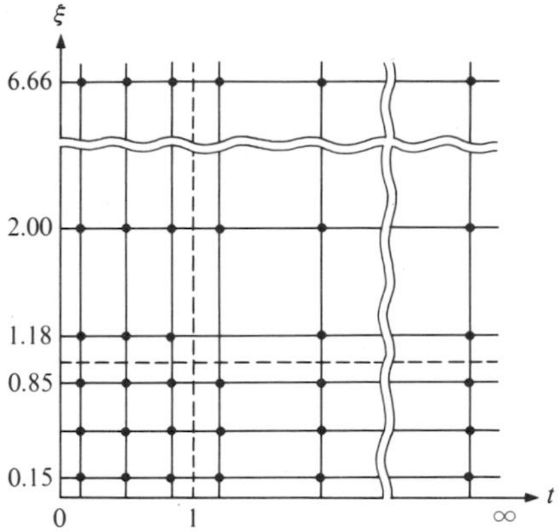

Fig. 7.1. Representative points

TABLE 7.1. Numerical solutions

τ	t	$x(t)$
0.8973	1.1145	-1.79×10^{-10}
0.5938	1.6841	-7.87×10^{-10}
0.4062	2.4615	-4.13×10^{-10}
0.1029	9.7371	1.04×10^{-10}

$$X(\tau) - \int_0^1 \exp\left[-2\left(\frac{1}{\tau} + \frac{1}{\xi}\right)\right] X(\zeta) \frac{d\zeta}{\zeta^2} = \frac{\tau}{\tau + 1}.$$

Calculating this by taking four representative points, we obtain the following simultaneous equations by Tschebyscheff's formula:

$$4.7x_1 + 3.4x_2 + 1.5x_3 + 1.2x_4 = 0.9 \times 10^{10},$$
$$11.1x_1 + 7.1x_2 + 3.2x_3 + 2.4x_4 = 0.3 \times 10^4,$$
$$4.6x_1 + 3.4x_2 + 1.5x_3 + 1.1x_4 = 0.4 \times 10^3,$$
$$14.4x_1 + 10.5x_2 + 5.4x_3 + 3.6x_4 = 0.5 \times 10^2.$$

The numerical solutions on the representative points are as given in Table 7.1.

7.2. Volterra-type singular integral equation

When in the Volterra-type integral equation of the second kind the integral domain becomes infinitely large, we have

$$x(t) - \int_{-\infty}^{t} K(t, \xi)x(\xi)d\xi = f(t), \qquad -\infty < t < +\infty. \tag{7.22}$$

Here $K(t, \xi)$ becomes zero in the half-plane $t < \xi$.† The change of variables gives

$$x(\xi) - \int_{\xi}^{\infty} K(t, \xi)x(t)dt = f(\xi);$$

that is,

$$x(t) - \int_{t}^{\infty} K(\xi, t)x(\xi)d\xi = f(t). \tag{7.23}$$

This kind of integral equation is called a *singular, Volterra-type, integral equation of the second kind.*

In this case, we express the iterated kernel made by the convolution of thc first kind as

$$\overset{*}{K}{}^2, \overset{*}{K}{}^3, \ldots,$$

where we have

$$\overset{*}{K}{}^n \equiv \int_{-\infty}^{t} \overset{*}{K}{}^{n-1}(t, \tau)K(\tau, \xi)d\tau = \overset{*}{K}{}^{n-1} * K, \tag{7.24}$$

or

$$\overset{*}{K}{}^n \equiv \int_{t}^{\infty} \overset{*}{K}{}^{n-1}(t, \tau)K(\tau, \xi)d\xi. \tag{7.25}$$

As with the method of solution by iterated kernel in Chapter 2, when the series of extended iterated kernels

$$K + \overset{*}{K}{}^2 + \overset{*}{K}{}^3 + \cdots$$

is absolutely and uniformly convergent, substitute $G(t, \xi)$, and if

$$x(t) = f(t) - \int_{-\infty}^{t} G(t, \xi)f(\xi)d\xi \tag{7.26}$$

or

$$x(t) = f(t) - \int_{t}^{\infty} G(t, \xi)f(\xi)d\xi \tag{7.27}$$

converges (uniformly bounded), these are the solutions of (7.22), (7.23).

† We extend the assumption of the boundedness of K and the normality of discontinuous points in the infinite domain as we did in § 2.1.

(1) *The case of convolution type.* For the equation

$$x(t) - \int_{-\infty}^{t} K(t-\xi)x(\xi)d\xi = f(t), \tag{7.28}$$

if the kernel is

$$K(t-\xi) = e^{-(m+\varepsilon)(t-\varepsilon)}k(t-\xi), \tag{7.29}$$

construct the kernel of the solution $G(t-\xi)$ for K on the lower limit of integration a, which is read as

$$G(t-\xi) = e^{-(m+\varepsilon)(t-\varepsilon)}g(t-\xi), \tag{7.30}$$

where g is the kernel of the solution of k.

Next, put

$$|k(t-\xi)| < m, 0 < t-\xi < \infty, \tag{7.31}$$

and then we have

$$|\overset{*}{k}{}^{n}| < \frac{m^n(t-\xi)^{n-1}}{(n-1)!}.$$

From § 1.4 we have

$$|g| < me^{m(t-\xi)}; \tag{7.32}$$

then we get

$$|K| < me^{-(m+\varepsilon)(t-\xi)}, \ |G| < me^{-\varepsilon(t-\xi)}. \tag{7.33}$$

Here, letting $a \to -\infty$, we have

$$\int_{-\infty}^{t} |K(t-\xi)|d\xi < \frac{m}{m+\varepsilon} < 1, \qquad \int_{-\infty}^{t} |G(t-\xi)|d\xi < \frac{m}{\varepsilon}. \tag{7.34}$$

If $f(t)$ is bounded in $-\infty < t \leqq b$, the right-hand side of

$$x(t) = f(t) - \int_{-\infty}^{t} G(t-\xi)f(\xi)d\xi \tag{7.35}$$

is uniformly convergent, and (7.35) is the solution of the integral equation.

We now show that this solution is unique. Suppose that there are two bounded solutions $x_1(t)$, $x_2(t)$; then $x_1 - x_2 = \varphi$ satisfies the homogeneous integral equation

$$\varphi(t) = \int_{-\infty}^{t} K(t-\xi)\varphi(\xi)d\xi, \tag{7.36}$$

which is bounded, $|\varphi| < M$. From (7.34) we have

$$|\varphi| < M\frac{m}{m+\varepsilon};$$

then substituting successively we can prove that

$$|\varphi| < M\left(\frac{m}{m+\varepsilon}\right)^n.$$

As $n \to \infty$ we have

$$|\varphi| = 0. \tag{7.37}$$

Hence we get $x_1 = x_2$, and thus the uniqueness of the solution is proved.

In order for the solution to be unique we require the assumption of boundedness. For example, in the homogeneous integral equation

$$x(t) - A\int_{-\infty}^{t} e^{-B(t-\xi)}x(\xi)d\xi, \tag{7.38}$$

in which A, B are arbitrary constants, multiply both sides by e^{Bt} and differentiate the result in terms of t, then we have

$$x^{(1)} + (B - A)x = 0. \tag{7.34}$$

(Recall that $x^{(1)} = dx/dt$.) Hence, the solution is

$$x(t) = Ce^{(A-B)t}. \tag{7.40}$$

On putting (7.40) into (7.38), we find that it has a unique solution, $x(t) \equiv 0$ in $A < 0$, and an infinite number of solutions, $x(t) = Ce^{(A-B)t}$ in $A > 0$. (When $A - B < 0$ these solutions become infinitely large as $t = -\infty$.)

(2) *The method of solution by variable transform.* We can make an integral interval bounded by the permutation of variables. That is, put

$$1/t = \tau, \quad 1/\xi = \zeta, \quad x(1/\tau) = X(\tau), \quad f(1/\tau) = F(\tau), \tag{7.41}$$

and (7.22) becomes

$$X(\tau) + \int_0^{\tau} K(1/\tau, 1/\zeta)\frac{X(\zeta)}{\zeta^2}d\zeta = F(\tau), \tag{7.42}$$

which is generally a case of the singular kernel.

The method of solution for the singular kernel requires the following theorem, which is given without proof (see Question 8).

THEOREM 7.1

In the singular integral equation (7.22) or (7.23), *if*

(i) $f(t)$, $K(t, \xi)$ *are bounded and continuous,*

(ii) $\int_{-\infty}^{t} |K(t, \xi)|\,d\xi$ *exists, and*

(iii) *there exists a value* t_0 *for which*

$$\int_{-\infty}^{t} |K(t, \xi)|\,d\xi < N < 1$$

always holds for an arbitrary value in $t < t_0$, *then the successive approximate solutions converge, and the solution exists and is unique.* (Question 8)

For example, in (7.38) the three conditions are satisfied when $B > 0$, $|A/B| < 1$. In $A < 0$ the solution is $x(t) \equiv 0$ and this is unique, while in $A > 0$ the solution is not unique, yet since $|A/B| < 1$, we have $A - B < 0$, which is not bounded. Thus there is but one unique and bounded solution.

Example 7.3

In the singular integral equation

$$x(t) + \int_{-\infty}^{t} \frac{1}{\xi^2}\left(\frac{1}{t} - \frac{1}{\xi}\right)x(\xi)d\xi = \frac{1}{t}, \tag{7.43}$$

if we permute the variables as in (7.41) then

$$X(\tau) - \int_{0}^{\tau} \zeta^2 \frac{(\tau - \zeta)}{\zeta^2} X(\zeta)d\zeta = \tau;$$

then the solution is

$$X(\tau) = \frac{1}{2}(e^{\tau} - e^{-\tau}),$$

and hence we have

$$x(t) = \frac{1}{2}(e^{1/t} - e^{-1/t}) = \sinh 1/t.$$

7.3. Simultaneous integral equations

We use the term *simultaneous integral equations* when we have functions that satisfy a number of integral equations with more than two unknown functions at the same time. They are classified into those of the second kind and of the first kind according to whether they have a disturbance function or not, and into Fredholm or Volterra type according to the type of integral involved. So that an integral equation may be solved, it must be linearly independent, and the number of equations must agree with the number of unknown functions.

In the case of the second kind, we can turn those of Fredholm type into the case of $[F_2]$ by enlarging the domain of the kernel, and in those of Volterra type we can extend the method of solution by the iterated kernel. As for those of the first kind, we can make them into the form of the second kind by the procedure described in § 2.5 and solve them using the method mentioned above.

1. *Simultaneous Fredholm-type integral equations of the second kind.* In the case of Fredholm type, we can turn the simultaneous equatinns very neatly into an integral equation of the second kind. In the following we explain the case when there are two unknown functions. Here we can take (0, 1) for the integration interval $I(a, b)$ without loss of generality.

In the simultaneous integral equations

$$\left.\begin{aligned} x_1(t) - \lambda\int_0^1 K_{11}(t, \xi)x_1(\xi)d\xi - \lambda\int_0^1 K_{12}(t, \xi)x_2(\xi)d\xi = f_1(t), \\ x_2(t) - \lambda\int_0^1 K_{21}(t, \xi)x_1(\xi)d\xi - \lambda\int_0^1 K_{22}(t, \xi)x_2(\xi)d\xi = f_2(t), \end{aligned}\right\} \quad (7.44)$$

the disturbance functions $f_1(t)$, $f_2(t)$ on the right-hand side are given in $I(0, 1)$, and the kernel $K_{ij}(t, \xi)$, in $S(0 \leqq t \leqq 1,\ 0 \leqq \xi \leqq 1)$.† We are going to decide the unknown functions $x_1(t)$, $x_2(t)$ within this domain.

Define the new kernel as in Fig. 7.2:

$$K(t, \xi) = \begin{cases} K_{11}(t, \xi), & 0 \leqq t < 1, \quad 0 \leqq \xi < 1, \\ K_{21}(t-1, \xi), & 1 \leqq t \leqq 2, \quad 0 \leqq \xi < 1, \\ K_{12}(t, \xi-1), & 0 \leqq t < 1, \quad 1 \leqq \xi \leqq 2, \\ K_{22}(t-1, \xi-1), & 1 \leqq t \leqq 2, \quad 1 \leqq \xi \leqq 2, \end{cases} \quad (7.45)$$

then, let

† In (7.44) λ can take the same value throughout without loss of generality; for if λ is not the same we have only to let λ be the overlapping group factor and put the other factors into the kernel K_{ij}.

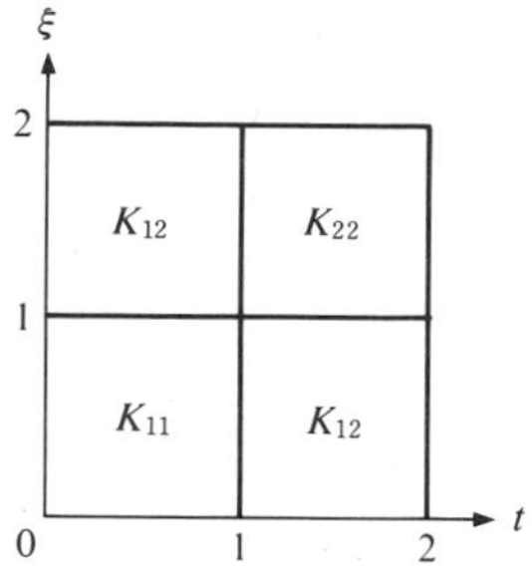

Fig. 7.2. Extension of the kernel

$$f(t) = \begin{cases} f_1(t), & 0 \leqq t < 1, \\ f_2(t-1), & 1 \leqq t \leqq 2 \end{cases}$$

be the new disturbance function and

$$x(t) = \begin{cases} x_1(t), & 0 \leqq t < 1, \\ x_2(t-1), & 1 \leqq t \leqq 2 \end{cases}$$

be the new unknown function, and then the simultaneous integral equations (7.44) can be replaced by a single integral equation:

$$x(t) - \lambda \int_0^2 K(t, \xi)x(\xi)d\xi = f(t) \tag{7.46}$$

(see § 6.8). Here the kernel $K(t, \xi)$ may not belong to the functional space $\mathscr{R}$ in the whole domain in Fig. 7.2, yet it belongs to $\mathscr{R}$ in each of the four subdomains. Therefore, it is bounded in the whole domain and its discontinuous points distribute regularly, except that it becomes discontinuous on the lines $t = 1$, $\xi = 1$ parallel to the coordinates axis. Further, $f(t)$ is also continuous except on $t = 1$. Thus, as in Theorem 2.1, we can prove that there exist continuous unique solutions $x(t)$. Therefore, any solution derived from this becomes a continuous unique solution in $I(0, 1)$.

7.3.1. The method of successive approximations

We have the following methods of solution for simultaneous integral equations.

Put

$$\left.\begin{aligned} x_1 &= x_{11} + x_{13} + x_{15} + \cdots, \\ x_2 &= x_{21} + x_{23} + x_{25} + \cdots, \end{aligned}\right\} \tag{7.47}$$

where x_{11} is supposed to satisfy

$$x_{11} - \lambda K \circ x_{11} = f_1.$$

Since this is an ordinary [F_2], we can solve it.

Put

$$x_1 = x_{11} + x_{12},$$

Then (7.44) becomes

$$\left.\begin{aligned} x_{12} - \lambda K_{11} \circ x_{12} - \lambda K_{12} \circ x_2 &= 0, \\ x_2 - \lambda K_{21} \circ x_{12} - \lambda K_{22} \circ x_2 &= f_2 + \lambda K_{21} \circ x_{11}. \end{aligned}\right\} \tag{7.48}$$

Then let x_{21} be the solution of

$$x_{21} - \lambda K_{22} \circ x_{21} = f_2 + \lambda K_{21} \circ x_{11}.$$

As x_{11} has already been solved, the right-hand side is a known function; therefore we can solve it since it is an ordinary [F_2].

Put

$$x_2 = x_{21} + x_{22},$$

and we have

$$\left.\begin{aligned} x_{12} - \lambda K_{11} \circ x_{12} - \lambda K_{12} \circ x_{22} &= \lambda K_{12} \circ x_{21}, \\ x_{22} - \lambda K_{21} \circ x_{12} - \lambda K_{22} \circ x_{22} &= 0. \end{aligned}\right\} \tag{7.49}$$

Then put

$$x_{12} = x_{13} + x_{14}$$

and let x_{13} be the solution of

$$x_{13} - \lambda K_{11} \circ x_{13} = \lambda K_{12} \circ x_{21}.$$

Then we have

$$\left.\begin{aligned} x_{14} - \lambda K_{11} \circ x_{14} - \lambda K_{12} \circ x_{22} &= 0, \\ x_{22} - \lambda K_{21} \circ x_{14} - \lambda K_{22} \circ x_{22} &= \lambda K_{21} \circ x_{13}. \end{aligned}\right\} \tag{7.50}$$

Then again put

$$x_{22} = x_{23} + x_{24}$$

and make x_{23} the solution of

$$x_{23} - \lambda K_{22} \circ x_{23} = \lambda K_{21} \circ x_{13}.$$

Thus solve the two [F_2]:

$$\left.\begin{aligned} x_{1j} - \lambda K_{11} \circ x_{1j} &= \lambda K_{12} \circ x_{2i}, \\ x_{2k} - \lambda K_{22} \circ x_{2k} &= \lambda K_{21} \circ x_{1j} \end{aligned}\right\} \tag{7.51}$$

generally in turn and the solution of (7.44) can be got in the form of (7.47), provided i, j, k are three contiguous odd numbers.

2. *Simultaneous Volterra-type integral equations of the second kind.* In the case of Volterra type, we can extend the method of solution by the iterated kernel. That is, we have the following theorem as an extension of Theorem 2.2.

THEOREM 7.2

In the simultaneous Volterra-type integral equation of the second kind

$$x_i(t) - \int_0^t \sum_1^n {}_j K_{ij}(t, \xi) x_j(\xi) d\xi = f_i(t) \qquad (i, j = 1, 2, \cdots, n), \tag{7.52}$$

if the kernel K_{ij} belongs to the functional space $\mathscr{R}$ in T, the necessary and sufficient condition for the continuous solution to exist is that $f_i(t)$ is continuous in I. The solution is

$$x_i(t) = f_i(t) - \int_0^t \sum_1^n {}_j G_{ij}(t, \xi) f_j(\xi) d\xi, \tag{7.53}$$

and is unique. Now the kernels of the solution G_{ij} are

$$G_{ij}(t) = - \sum_1^\infty {}_h \overset{*}{K}_{ij}{}^h, \tag{7.54}$$

$$\left.\begin{aligned} \overset{*}{K}_{ij} &= K_{ij}, \\ \overset{*}{K}_{ij}{}^h &= \sum_1^n {}_r \overset{*}{K}_{ir}{}^s * \overset{*}{K}_{rj}{}^{h-s} \qquad (h = 2, 3, \cdots). \end{aligned}\right\} \tag{7.55}$$

Proof.(i) *Convergence.* There is no connection between the iterated kernel $\overset{*}{K}{}^h$ and s, where $s = 1, 2, \ldots, h-1$. Put $|K_{ij}| \leqq M$ in T and we have

$$|\overset{*}{K}{}^h| < n^{h-1} M^h \frac{|t-\xi|^{h-1}}{(h-1)!};$$

therefore all G_{ij} are uniformly convergent. Hence (7.54) defines n^2 bounded functions G_{ij}.

(ii) *Reciprocal theorem.* We may prove that

$$K_{ij} + G_{ij} = \sum_1^n {}_r K_{ir} * G_{rj} = \sum_1^n {}_r G_{ir} * K_{rj} \tag{7.56}$$

and

$$K_{ij} = -\sum_{1}^{\infty}{}_h \overset{*}{G}_{ij}{}^h.$$

(iii) *Inversion.* Put (7.53) into x_i on the left-hand side of (7.52) and we get

$$f_i - \sum_j (G_{ij} + K_{ij}) * f_j + \sum_j K_{ij} * \sum_r G_{jr} * f_r,$$

while

$$\sum_j K_{ij} * \sum_r G_{jr} * f_r = \sum_j (\sum_r K_{ir} * G_{rj}) * f_j;$$

this then becomes f_j by the reciprocal theorem. Thus we have proved that (7.53) is the solution of (7.52).

(iv) *Uniqueness.* Suppose that (7.52) has the form

$$x_i(t) = f(t) + \int_0^t \sum_j K_{ij}(t, \xi) x_j(\xi) d\xi,$$

then put this expression again into $x_j(\xi)$ on the right-hand side, and repeat the process successively, until

$$x_i(t) = f(t) - \int_0^t G_{ij}(t, \xi) f(\xi) d\xi.$$

Hence uniqueness has been proved.

7.3.2 Simultaneous Volterra-type integral equations of the first kind

In

$$\int_0^t \sum_j K_{ij}(t, \xi) x_j(\xi) d\xi = f_i(t) \; (i, j = 1, 2, \ldots, n), \tag{7.57}$$

if K_{ij}, f_i are differentiable, then

$$\sum K_{ij}(t, t) x_j(t) + \int_0^t \sum_j \frac{\partial K_{ij}(t, \xi)}{\partial t} x_j(\xi)\, d\xi = f_i^{(1)}(t), \tag{7.58}$$

Now, for the determinant $[K_{ij}(t, \xi)]$ or

$$D(t, \xi) = \begin{vmatrix} K_{11} & K_{12} & \ldots & K_{1n} \\ \ldots & \ldots & \ldots & \ldots \\ K_{n1} & K_{n2} & \ldots & K_{nn} \end{vmatrix},$$

where if $D(t, t)$ does not become zero in I, we can express (7.58) in terms of simultaneous equations of the second kind like (7.52) by solving it for x_j.

7.4. The case when the upper and lower limits of an integral are indeterminate (Volterra type)

(1) In a Volterra-type integral equation of the second kind, when both the upper and lower limits of the integral in the regular part are variables we have

$$x(t) - \int_{-t}^{t} K(t, \xi)x(\xi)d\xi = f(t), \tag{7.59}$$

provided the disturbance function $f(t)$ is given by $t(-b, b)$ and the kernel is given in the domain $-b \leqq \xi \leqq b, \quad -t \gtreqless \xi \gtreqless t$, as in Fig. 7.3.

In (7.59) divide the integral in the regular part into two and transform the variable t to $-t$, and ξ to $-\xi$, then we have

$$x(t) - \int_{0}^{t} K(t, \xi)x(\xi)d\xi - \int_{0}^{t} K(t, -\xi)x(-\xi)d\xi = f(t). \tag{7.60}$$

Here put $-t$ in place of t and we have

$$x(-t) - \int_{0}^{-t} K(-t, \xi)x(\xi)d\xi - \int_{0}^{-t} K(-t, -\xi)x(-\xi)d\xi = f(-t).$$

Transform $-\xi$ into ξ again and this leads us to

$$x(-t) + \int_{0}^{t} K(-t, \xi)x(\xi)d\xi + \int_{0}^{t} K(-t, -\xi)x(-\xi)d\xi = f(-t), \tag{7.61}$$

in which we may put $0 < t < b$.

Now for $0 < t < b$, put

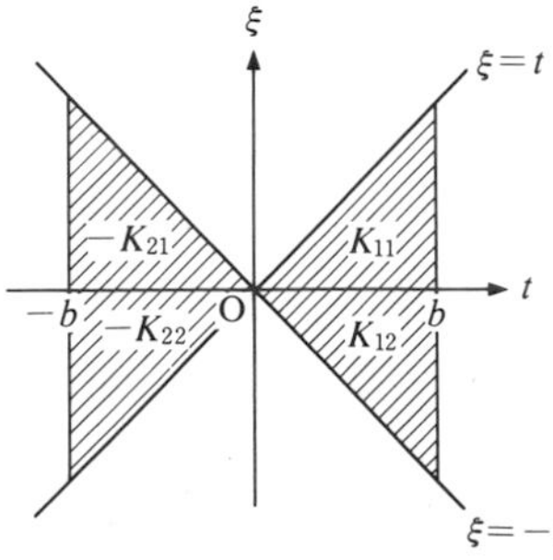

Fig. 7.3. Extension of the kernel

$$x(t) = x_1(t), \qquad x(-t) = x_2(t),$$
$$f(t) = f_1(t), \qquad f(-t) = f_2(t),$$

and for $0 < \xi < t < b$,

$$K(t, \xi) = K_{11}(t, \xi), \qquad K(t, -\xi) = K_{12}(t, \xi),$$
$$-K(-t, \xi) = K_{21}(t, \xi), \qquad -K(-t, -\xi) = K_{22}(t, \xi);$$

then (7.60) and (7.61) become a set of simultaneous integral equations of the second kind:

$$\left.\begin{aligned} x_1(t) - \int_0^t K_{11}(t, \xi)x_1(\xi)d\xi - \int_0^t K_{12}(t, \xi)x_2(\xi)d\xi &= f_1(t), \\ x_2(t) - \int_0^t K_{21}(t, \xi)x_1(\xi)d\xi - \int_0^t K_{22}(t, \xi)x_2(\xi)d\xi &= f_2(t). \end{aligned}\right\} \tag{7.62}$$

Now suppose that all the functions we are treating here are bounded and continuous. Then if the kernels of the solution are G_{ij} we have

$$\left.\begin{aligned} x_1(t) &= f_1(t) - \int_0^t G_{11}(t, \xi)f_1(\xi)d\xi - \int_0^t G_{12}(t, \xi)f_2(\xi)d\xi, \\ x_2(t) &= f_2(t) - \int_0^t G_{21}(t, \xi)f_1(\xi)d\xi - \int_0^t G_{22}(t, \xi)f_2(\xi)d\xi. \end{aligned}\right\} \tag{7.63}$$

Let

$$G_{11}(t, \xi) = G(t, \xi), \qquad G_{12}(t, \xi) = G(t, -\xi),$$
$$G_{21}(t, \xi) = -G(-t, \xi), \qquad G_{22}(t, \xi) = -G(-t, -\xi)$$

be the function $G(t, \xi)$ defined by $-t \gtreqless \xi \gtreqless t$, $-b < \xi < b$; then using the above $f(t)$ (7.63) becomes

$$x(t) = f(t) - \int_{-t}^t G(t, \xi)f(\xi)d\xi. \tag{7.64}$$

(2) Consider the integral equation

$$x(t) - \int_{\alpha t}^t K(t, \xi)x(\xi)d\xi = f(t), \ |\alpha| < 1 \tag{7.65}$$

in the same manner. If α is positive, we may take it as a case of

$$x(t) - \int_0^t K(t, \xi)x(\xi)d\xi = f(t), \qquad [\mathrm{V}_2]$$

the kernel is $K(t, \xi) = 0$ when $\xi < \alpha t$ $(t > 0)$.

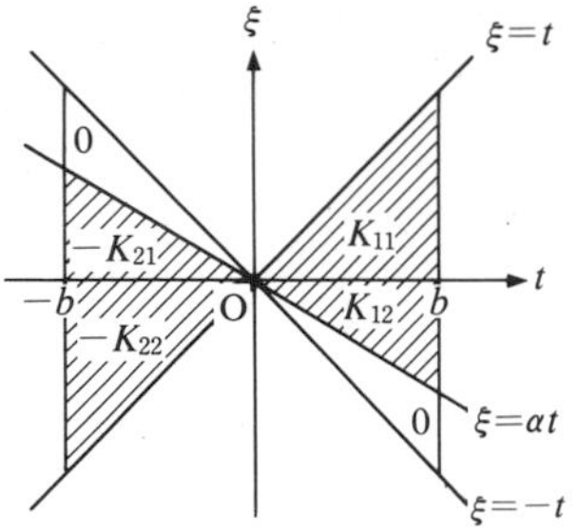

Fig. 7.4. Extension of the kernel

If α is negative, we may think of the kernel in the form

$$K(t,\ \xi) = 0\begin{cases} -t < \xi < \alpha t & (t > 0), \\ \alpha t < \xi < -t & (t < 0) \end{cases}$$

in (7.59) as is shown in Fig. 7.4. In this case, the previously given K_{12}, K_{21} become

$$K_{12} = K_{21} = 0, \quad -\alpha t < \xi < t.$$

Accordingly, their kernels become zero in this domain. This is clear, for example, since we have

$$\overset{*}{K}_{12}{}^{i}(t,\ \xi) = \sum_{1}^{2}{}_{r} \int_{\xi}^{t} \overset{*}{K}_{1r}{}^{i-1}(t,\ \tau)\overset{*}{K}_{r2}(\tau,\ \xi)d\tau.$$

Thus we can see that G_{12}, G_{21} become zero in this domain for the kernel of the solution, too, and then lastly we obtain

$$x(t) = f(t) - \int_{\alpha t}^{t} G(t,\ \xi)x(\xi)d\xi. \tag{7.66}$$

Further, we have

$$x(t) - \int_{\alpha t}^{\beta t} K(t,\ \xi)x(\xi)d\xi = f(t)\ (|\alpha| < 1,\ |\beta| < 1), \tag{7.67}$$

or, in a more generalized form,

$$x(t) - \int_{Z_1(t)}^{Z_2(t)} K(t,\ \xi)x(\xi)d\xi = f(t), \tag{7.68}$$

where $Z_1(t)$, $Z_2(t)$ are defined in $t(-b,\ b)$, and $|Z_1(t)| < |t|$, $|Z_2(t)| < |t|$. We can treat these cases using the methods explained above.

7.5. Integral equations with multiple integrals

In a given function, when its unknown function has more than two variables, the regular part becomes a multiple integral. This is called an *integral equation with a multiple integral.* Such equations are classified into Volterra type and Fredholm type, and in each type there are those of the first kind, of the second kind, homogeneous ones, and others. In order to solve the integral equations of the second kind, we can directly apply the method of solution by iterated kernel or by determinant given in Chapter 2. In the case of Volterra-type of the first kind, it is made into the form of the second kind. Among differential equations the difference between ordinary differential equations and partial differential equations is great and the latter are quite difficult to solve. But for integral equations, we have no such difficulty, even when there are many variables. This indicates the advantage of solving a problem when it is expressed in terms of an integral equation with a multiple integral rather than treating a partial differential equation.

7.5.1. *Integral equations of the second kind with multiple integrals, Fredholm type*

Let M and P be two points in the domain V in multidimensional space. Let $x(M)$, $f(M)$ be given as functions of the position of point M in V, of which $x(M)$ is unknown and $f(M)$ is known. Also, there is a function $K(M, P)$ which is determined by the position of M and P. When an infinitesimal minor element (line element, surface element, volume element, etc.) in the neighbourhood of the point P in the domain V is given by dP,

$$x(M) - \lambda \int_V K(M, P)x(P)dP = f(M) \tag{7.69}$$

is an integral equation on $x(M)$. Here we see that the regular part is the multiple integral of the coordinates. Then we can directly extend Fredholm's theory. For example, when Fredholm's determinant is

$$K\begin{pmatrix} M_1 & M_2 & \dots & M_\nu \\ P_1 & P_2 & \dots & P_\nu \end{pmatrix} = \begin{vmatrix} K(M_1, P_1) & K(M_1, P_2) & \dots & K(M_1, P_\nu) \\ \dots & \dots & \dots & \dots \\ K(M_\nu, P_1) & K(M_\nu, P_2) & \dots & K(M_\nu, P_\nu) \end{vmatrix}, \tag{7.70}$$

we have

$$\Delta(\lambda) = 1 + \sum_{1}^{\infty}{}_{\nu}(-1)^{\nu}\frac{\lambda^{\nu}}{\nu!}\int_{V}\int_{V}\cdots\int K\begin{pmatrix} M_1 & M_2 & \ldots & M_\nu \\ P_1 & P_2 & \ldots & P_\nu \end{pmatrix} dP_1\, dP_2 \ldots dP_\nu. \tag{7.71}$$

In this manner, when $\Delta(\lambda) \neq 0$, the kernel of the solution becomes

$$\Gamma(M, P; \lambda) = -\Delta\begin{pmatrix} M \\ P \end{pmatrix}; \lambda\Big)/\Delta(\lambda), \tag{7.72}$$

and the solution is

$$x(M) = f(M) - \lambda\int_{V}\Gamma(M, P; \lambda)f(P)dP. \tag{7.73}$$

We make a similar extension in the case when λ is an eigenvalue.

For example in the integral equation

$$u(x, y) - \lambda\int_{0}^{1}d\xi\int_{0}^{1}d\eta K(x, y\,|\,\xi, \eta)u(\xi, \eta) = f(x, y),$$

when

$$K\begin{pmatrix} x_1, y_1; x_2, y_2; \ldots; x_\nu, y_\nu \\ x_1, y_1; x_2, y_2; \ldots; x_\nu, y_\nu \end{pmatrix} = \begin{vmatrix} K(x_1, y_1\,|\,x_1, y_1) & K(x_1, y_1\,|\,x_2, y_2) & \ldots & K(x_1, y_1\,|\,x_\nu, y_\nu) \\ \ldots & \ldots & \ldots & \ldots \\ K(x_\nu, y_\nu\,|\,x_1, y_1) & K(x_\nu, y_\nu\,|\,x_2\, y_2) & \ldots & K(x_\nu, y_\nu\,|\,x_\nu, y_\nu) \end{vmatrix},$$

we obtain

$$\Delta(\lambda) = 1 + \sum_{1}^{\infty}{}_{\nu}(-1)^{\nu}\frac{\lambda^{\nu}}{\nu!}\int_{0}^{1}\int_{0}^{1}\cdots\int_{0}^{1} K\begin{pmatrix} x_1, y_1; \ldots; x_\nu, y_\nu \\ x_1, y_1; \ldots; x_\nu, y_\nu \end{pmatrix} dx_1dy_1 \ldots dx_\nu dy_\nu.$$

In the case of simultaneous integral equations with multiple integrals we have only to arrange them into one equation using the method explained in the previous sections.

7.5.2. *Integral equations of the second kind with multiple integrals, Volterra type*

When the unknown function $x(t_1, t_2, \ldots, t_n)$ is a function with n variables, the integral equation of the second kind becomes

$$x(t_1, t_2, \ldots, t_n) - \int_0^{t_1} d\xi_1 \int_0^{t_2} d\xi_2 \ldots \int_0^{t_n} K(t_1, t_2, \ldots, t_n | \xi_1, \xi_2, \ldots, \xi_n) \times x(\xi_1, \xi_2, \ldots, \xi_n) d\xi_n = f(t_1, t_2, \ldots, t_n). \tag{7.74}$$

We generalize it by extending the convolution of the first kind to

$$f * g = \int_{t_1}^{\xi_1} d\tau_1 \ldots \int_{t_n}^{\xi_n} f(t_1, t_2, \ldots, t_n | \tau_1, \tau_2, \ldots, \tau_n) \times g(\tau_1, \tau_2, \ldots, \tau_n | \xi_1, \xi_2, \ldots, \xi_n) d\tau_n, \tag{7.75}$$

and the kernel of the solution is given by the following expression:

$$G(t_1, t_2, \ldots, t_n | \xi_1, \xi_2, \ldots, \xi_n) = \sum_1^\infty {}_h\overset{*}{K}{}^h(t_1, t_2, \ldots, t_n | \xi_1, \xi_1, \ldots, \xi_n). \tag{7.76}$$

If we put

$$\left.\begin{aligned} \overset{*}{K}(t_1, t_2, \ldots, t_n | \xi_1, \xi_2, \ldots, \xi_n) &= K(t_1, t_2, \ldots, t_n | \xi_1, \xi_2, \ldots, \xi_n), \\ \overset{*}{K}{}^2(t_1, t_2, \ldots, t_n | \xi_1, \xi_2, \ldots, \xi_n) &= \overset{*}{K} * K, \\ \ldots, & \\ \overset{*}{K}{}^h(t_1, t_2, \ldots, t_n | \xi_1, \xi_2, \ldots, \xi_n) &= \overset{*}{K}{}^{(h-1)} * K, \end{aligned}\right\}$$

the solution is

$$x(t_1, t_2, \ldots, t_n) = f(t_1, t_2, \ldots, t_n) - \int_0^{t_1} d\xi_1 \int_0^{t_2} d\xi_2 \cdots \int_0^{t_n} G(t_1, t_2, \ldots, t_n | \xi_1, \xi_2, \ldots, \xi_n) \times f(\xi_1, \xi_2, \ldots, \xi_n) d\xi_1, d\xi_2, \ldots, d\xi_n. \tag{7.77}$$

This can be proved for convergence, reciprocity, and inversion, as in Theorem 2.2

7.5.3. Integral equations of the first kind with multiple integrals, Volterra type

In the case of the first kind its treatment is a little complex compared with that of the second kind, therefore we explain here the case *with* two variables. For example, take

$$\int_0^{t_1} d\xi_1 \int_0^{t_1} K(t_1, t_2 | \xi_1, \xi_2) x(\xi_1, \xi_2) d\xi_2 = f(t_1, t_2). \tag{7.78}$$

The same rule applies in general. The plan of the method of solution is to differentiate in t_1, t_2 and re-form the result into that of the second kind. After differentiation, the equation (7.78) becomes

$$K(t_1, t_2 | t_1, t_2)x(t_1, t_2)$$
$$+\int_0^{t_1} x(\xi_1, t_2)\frac{\partial K}{\partial t_1}(\xi_1, t_2 | t_1, t_2)d\xi_1 + \int_0^{t_2} x(t_1, \xi_2)\frac{\partial K}{\partial t_2}(t_1, \xi_2, t_1 | t_2)d\xi_2$$
$$+\int_0^{t_1} d\xi_1 \int_0^{t_2} x(\xi_1, \xi_2)\frac{\partial^2 K}{\xi t_1 \partial t_2}(\xi_1, \xi_2 | t_1, t_2)d\xi_2 = \frac{\partial^2 f}{\partial t_1 \partial t_2}.$$

This equation becomes $f(t_1, t_2) = 0$ when $t_1 = t_2 = 0$ (this is of course necessary), then further when

$$\left(\frac{\partial f}{\partial t_1}\right)_{t_2=0} = \left(\frac{\partial f}{\partial t_2}\right)_{t_1=0} = 0$$

it coincides with (7.78).

Suppose that $K(t_1, t_1 | t_1, t_2) \not\equiv 0$; then if we put

$$\frac{\partial K}{\partial t_1}\Big/K \equiv F, \quad \frac{\partial K}{\partial t_2}\Big/K \equiv G, \quad \frac{\partial^2 K}{\partial t_1 \partial t_2}\Big/K \equiv H,$$

we get

$$x(t_1, t_2) + \int_0^{t_1} x(\xi_1, t_2)F(\xi_1; t_1, t_2)d\xi_1 + \int_0^{t_2} x(t_2, \xi_2)G(\xi_n; t_1, t_2)d\xi_2$$
$$+\int_0^{t_1} d\xi_1 \int_0^{t_2} x(\xi_1, \xi_2)H(\xi_1, \xi_2; t_1, t_2)d\xi_2 = h(t_1, t_2).$$

In this equation we find that all the simple and the multiple integrals contain unknown functions, and we must, therefore, remove the terms of the simple integral. To do this, we use the method of successive approximations. Remove all the terms except two on the right to on the left-hand side and express it by $T(t_1, t_2)$; then we have

$$x(t_1, t_2) + \int_0^{t_1} x(\xi_1, t_2)F(\xi_1; t_1, t_2)d\xi_1 = T(t_1, t_2).$$

Treat t_2 as a parameter and $T(t_1, t_2)$ as known, then the above expression becomes an integral equation of the second kind, the solution of which is expressed by the reciprocal function $\mathfrak{F}$ of F in

$$x(t_1, t_2) = T(t_1, t_2) + \int_0^{t_1} T(\xi_1, t_2)\,\mathfrak{F}(\xi_1; t_1, t_2)d\xi_1.$$

Put the original form in T, and we get an integral equation of the same form as the original one, yet here we find one simple integral has been dropped off as

$$x(t_1, t_2) + \int_0^{t_2} x(t_1, \xi_2)G_1(\xi_2; t_1, t_2)d\xi_2$$
$$+ \int_0^{t_1} d\xi_1 \int_0^{t_2} H_1(\xi_1, \xi_2; t_1, t_2)x(\xi_1, \xi_2)d\xi_2 = h_1(t_1, t_2).$$

Repeat, and finally we obtain

$$x(t_1, t_2) + \int_0^{t_1} d\xi_1 \int_0^{t_2} H_2(\xi_1, \xi_2; t_1, t_2)x(\xi_1, \xi_2)d\xi_2 = h_2(t_1, t_2),$$

which is the form we have already treated.

When there are n variables, we may introduce an expression containing simple integrals and integral equations with as many as n-tupl multiple integrals by differentiating once for each variable. Low-degree integrals may also be removed in the same way.

Exercises

1. Solve the following singular integral equations of the second kind:

(a) $x(t) - \int_{-\infty}^{t} e^{-2(t-\xi)}(t - \xi)x(\xi)d\xi = te^t,$

(b) $x(t) + \int_{-\infty}^{t} \frac{1}{\xi^2}\left(\frac{1}{t} - \frac{1}{\xi}\right)x(\xi)d\xi = \frac{1}{t} + e^t\left(\frac{1}{t^4} + \frac{1}{t^3}\right)$

(c) $x(t) - \int_{-\infty}^{\infty} t \sin \xi\, x(\xi)d\xi = 2e^{-t} + 3t,$

(d) $x(t) - \int_0^{\infty} te^{-\xi}x(\xi)d\xi = \sin 2t + \frac{1}{2}\, t,$

(e) $x(t) - \int_0^{\infty} \frac{tx(\xi)}{\xi(1 + \xi)^4}\, d\xi = t^2 - \frac{t}{6},$

(f) $x(t) - \int_1^{\infty} e^{-(t-\xi)}\, x(\xi)d\xi = e^{-t} + t + 1,$

(g) $x(t) - \int_1^{\infty} t\xi x(\xi)d\xi = 2e^{-t} + e^{-1} + 1.$

2. Solve the following convolution-type singular integral equations:

(a) $x(t) - \int_0^{\infty} (t - \xi)x(\xi)d\xi = t + 1,$

(b) $x(t) - \int_0^{\infty} e^{-(t-\xi)^2}x(\xi)d\xi = \frac{1}{t + 1},$

(c) $x(t) - \int_{-\infty}^{\infty} e^{-(t-\xi)^2} x(\xi) d\xi = \sin t/t.$

3. Solve the following simultaneous integral equations:

(a) $$\left.\begin{aligned} x_1(t) + x_2(t) + \int_0^1 (t+\xi) x_1(\xi) d\xi &= at, \\ x_1(t) - x_2(t) + \int_0^1 t\xi x_2(\xi) d\xi &= bt \end{aligned}\right\}$$

(b) $$\left.\begin{aligned} x_1(t) + x_2(t) - \int_0^t (t-\xi) x_1(\xi) d\xi &= at, \\ x_1(t) - x_2(t) - \int_0^t (t-\xi)^2 x_2(\xi) d\xi &= bt^2 \end{aligned}\right\}$$

(c) $$\left.\begin{aligned} x_1(t) + x_2(t) - \int_0^t (t-\xi) x_1(\xi) d\xi &= at, \ 0 < t < 1, \\ x_1(t) - x_2(t) - \int_0^1 t\xi x_2(\xi) d\xi &= bt. \end{aligned}\right\}$$

4. Solve the following Volterra-type integral equations:

(a) $x(t) - \int_{-t}^{t} x(\xi) d\xi = t + 1,$

(b) $x(t) - \int_{-t}^{t} (t\xi + \xi) x(\xi) d\xi = t^2 + t + 1,$

(c) $x(t) - \int_{-t}^{2t} t\xi x(\xi) d\xi = 2t + 3.$

5. Solve the following integral equations with multiple integrals:

(a) $x(t_1, t_2) - \int_0^{t_1} d\xi_1 \int_0^{t_2} (t_1\xi_2 + t_2\xi_1) x(\xi_1, \xi_2) d\xi_2 = at_1 + bt_2,$

(b) $u(x, y) - \lambda \int_0^1 d\xi \int_0^1 d\eta u(\xi, \eta) = x + y + c,$

(c) $u(x, y) - \lambda \int_0^1 d\xi \int_0^1 (x\xi + y\eta + 1) u(\xi, \eta) d\eta = x + y + c.$

6. Solve the following integral equations using the numerical method of solution:

(a) $x(t) - \int_0^{\infty} e^{-2(t+\xi)} x(\xi) d\xi = 1/(t^2 + 1),$

(b) $x(t) - \int_{-\infty}^{\infty} e^{-(t-\xi)^2} x(\xi) d\xi = 1/(t^2 + 1).$

7. Study the reciprocal theorem on the kernel $K(x, y|\xi, \eta)$ and the kernel of the solution $\Gamma(x, y|\xi, \eta; \lambda)$.

8. Prove Theorem 7.1.

9. Solve Example 7.1 by the numerical method of solution.

10. In the Fredholm-type integral equation of the second kind with multiple integral

$$x(M) - \lambda \int_V K(M, P)x(P)dP = f(M),$$

when the kernel of the solution is $\Gamma(M, P; \lambda)$, prove that the reciprocal theorem

$$K(M, P) + \Gamma(M, P; \lambda) = \lambda \int_V K(M, Q)\Gamma(Q, P; \lambda)dQ$$

$$= \lambda \int_V \Gamma(M, Q; \lambda)K(Q, P)dQ$$

holds (see § 2.2).

11. In the Fredholm-type homogeneous integral equation

$$x(M) = \lambda \int_V K(M, P)x(P)dP,$$

when λ is the eigenvalue of index 1 and the eigenfunction is properly selected in P, prove that

$$\varphi(M) = \Delta\left(\begin{matrix} M \\ P' \end{matrix}; \lambda\right).$$

Then what will become of the eigenfunctions when the index of the eigenvalue is m? (see § 2.3).

12. Write down the conditions for the existence of a solution when λ is an eigenvalue in an integral equation of the second kind with multiple integral, Fredholm type:

$$x(M) - \lambda \int_V K(M, P)x(P)dP = f(M)$$

(see § 2.4).

13. Given a regular cube V of a side length 2, let M, P denote two arbitrary points in the interior and let the kernel $K(M, P)$ represent the distance between M and P. Further, it $f(M)$ is equal to the cube (or third power) of the distance between the centre O of the cube and the point M, then formulate a Fredholm-type integral equation of the second kind with triple integrals, the unknown function of which is the function $x(M)$ of the position of an arbitrary point M.

14. Find Fredholm's minor $\Delta\begin{pmatrix} x, y \\ \xi, \eta \end{pmatrix}; \lambda)$ on the integral equation

$$u(x, y) - \lambda \int_0^1 d\xi \int_0^1 d\eta\, K(x, y \mid \xi, \eta) u(\xi, \eta) = f(x, y).$$

15. Devise the numerical method of solution for the above integral equation. Then, when

$$u(x_i, y_i) = u_{ij}, \quad K(x_i, y_j \mid \xi_m, \eta_n) = k_{ijmn}, \quad f(x_i, y_j) = f_{ij}$$

denote the values at the lattice points, prove that we can introduce the simultaneous equations in u_{ij}. Solve Question 5 numerically in this way and compare the results with the exact solutions.

References

Churchill, R.V. (1941) *Fourier Series and Boundary Value Problems* (McGraw, New York) Formula 7.

8 Singular kernels

So far, we have restricted the kernel to be bounded in the domain; now we shall study the case when this restriction is removed. In applications we often see that the kernel becomes infinitely large. In such cases we may as well extend Fredholm's or Volterra's theories. Yet the methods to be used vary according to the mathematical properties of the kernel in the neighbourhood of the singular point, and many studies can be made of different cases. Most of these topics are complex and extend far from the general theories; therefore, we shall explore only those which are simple and easily applicable.

When the kernel takes a special form, operational calculus can be employed, and we shall also touch on the techniques of numerical calculation.

8.1. Fredholm-type integral equations

If the poles of a kernel $K(t, \xi)$ are located in the interior of its closed domain S or T we call it a *singular kernel.* When this sort of singular kernel stands on a line parallel to the coordinate axis, for example, if the poles stand on $t = t_0$, then for an arbitrary function $f(t)$ the integral

$$\int_a^b K(t, \xi) f(\xi) d\xi$$

diverges on $t = t_0$, and the singular points never vanish by convolution.

Studies have been made on the methods of solving integral equations with singular kernels in cases when the integral

$$\int_a^b \int_a^b K(t, \xi) dt\, d\xi$$

or

$$\int_a^b \int_a^b |K(t, \xi)|^\alpha dt\, d\xi, \qquad 1 < \alpha < 2$$

exists, or the integral

$$\int_a^b |K(t, \xi)| d\xi$$

is bounded, or divergent, but we shall leave minute explanations about these cases, and go on with a few special cases.

(1) *Successive method of substitution.*

THEOREM 8.1

For the Fredholm-type integral equation of the second kind:

$$x(t) - \lambda \int_a^b K(t, \xi) x(\xi) d\xi = f(t), \qquad [F_2]$$

if

$$\int_a^b |K(t, \xi)| d\xi \leqq q(b - a) \qquad (8.1)$$

for a positive number q almost everywhere in I, iterated kernels exist almost everywhere and

$$-\{K + \lambda \mathring{K}^2 + \lambda^2 \mathring{K}^3 + \cdots\}$$

converges for $|\lambda| < 1/q$ *almost everywhere in I. Let* $\Gamma(t, \xi; \lambda)$ *denote this expression, then when the integral*

$$\int_a^b |K(t, \xi)| f(\xi) d\xi \qquad (8.2)$$

is bounded, the solution $x(t)$ *is*

$$x(t) = f(t) - \lambda \int_a^b \Gamma(t, \xi; \lambda) f(\xi) d\xi.$$

Proof. See (Hille and Tamarkin 1930).

This theorem holds if we put

$$\int_a^b |K(t, \xi)| dt \leqq q(b - a) \qquad (8.3)$$

instead of the condition (8.1) and assume that

$$\int_a^b f(\xi) d\xi \qquad (8.4)$$

instead of (8.2) is bounded.

When $f(t)$ is not continuous, the solution $x(t)$ is generally not continuous, but if $f(t)$ is continuous the conditions (8.2), (8.4) are not necessary. This time, the solution is also continuous.

(2) *Extension of Fredholm's method.* By Fredholm's method of solution, the kernel of the solution becomes

$$\Gamma(t, \xi; \lambda) = -\Delta\binom{t}{\xi}; \lambda\Big) \Big/ \Delta(\lambda). \tag{8.5}$$

Then the iterated kernel

$$K(t, \xi), \mathring{K}^2(t, \xi), \ldots, \mathring{K}^{n-1}(t, \xi)$$

becomes discontinuous (infinitely large) at $t = \xi$, but if the terms after $\mathring{K}^n(t, \xi)$ are all bounded, that is, denoting by $D_n\left(\begin{matrix}t\\ \xi\end{matrix}; \lambda\right)$, $D_n(\lambda)$ the expression $\Delta\left(\begin{matrix}t\\ \xi\end{matrix}; \lambda\right)$, $\Delta(\lambda)$ in which $K(t, t), \mathring{K}(t, t), \ldots, \mathring{K}^{n-1}(t, t)$ are equated to zero, the kernel of the solution of $[F_2]$ becomes

$$\Gamma(t, \xi; \lambda) = -D_n\left(\begin{matrix}t\\ \xi\end{matrix}; \lambda\right) \Big/ D_n(\lambda). \tag{8.6}$$

We may solve $[F_2]$ using this method (Poincaré 1910).

(3) *Bounded and continuous iterated kernel* $\mathring{K}^n$.

When $\mathring{K}^n$ is bounded and continuous, $\mathring{K}^{n+1}$, ... are bounded and continuous, then if $|\lambda|$ is sufficiently small,

$$-(K + \lambda\mathring{K}^2 + \lambda^2\mathring{K}^3 + \ldots + \lambda^{n-1}\mathring{K}^n + \ldots) \tag{8.7}$$

defines $\Gamma(t, \xi; \lambda)$. From this we obtain the reciprocal theorem:

$$\Gamma + K = \lambda\Gamma \circ K = \lambda K \circ \Gamma.$$

Hence, Γ is the kernel of the solution (see § 2.2).

Let $\Delta_n(\lambda)$ be the determinant of $\mathring{K}^n$, then from Fredholm's formula the reciprocal function for $\mathring{K}^n$ is given by

$$\Gamma_n(t, \xi; \lambda) = -\frac{\Delta_n\left(\begin{matrix}t\\ \xi\end{matrix}; \lambda\right)}{\Delta_n(\lambda)}, \tag{8.8}$$

where $\Delta_n(\lambda)$ is the determinant of $\mathring{K}^n$. In addition, we have

$$\Gamma_n = -(\mathring{K}^n + \lambda\mathring{K}^{2n} + \cdots). \tag{8.9}$$

If we compare (8.7) with (8.9), and when $|\lambda|$ is small put

$$L = K + \lambda\mathring{K}^2 + \cdots + \lambda^{n-2}\mathring{K}^{n-1},$$

then we get

$$\begin{aligned}\Gamma &= -L - \lambda^{n-1}(\mathring{K}^n + \lambda\mathring{K}^{n+1} + \lambda^2\mathring{K}^{n+2} + \cdots) \\ &= -L + \lambda^{n-1}\Gamma_n(t, \xi; \lambda^n) - \lambda^n(\mathring{K}^{n+1} + \lambda\mathring{K}^{n+2} + \cdots) \\ &= -L + \lambda^{n-1}\Gamma_n(t, \xi; \lambda^n) + \lambda^n L(\lambda) \circ \Gamma_n(\lambda^n). \qquad (8.10)\end{aligned}$$

In this way the reciprocal function Γ of K is obtained from the reciprocal function Γ_n of $\mathring{K}^n$.

From (8.10) we know that the poles of $\Gamma(\lambda)$ are at the zero points of $\Delta_n(\lambda^n)$. But every zero point of $\Delta_n(\lambda^n)$ does not necessarily become a pole of Γ.

When indeed $\Delta(\lambda)$ exists, we directly obtain

$$\Delta_n(\lambda^n) = \Delta(\lambda)\Delta(\omega\lambda) \cdots \Delta(\omega^{n-1}\lambda),$$

where ω is the n-tuple root of 1, and the roots of $\Delta_n(\lambda^n)$ contain not only the root λ_i of $\Delta(\lambda)$, but also the roots of

$$\Delta(\omega^k\lambda), \qquad k = 1, 2, \dots, n-1.$$

Suppose that $\lambda = c$ is the pole of the kernel of the solution $\Gamma(t, \xi; \lambda)$, and φ is an eigenfunction; then we have

$$\varphi \circ (E - cK) = 0.$$

Multiply this by $(E + cL(c))$ from the right and we get

$$\varphi \circ (E - cK + cL - c^2K \circ L) = 0, \qquad \varphi \circ (E - c^n\mathring{K}^n) = 0.$$

Hence, the eigenfunction of K becomes the eigenfunction of $\mathring{K}^n$ for c^n. When $\mathring{K}^n$ is continuous, the eigenfunctions are also continuous and finite.

In this manner, the method of solving the singular kernel is simplified to Fredholm's method of solution for cases when the kernel is bounded and continuous.

8.1.1. *The kernel* $k(t, \xi)/|t - \xi|^\alpha$

When the kernel takes the form

$$K(t, \xi) = \frac{k(t, \xi)}{|t - \xi|^\alpha} \qquad (0 < \alpha < 1)$$

and $k(t, \xi)$ belongs to $\mathscr{R}$ in D, iterated kernels of more than a certain degree become bounded. Now let $k_1(t, \xi)$, $k_2(t, \xi)$ be two functions which belong to $\mathscr{R}$; then the convolution of

$$K_1(t, \xi) = \frac{k_1(t, \xi)}{|t - \xi|^{\alpha_1}}, \qquad K_2(t, \xi) = \frac{k_2(t, \xi)}{|t - \xi|^{\alpha_2}};$$
$$0 < \alpha_1, \alpha_2 < 1,$$

that is,

$$K_1 \circ K_2 = \int_a^b \frac{k_1(t, \tau) k_2(\tau, \xi)}{|t - \tau|^{\alpha_1} |\tau - \xi|^{\alpha_2}} d\tau,$$

has meaning, and in D, when

$$|k_1| \leqq m_1, \ |k_2| \leqq m_2$$

holds, we have

$$|K_1 \circ K_2| \leqq m_1 m_2 \int_a^b \frac{1}{|t - \tau|^{\alpha_1} |\tau - \xi|^{\alpha_2}} d\tau.$$

For the integral on the right-hand side when $\alpha_1 + \alpha_2 - 1 < 0$, if $t < \xi$, we have

$$\int_a^t \frac{1}{|t - \tau|^{\alpha_1} |\tau - \xi|^{\alpha_2}} d\tau < \int_a^t \frac{1}{|t - \tau|^{\alpha_1 + \alpha_2}} d\tau$$
$$= \frac{|t - a|^{-(\alpha_1 + \alpha_2 - 1)}}{\alpha_1 + \alpha_2 - 1},$$

$$\int_t^\xi \frac{1}{|t - \tau|^{\alpha_1} |\tau - \xi|^{\alpha_2}} d\tau$$
$$< \operatorname{Max}\left\{ \int_t^\xi \frac{1}{|\tau - \xi|^{\alpha_1 + \alpha_2}} d\tau, \ \int_t^\xi \frac{1}{|t - \tau|^{\alpha_1 + \alpha_2}} d\tau \right\}$$
$$= \frac{-1}{\alpha_1 + \alpha_2 - 1} \operatorname{Max}\{ |\tau - a|^{-(\alpha_1 + \alpha_2 - 1)}, \ |\tau - b|^{-(\alpha_1 + \alpha_2 - 1)} \},$$

$$\int_\xi^b \frac{1}{|t - \tau|^{\alpha_1} |\tau - \xi|^{\alpha_2}} < \int_\xi^b \frac{1}{|\tau - \xi|^{\alpha_1 + \alpha_2}} d\tau$$
$$= \frac{|b - \xi|^{-(\alpha_1 + \alpha_2 - 1)}}{-(\alpha_1 + \alpha_2 - 1)},$$

then consequently all these integrals become bounded, and therefore $K_1 \circ K_2$ is bounded.

On the other hand, if $\alpha_1 + \alpha_2 - 1 > 0$, we change variables so that

$$\frac{t - \tau}{t - \xi} = \eta,$$

which gives

$$\int_a^b \frac{d\tau}{|t-\tau|^{\alpha_1}|\tau-\xi|^{\alpha_2}} < \frac{1}{|t-\xi|^{\alpha_1+\alpha_2-1}} \int_{-\infty}^{\infty} \frac{d\eta}{|1-\eta|^{\alpha_2}|\eta|^{\alpha_1}}.$$

The integral on the right-hand side is clearly bounded.

For the kernel

$$K(t, \xi) = \frac{k(t, \xi)}{|t-\xi|^{\alpha}},$$

we carry out integrated kernels $\mathring{K}^2, \mathring{K}^3, \ldots, \mathring{K}^n, \ldots$ While the degree of pole $\dfrac{f(t, \xi)}{|t-\xi|^{\nu}}$ at $t = \xi$ is ν, so the degrees here are $2\alpha - 1$, $3\alpha - 2$, $\cdots$, $n\alpha - n + 1$ respectively. Therefore, when

$$n\alpha - n + 1 < 0 \quad \text{i.e.} \quad n > \frac{1}{1-\alpha},$$

$\mathring{K}^n$ is bounded and each of $\mathring{K}^{n+1}, \mathring{K}^{n+2}, \ldots$ is bounded. In particular, when $0 < \alpha < 1/2$, $\mathring{K}^2$ is bounded. Then we equate all the terms containing $K(\xi, \xi)$ and so on in the expression $\Delta(\lambda)$, $\Delta\left(\begin{matrix} t \\ \xi \end{matrix}; \lambda\right)$ to zero, and all the main-diagonal elements, such as, $K(\xi_i, \xi_i)$, in the determinants

$$K\begin{pmatrix} \xi_1 & \cdots & \xi_\nu \\ \xi_1 & \cdots & \xi_\nu \end{pmatrix} \quad \text{and} \quad K\begin{pmatrix} t & \xi_1 & \xi_2 & \cdots & \xi_\nu \\ \xi & \xi_1 & \xi_2 & \cdots & \xi_\nu \end{pmatrix}$$

to zero; then the kernel of the solution becomes

$$\Gamma(t, \xi; \lambda) = -\Delta\left(\begin{matrix} t \\ \xi \end{matrix}; \lambda\right) \Big/ \Delta(\lambda),$$

and the solution is

$$x(t) = f(t) - \lambda \int_a^b \Gamma(t, \xi; \lambda) f(\xi) d\xi.$$

8.1.2. *Applications*

Many applications of mathematical physics deal with these sorts of singular kernels. For example, in § 7.5.1 we have

$$x(M) - \lambda \int_V K(M, P) x(P) dP = f(M)$$

where V is a bounded domain in p-dimensional Euclidian space. When the kernel denotes the function of distance $\overline{MP}$

$$K(M, P) = \frac{k(M, P)}{(\overline{MP})^{\alpha}} \qquad (0 < \alpha < p)$$

and $k(M, P)$ is bounded, then consider $M(t_1, t_2, \ldots, t_p)$, $P(\xi_1, \xi_2, \ldots, \xi_p)$ when

$$\overline{MP} = ((t_1 - \xi_1)^2 + (t_2 - \xi_2)^2 + \cdots + (t_p - \xi_p)^2)^{\frac{1}{2}}.$$

Let n be the first integer larger than $1/(1 - (\alpha/p))$; then all the iterated kernels bigger than K^n are bounded, and hence the above theorem can be applied.

8.2. Fredholm-type integral equations of the third kind

In the Fredholm-type integral equation of the third kind:

$$A(t)x(t) - \lambda \int_a^b K(t, \xi)x(\xi)d\xi = f(t), \qquad [F_3]$$

to facilitate the discussion, we shall suppose that $A(t)$ has a simple root in $I(a, b)$ and put

$$A(t)x(t) \equiv X(t);$$

then $[F_3]$ becomes an integral equation in $X(t)$:

$$X(t) - \lambda \int_a^b \frac{K(t, \xi)}{A(\xi)} X(\xi)d\xi = f(t),$$

which results in the case of the singular kernel.†

Picard (1911), supposing that I has only a single root $t = t_0$, studied the case when $K(t, \xi), A(t), f(t)$ are analytic in the domain D, surrounded by the line (a, b) and the simple curve C. The result runs as follows.

Fredholm's theory applies as it is if we eliminate the minor interval $(t_0 - \varepsilon, t_0 + \eta)$ containing the discontinuous point t_0. Next, let ε and η converge to zero at the same time such that $\varepsilon/\eta = c$ (where c is constant); then we may prove that $X(t)$ converges to a certain constant value that is related only to

† In fact it becomes an ordinary $[F_2]$, if $A(t)$ has no zero points in I.

$$\lim \log \frac{\eta}{\varepsilon}.$$

Therefore, we obtain the solution $x(t)$, which is related to $\lim \log(\eta/\varepsilon)$ with regard to $[F_3]$ and has t_0 as its simple pole. If the residual is zero then $[F_3]$ has a continuous solution. Also, the condition is such that the solution has no connection with C, which shows that the value of λ agrees with a root of a certain equation.

Hence, if λ takes this special value, the solution of $[F_3]$ has no connection with C and is continuous.

8.3. Integral equation which takes Cauchy's principal value

At times, in the integral in the regular part of an integral equation with singular kernel, there are cases where we calculate by taking Cauchy's principal value. These cases are often found in applications in engineering and mathematical physics.

For example, in $[F_3]$ of the previous section, if $A(t)$ contains the factor $t - t_0$, then to calculate the integral we eliminate the interval $(t_0 - \varepsilon, t_0 + \varepsilon)$ and then evaluate the limit as $\varepsilon \to 0$. But, as explained above, since the solution under a certain condition is independent of taking the interval containing the singular point and of its limiting process, to facilitate the calculation we introduce the solution taking the regular part as a Cauchy integral.

Generally, for the two functions

$$K(t, \xi) = \frac{k(t)}{t - \xi}, \qquad L(t, \xi) = \frac{l(t)}{t - \xi}, \tag{8.11}$$

when $k(t)$, $l(t)$ are continuous in I, extend the convolution of the second kind into

$$K \circ L = ⨍_a^b \frac{k(t)}{t - \tau} \frac{l(\tau)}{\tau - \xi} d\tau, \tag{8.12}$$

in which the principle of combination does not hold, and generally it comes to

$$L \circ (K \circ f) \neq (L \circ K) \circ f. \tag{8.13}$$

Here $⨍$ means take Cauchy's principal value. Then let

$$D \equiv (L \circ K) \circ f - L \circ (K \circ f) \tag{8.14}$$

be the remainder, and, according to Poincaré, the formula

$$\oint_a^b \frac{d\xi}{t-\xi} d\xi \oint_a^b \frac{F(t, \xi, \eta)}{\xi - \eta} d\eta = \oint_a^b d\eta \oint_a^b F(t, \xi, \eta) \frac{d\xi}{(t-\xi)(\xi-\eta)} - \pi^2 F(t, t, t) \tag{8.15}$$

holds. As for the above $k(t)/(t-\xi)$, $l(t)/(t-\xi)$, we get

$$D = \pi^2 l(t) k(t) f(t). \tag{8.16}$$

Multiply both sides of the integral equation which takes Cauchy's principal value

$$x - \lambda K \circ x = f, \qquad K = k/(t-\xi) \tag{8.17}$$

by λK from the left:

$$\lambda K \circ x - \lambda^2 K \circ (K \circ x) = \lambda K \circ f. \tag{8.18}$$

From (8.16) this becomes

$$K \circ (K \circ x) = \mathring{K}^2 \circ x + \pi^2 k^2 x. \tag{8.19}$$

Therefore, using (8.17), (8.19) we find that (8.18) becomes

$$(E + \lambda^2 \pi^2 k^2) \circ x - \lambda^2 \mathring{K}^2 \circ x = f + \lambda K \circ f;$$

that is,

$$[1 + \lambda^2 \pi^2 k^2(t)] x(t) - \lambda^2 \oint_a^b \mathring{K}^2(t, \xi) x(\xi) d\xi = f(t) + \lambda \oint_a^b K(t, \xi) f(\xi) d\xi. \tag{8.20}$$

Since $\mathring{K}^2$ no longer has a singular point, the integral on the left-hand side is an ordinary definite integral; and when $[1 + \lambda^2 \pi^2 k^2(t)]$ does not become zero in I, (8.20) is nothing but an integral equation of the second kind, [F_2].

Now put

$$x - \lambda K \circ x - f \equiv X;$$

then (8.20) becomes

$$X + \lambda K \circ X = 0.$$

Hence we can see easily that the solution of (8.17) satisfies (8.20).

Example 8.1. *Villat's example*
If

$$K(t, \xi) = \frac{1}{2\pi} \cot \frac{t-\xi}{2}, \quad I(0, 2\pi), \tag{8.21}$$

we have

$$\mathring{K}^2 = 1/2\pi$$

Then (8.20) becomes

$$x(t)(1+\lambda^2) - \frac{\lambda^2}{2\pi} \int_0^{2\pi} x(\xi)d\xi = f(t) + \lambda \int K(t, \xi)f(\xi)d\xi. \tag{8.22}$$

Let $F(t)$ be the right-hand side, which is a known function, and the solution is

$$x(t) = \frac{F(t)}{1+\lambda^2} + C, \tag{8.23}$$

C a constant. Put this into the above expression again and C can be determined as

$$C = \frac{\lambda^2}{2\pi} \int_0^{2\pi} F(\xi)d\xi. \tag{8.24}$$

As for the integral equation of the first kind,

$$\frac{1}{2\pi} \int_0^{2\pi} \cot \frac{t-\xi}{2} x(\xi)d\xi = f(t), \tag{8.25}$$

convolute the kernels and we get

$$-x(t) + \frac{1}{2\pi} \int_0^{2\pi} x(\xi)d\xi = \frac{1}{2\pi} \int_0^{2\pi} \cot \frac{t-\xi}{2} f(\xi)d\xi,$$

which produces

$$x(t) = -\frac{1}{2\pi} \int_0^{2\pi} \cot \frac{t-\xi}{2} f(\xi)d\xi + C, \tag{8.26}$$

where C is a constant. Put this into the original integral equation and the condition

$$\int_0^{2\pi} f(\xi)d\xi = 0$$

is obtained. Then if we make

$$\frac{1}{2\pi} \int_0^{2\pi} \cot \frac{t-\xi}{2} x(\xi)d\xi = f(t) - \frac{1}{2\pi} \int_0^{2\pi} f(\xi)d\xi, \tag{8.27}$$

(8.26) becomes the general solution of (8.27).

This kind of method of solution can be extended to the case containing multiple integrals. For example, in the integral equation

$$x(M) - \lambda \oint_V K(M, P)x(P)dP = f(M), \tag{8.28}$$

when the kernel $K(M, P)$ is determined by the two points M, P in the domain V of the p-dimensional space, and becomes infinitely large with the order MP^{-p} when $M \to P$, $\oint_V$ is defined as the limit of the integral over the domain V' from which an m-dimensional small spherical domain with centre P is excluded, and the radius of the sphere is made to converge to zero; that is, $V' \to V$. Here again, if

$$D = \int_V L \circ K(M, P)f(P)dP - \int_V M(M, P)dP \int_V K(P, Q)f(Q)dQ, \tag{8.29}$$

the result of the calculation reads

$$D = A(P)f(P), \tag{8.30}$$

where $A(P)$ is a function of P which is connected with only the singular parts of K, L. Then multiply (8.28) by L from the left and like (8.20) it becomes

$$\begin{aligned} x(M)[1 + \lambda^2 A(M; \lambda)] - \lambda \oint_V T(M, P; \lambda)x(P)dP \\ = f(M) + \oint_V L(M, P; \lambda)f(P)dP. \end{aligned} \tag{8.31}$$

Here T turns out to be

$$T(M, P; \lambda) = K - L + \lambda L \circ K. \tag{8.32}$$

Giraud (1934) showed that we can choose L so that T may no longer have a pole of order $\overline{MP}^{-p}$. Since (8.31) is an ordinary integral equation, in so far as λ does not make $[1 + \lambda^2 A(M; \lambda)]$ become zero, its solution is

$$x(M) = \frac{f(M)}{1 + \lambda^2 A(M, \lambda)} - \lambda \oint_V \Gamma(M, P; \lambda)f(P)dP. \tag{8.33}$$

We may prove that only if (8.33) does not take an eigenvalue it may serve as the unique solution of (8.28). If λ becomes the pole of Γ, the homogeneous integral equation corresponding to (8.28) has an eigenfunction not identically zero, and the general solution of the homogeneous integral equation corresponding to (8.31) is expressed as a multiple of this eigenfunction. In this way we may extend and solve Fredholm's theories. But since the calculation is considerably complicated we have shown here only an outline of the plan of solving the equation and the result.

8.4. Volterra-type integral equations

The most important case among the Volterra types is when the kernel becomes infinitely large at $t = \xi$. That is, when the kernel contains a factor such as $(t-\xi)^{\alpha-1}$, $0 < \alpha < 1$. In this case, the method of solution by the iterated kernel is applied.

(1) *Convolution of the singular kernel*

Definition

When the kernel $K(t, \xi)$ is defined in T and is expressed in the form

$$K(t, \xi) = \frac{(t-\xi)^{\alpha-1}}{\Gamma(\alpha)} k(t, \xi), \qquad \alpha > 0, \tag{8.34}$$

we say that the order of the kernel is α. If $k(t, \xi)$ is bounded and continuous, and $k(t, t) \not\equiv 0$, then $k(t, \xi)$ is called the *characteristic function* of the kernel, and $k(t, t)$ is called a *diagonal*; $\Gamma(\alpha)$ is a gamma function.

When $\alpha \geqq 1$, the kernel is bounded. Therefore, we must look at the case when $0 < \alpha < 1$. Let us examine the convolution of this kind of singular kernel of the first kind. Take the kernel of order β:

$$G(t, \xi) = \frac{(t-\xi)^{\beta-1}}{\Gamma(\beta)} g(t, \xi),$$

and calculate the convolution of the first kind:

$$K * G = \int_\xi^t \frac{(t-\tau)^{\alpha-1}(\tau-\xi)^{\beta-1}}{\Gamma(\alpha)\Gamma(\beta)} k(t, \tau) g(\tau, \xi) \mathrm{d}\tau. \tag{8.35}$$

If we change variables here, putting

$$\frac{t-\tau}{t-\xi} = \eta,$$

this makes the above expression

$$K * G = \frac{(t-\xi)^{\alpha+\beta-1}}{\Gamma(\alpha+\beta)} H(t, \xi), \tag{8.36}$$

with

$$H(t, \xi) = \frac{\Gamma(\alpha+\beta)}{\Gamma(\alpha)\Gamma(\xi)} \int_0^1 k(t, t+\eta(\xi - t))$$
$$\times g(t+\eta(\xi-t), \xi)\eta^{\alpha-1}(1-\eta)^{\beta-1} d\eta. \qquad (8.37)$$

Therefore the following properties are obtained.

(i) When an index is a positive number smaller than 1, all the operational laws about the convolution in § 1.4 hold.

(ii) The order of the convolution proves to be the sum of the original orders.

(iii) The diagonal of the convolution proves to be the product of the original function.†

(iv) If $|k| < m, |g| < n$ in T, we have

$$|K * G| < \frac{(t-\xi)^{\alpha+\beta-1}}{\Gamma(\alpha+\beta)} mn. \qquad (8.38)$$

(2) *Convolution of* 1 *of the first kind.*

According to § 1.4, when n is a positive integer we have

$$\overset{*}{1}{}^{n} = \frac{(t-\xi)^{n-1}}{(n-1)!} = \frac{(t-\xi)^{m-1}}{\Gamma(n)};$$

and for m, n,

$$\overset{*}{1}{}^{m} * \overset{*}{1}{}^{n} = \overset{*}{1}{}^{m+n}$$

holds, then when α is an arbitrary positive number we extend the definition into

$$\overset{*}{1}{}^{\alpha} = \frac{(t-\xi)^{\alpha-1}}{\Gamma(\alpha)}. \qquad (8.39)$$

(3) *Integral equation of the second kind or the case when the order of the kernel is* $\alpha (0 < \alpha < 1)$.

In the Volterra-type integral equation of the second kind:

$$x(t) - \int_0^t K(t, \xi) x(\xi) d\xi = f(t), \qquad (8.40)$$

when the kernel is given in the form (8.34), if $0 < \alpha < 1$, by (8.38) we have

† This is due to the mathematical property of the gamma function, that $\int_0^1 t^{\alpha-1}(1-\beta)^{\beta-1}dt = \Gamma(\alpha)\Gamma(\beta)/\Gamma(\alpha+\beta)$.

$$|\overset{*}{K}^n| < \frac{(t-\xi)^{n\alpha-1}}{\Gamma(n\alpha)} m^n.$$

Therefore, when n is an integer bigger than $1/\alpha$, the iterated kernel becomes bounded and continuous. Therefore,

$$-(K + \overset{*}{K}^2 + \overset{*}{K}^3 + \cdots + \overset{*}{K}^n + \cdots)$$

is uniformly absolutely convergent. Let this be G; then the reciprocal theorem

$$K + G = K * G = G * K$$

holds; then the solution of (8.40) becomes

$$x(t) = f(t) - \int_0^t G(t, \xi) f(\xi) d\xi. \tag{8.41}$$

(4) *Integral equation of the first kind for the case when the order of the kernel is* α $(0 < \alpha < 1)$.
In the Volterra-type integral equation of the first kind:

$$\int_0^t K(t, \xi) x(\xi) d\xi = f(t), \tag{8.42}$$

when the order of the kernel is α $(0 < \alpha < 1)$, multiply it by $\overset{*}{1}^{1-\alpha}$ from the left and we find that

$$K_1 * x = \overset{*}{1}^{1-\alpha} * f. \tag{8.43}$$

From (8.36), (8.37) we get

$$K_1 = \overset{*}{1}^{1-\alpha} * K = \frac{1}{\Gamma(\alpha)\Gamma(1-\alpha)} \int_0^1 \frac{k(t + \eta(\xi - t), \xi)}{\eta^\alpha \eta^{1-\alpha}} d\eta. \tag{8.44}$$

Now take a partial integral of

$$\overset{*}{1}^{1-\alpha} * f = \int_0^t \frac{(t-\tau)^{-\alpha}}{\Gamma(1-\alpha)} f(\tau) d\tau,$$

and when $f(t)$ is differentiable, it becomes

$$\left[-f(\tau)\frac{(t-\tau)^{1-\alpha}}{\Gamma(2-\alpha)}\right]_0^t + \int_0^t \frac{(t-\tau)^{1-\alpha}}{\Gamma(2-\alpha)} f^{(1)}(\tau) d\tau.$$

Then we have

$$\overset{*}{1}^{1-\alpha} * f = f(0)\frac{t^{1-\alpha}}{\Gamma(2-\alpha)} + \int_0^t \frac{(t-\tau)^{1-\alpha}}{\Gamma(2-\alpha)} f^{(1)}(\tau) d\tau. \tag{8.45}$$

Therefore, the integral equation (8.42) becomes [V₁]:

$$\int_0^t K_1(t, \xi)x(\xi)d\xi = \left[f(0)t^{1-\alpha} + \int_0^t (t-\xi)^{1-\alpha} f^{(1)}(\xi)d\xi \right] \Gamma(2-\alpha). \tag{8.46}$$

As the kernel here is non-singular, this is solved by the method of § 2.5.

If we differentiate (8.42) in t, it becomes

$$x(t)K(t, t) + \int_0^t K_t(t, \xi)x(\xi)d\xi = f^{(1)}(t),$$

which is solved without difficulty when $K(t, t) \neq 0$.

8.4.1. Abel's integral equation

The Volterra-type integral equation of the first kind with singular kernel:

$$\int_0^t \frac{x(\xi)}{(t-\xi)^\alpha} d\xi = f(t) \qquad (0 < \alpha < 1), \tag{8.47}$$

is called *Abel's integral equation*, where the kernel is $\Gamma(1-\alpha)\overset{*}{1}{}^{1-\alpha}$ in (8.42). Therefore, express (8.46) as

$$\Gamma(1-\alpha)\overset{*}{1}{}^{1-\alpha} * x = f,$$

and multiply both sides by $\Gamma(\alpha)\overset{*}{1}{}^{\alpha}$, which results in

$$1 * x = \frac{\Gamma(\alpha)}{\Gamma(\alpha)\Gamma(1-\alpha)} \overset{*}{1}{}^{\alpha} * f.$$

Now from

$$\Gamma(\alpha) \cdot \Gamma(1-\alpha) = \frac{\pi}{\sin \alpha\pi}$$

we obtain

$$x(t) = \frac{\sin \alpha\pi}{\pi} \frac{d}{dt} \int_0^t \frac{f(\xi)}{(t-\xi)^{1-\alpha}} d\xi.$$

Taking the partial integral we also have

$$x(t) = \frac{\sin \alpha\pi}{\pi} \left\{ \frac{f(0)}{t^{1-\alpha}} + \int_0^t \frac{f^{(1)}(\xi)}{(t-\xi)^{1-\alpha}} d\xi \right\}.$$

In particular, when $\alpha = 1/2$, this becomes

$$x(t) = \frac{1}{\pi}\frac{d}{dt}\int_0^t \frac{f(\xi)}{(t-\xi)^{1/2}}\,d\xi,$$

since $\Gamma(1/2) = \sqrt{\pi}$.

8.4.2. *Sonine's integral equation*

In the Volterra-type integral equation of the first kind with a singular kernel:

$$K * x = f,$$

when the kernel K takes the form

$$K = \overset{*}{1}{}^{\alpha} * (E + a_1\overset{*}{1} + a_2\,\overset{*}{1}{}^2 + \cdots), \tag{8.48}$$

we call the integral equation *Sonine's integral equation*. Here a_1, a_2, ... are constants.

Now for *a* series in z, if we determine the values of b_1, b_2, ... so that

$$(1 + a_1z + a_2z^2 + \cdots)(1 + b_1z + b_2z^2 + \cdots) = 1,$$

the result of multiplication of both sides of (8.48) by $\overset{*}{1}{}^{1-\alpha} * (E + b_1\overset{*}{1} + b_2\overset{*}{1}{}^2 + \cdots)$ from the right is

$$1 * x = f * \overset{*}{1}{}^{1-\alpha} * (E + b_1\overset{*}{1} + b_2\overset{*}{1}{}^2 + \cdots).$$

Then put

$$\overset{*}{1}{}^{1-\alpha} * (E + b_1\overset{*}{1} + b_2\overset{*}{1}{}^2 + \cdots) = S,$$

and we get

$$1 * x = f * S;$$

therefore, we obtain

$$x(t) = \frac{d}{dt}\int_0^t S(t-\xi)f(\xi)d\xi, \tag{8.49}$$

or

$$x(t) = f(0)S(t) + \int_0^t f^{(1)}(\xi)S(t-\xi)d\xi. \tag{8.49a}$$

8.4.3. *Logarithmic kernel*

In the Volterra-type integral equation of the first kind:

$$K * x = f,$$

when

$$K(t, \xi) = \log(t - \xi) + c \tag{8.50}$$

(c is a constant), the integral equation is called a *logarithmic kernel.* kind of singular kernel becomes infinitely large when $t = \xi$.

To find the inverse element of a logarithmic kernel, suppose that we have the following convolution:

$$\overset{*}{1}^{\alpha} * \overset{*}{1}^{\beta} = \overset{*}{1}^{\alpha+\beta}; \tag{8.51}$$

that is,

$$\int_{\xi}^{t} \frac{(t-\tau)^{\alpha-1}}{\Gamma(\alpha)} \frac{(\tau-\xi)^{\beta-1}}{\Gamma(\beta)} d\tau = \frac{(t-\xi)^{\alpha+\beta-1}}{\Gamma(\alpha+\beta)}.$$

Differentiate both sides by α and integrate them by β (represent the integral variable as η), and then we get

$$\int_{\xi}^{t} \frac{\partial}{\partial\alpha} \frac{(t-\tau)^{\alpha-1}}{\Gamma(\alpha)} d\tau \int_{\beta}^{\infty} \frac{(\tau-\xi)^{\eta-1}}{\Gamma(\eta)} d\eta = -\frac{(t-\xi)^{\alpha+\beta-1}}{\Gamma(\alpha+\beta)}.$$

Putting $\alpha = \beta = 1$ in particular, we get

$$\int_{\xi}^{t} [\log(t-\tau) + C] d\tau \int_{0}^{\infty} \frac{(\tau-\xi)^{\beta}}{\Gamma(\beta+1)} d\beta = -(t-\xi),$$

where η is replaced by $\beta + 1$. In this equation, C serves as Euler's constant; that is,

$$C = -\Gamma^{(1)}(1) = 0.57721\ldots.$$

Then if

$$G_1(t-\xi) = \int_{0}^{\infty} \frac{(\tau-\xi)^{\beta}}{\Gamma(\beta+1)} d\beta,$$

we get

$$K * G_1 = -\overset{*}{1}^{2}.$$

Therefore, on multiplying both sides of (8.50) by G_1 from the right we have

$$-\overset{*}{1}^{2} * x = G_1 * f.$$

Then since

$$x(t) = -\frac{d^2}{dt^2} \int_{0}^{t} G_1(t-\xi) f(\xi) d\xi,$$

or $f(0) = 0$, we obtain

$$x(t) = -f^{(1)}(0)G_1(t-\xi) - \int_0^t G_1(t-\xi)f^{(2)}(\xi)d\xi. \qquad (8.52)$$

8.5. Special method of solution

We can find the solution using the convolution theorem in the case when the regular part of an integral equation is of the convolution type of functional transformations. In particular, in the case of the singular kernel, if the transform is expressed simply, we can apply the theorem to every kind of integral equation in this chapter.

8.5.1. The use of the Laplace transform (an application of operational calculus)

In the case of the singular kernel $K(t, \xi) = (t-\xi)^{-\alpha}$ $(0 < \alpha < 1)$, the image function of $K(t)$ becomes

$$\mathrm{K}(s) = \Gamma(1-\alpha)s^{\alpha-1}$$

(Churchill (1941), formula(7)). Then in the image space, Abel's integral equation is

$$\Gamma(1-\alpha)s^{\alpha-1}X(s) = F(s).$$

Therefore, we have

$$X(s) = \frac{\Gamma(\alpha)}{\Gamma(\alpha)\Gamma(1-\alpha)}\frac{1}{s}s^{\alpha}F(s),$$

which produces

$$x(t) = \frac{\sin \alpha\pi}{\pi}\frac{d}{dt}\int_0^t \frac{f(\xi)}{(t-\xi)^{1-\alpha}}d\xi.$$

As in the case of Sonine's integral equation, the same method is applied in the following calculation. That is, when

$$K(t) = t^{-\alpha}(1 + a_1t + a_2t^2 + \cdots + a_nt^n) \qquad 0 < \alpha < 1$$

we have

$$K(s) = \Gamma(1-\alpha)s^{\alpha-n-1}[s^n + (1-\alpha)a_1s^{n-1} + (1-\alpha)(2-\alpha)a_2s^{n-2} + \cdots + (1-\alpha)(2-\alpha)\cdots(n-\alpha)a_n]$$

in the image space. Then, expressing the solution as

$$x = \frac{\sin \alpha\pi}{\pi} F_1 * f,$$

we get $F_1(t)$ in the form of

$$F_1(t) \equiv t^{\alpha-1} + \frac{z_1{}^{n-\alpha}}{A^{(1)}(z_1)} \Gamma_\alpha(z_1 t) + \frac{z_2{}^{n-\alpha}}{A^{(1)}(z_2)} \Gamma_\alpha(z_2 t) + \dots$$

$$+ \frac{z_n{}^{n-\alpha}}{A^{(1)}(z_n)} \Gamma_\alpha(z_n t),$$

where

$$A(z) \equiv z^n + (1-\alpha)a_1 z^{n-1} + (1-\alpha)(2-\alpha)a_2 z^{n-2} + \cdots$$
$$+ (1-\alpha)(2-\alpha)\cdots(n-\alpha)a_n.$$

Here $A^{(1)}(z_i)$ is the value at $z = z_i$ of its derivative in z, and $z_1, z_2, \dots, z_n$ are the simple roots of the algebraic equation $A(z) = 0$. Further, Γ_α is an incomplete gamma function:

$$\Gamma_\alpha(t) = \int_0^t \xi^{\alpha-1} e^{(-\xi)}\, d\xi$$

(see Question 9).

Next, in the case of the logarithmic kernel, we have

$$K(s) = -\frac{1}{s}(\log s + C)$$

(Churchill(1941), formula (95)), then we have

$$X = -\frac{s}{\log s + C} \cdot F = -s^2 \left\{ \frac{1}{s(\log s + C)} \cdot F \right\}.$$

Put

$$G_1 = \frac{1}{s(\log s + C)},$$

which makes

$$G_1 = \int_0^\infty \frac{1}{s^{\beta+1}} d\beta;$$

while since $t^\beta/\Gamma(\beta+1) \supset 1/s^{\beta+1}$, we have

$$G_1(t) = \int_0^\infty \frac{t^\beta}{\Gamma(\beta+1)} d\beta,$$

and this agrees with the result obtained in § 8.4.

8.6. Method of approximate solution of the integral equation with a singular kernel

In the case of the singular kernel, the method of simultaneous equations by an approximate calculation using representative points is inadequate as its coefficients may diverge, and its precision becomes very low. In such a case, we separate the integral in the neighbourhood of the singular point of the kernel and applying a sort of mean-value theorem to the interval, we introduce simultaneous equations. That is, in the case of the integral equation

$$x(t) - \lambda \int_a^b K(t, \xi) x(\xi) d\xi = f(t),$$

when the kernel $K(t, \xi)$ becomes infinitely large at $\xi = c \quad (a < c < b)$, let

$$x(t) - \lambda \int_a^b K_1(t, \xi) d\xi - \lambda x(c') \int_{c-\varepsilon}^{c+\varepsilon} K(t, \xi) d\xi = f(t),$$
$$c - \varepsilon < c' < c + \varepsilon,$$

where we put

$$K_1(t, \xi) = \begin{cases} K(t, \xi) & (\xi < c - \varepsilon,\ \xi > c + \varepsilon), \\ 0 & (c - \varepsilon < \xi < c + \varepsilon). \end{cases}$$

Here we approximate the integral in the second term on the left-hand side by the method of representative points. Thus we can introduce simultaneous equations with $n + 1$ unknown quantities as the values of the unknown functions at $n + 1$ coordinate points of n representative points and $t = c'$.

In addition to this, we may well consider carrying out a numerical calculation by substituting some proper approximate expression for the singularity of the kernel at $x = c$, or else solving by approximating the kernel so that we may use the special method of solution in § 8.5 for the case of convolution type. It is desirable, if possible, to measure out the method of solution with an approximate kernel, since the numerical method of solution is hard to deal with, and we can only expect low precision from it.

Exercises

1. Solve the following integral equations:

$$\text{(a)} \int_0^t \frac{x(\xi)}{(t-\xi)^{\frac{1}{2}}} d\xi = t,$$

$$\text{(b)} \int_0^t \frac{x(\xi)}{(t-\xi)^{\frac{1}{3}}} d\xi = t + \sqrt{t},$$

$$\text{(c)} \; x(t) - \int_0^t \frac{x(\xi)}{(t-\xi)^{\frac{1}{2}}} d\xi = t \sin t,$$

$$\text{(d)} \left. \begin{aligned} x_1(t) + x_2(t) - \int_0^t \frac{x_1(\xi)}{(t-\xi)^{\frac{1}{2}}} d\xi &= t, \\ x_1(t) + 2x_2(t) - \int_0^t \frac{x_2(\xi)}{(t-\xi)^{\frac{1}{3}}} d\xi &= t \sin t \end{aligned} \right\}$$

$$\text{(e)} \; x(t) - \int_0^t K(t-\xi)x(\xi)d\xi = e^{-t}\sin t,$$

$$K(t) = (1 + 2t + 3t^2)/t^{\frac{1}{3}}.$$

2. Solve the following integral equation by the method of successive approximations:

$$\text{(a)} \; x(t) - \int_0^1 \frac{x(\xi)}{(t-\xi)^{\frac{1}{2}}} d\xi = \sqrt{t} + (t-1)^{\frac{1}{2}},$$

$$\text{(b)} \; x(t) - \int_0^1 \frac{x(\xi)}{t\sqrt{\xi}} d\xi = 4t^{-1} + \sqrt{t},$$

$$\text{(c)} \; x(t) - \int_0^1 \frac{\xi x(\xi)}{t^3} d\xi = 2t^{-3}.$$

3. Solve the following integral equation:

$$x(t_1, t_2) - \int_0^1 d\xi_1 \int_0^1 d\xi_2 \frac{x(t_1, t_2)}{((t_1 - \xi_1)^2 + (t_2 - \xi_2{}^2))^{\frac{1}{2}}} = 1.$$

4. Solve the following integral equation of the second kind:

$$(t-1)x(t) - \int_0^2 (t\xi - t - \xi + 1)x(\xi)d\xi = t^2 - 1.$$

5. Solve the following integral equations:

$$\text{(a)}\ x(t) - \int_0^1 \frac{x(\xi)}{(t-\xi+1)^{\frac{1}{2}}} d\xi = e^t,$$

$$\text{(b)}\ x(t) - \int_0^1 \frac{x(\xi)}{(t-\xi)^{\frac{1}{3}}} d\xi = t,$$

$$\text{(c)}\ x(t) - \int_0^1 \frac{(t+\xi)x(\xi)}{(t-2\xi)^{\frac{1}{2}}} d\xi = \sqrt{t} + 1.$$

6. Solve the following integral equation:

$$x(t) - \int_0^t K(t, \xi)x(\xi)d\xi = f(t),$$

where $K(t, \xi) = \log(t-\xi) + C$.

7. Solve **1.** (a), (b) and **2.** (a), (b) by the numerical method of solution, and compare the results with exact solutions.

8. Solve the following integral equations:

$$\text{(a)}\ \int_0^t \log(t-\xi)x(\xi)d\xi = t,$$

$$\text{(b)}\ \int_0^t K(t, \xi)x(\xi)d\xi = t,$$

$$K(t, \xi) = \frac{a + b(t-\xi)}{(t-\xi)^{1/2}},$$

$$\text{(c)}\ x(t) - \int_0^t K(t, \xi)x(\xi)d\xi = e^t,$$

with the same kernel as in (b)

9. In the method of solution of Sonine's integral equation in § 8.5, what will become of the solution when $A(z) = 0$ has multiple roots? [Kondo]

10. Show the solutions of the following integral equations if their regular parts take Cauchy's principal value:

$$\text{(a)}\ x(t) - \int_{-1}^1 \frac{x(\xi)}{\xi} d\xi = 1,$$

$$\text{(b)}\ x(t) - \int_{-1}^1 \frac{x(\xi)}{|t-\xi|} d\xi = 1.$$

11. Solve the integral equation

$$\int_t^\infty \frac{x(\xi)}{\sqrt{(\xi - t)}} d\xi = f(t)$$

[Reduce to the Abel's equation by putting

$$\xi = 1/\eta \text{ and } t = 1/\tau \,]$$

12. Show that the integro differential equation of a lifting line yields

$$x(t) - \lambda \int_{-b/2}^{b/2} \frac{x^{(1)}(\xi)}{t - \xi}\, d\xi = f(t)$$

where we have

$$\lambda = -1/8\pi \, a_0 c, \quad f(t) = 1/2 a_0 c v_0 \alpha_0.$$

In the study of three dimensional wing theory, L. Prandtle (1875–1953) substituted a wing by a line of circulation $x(t)$, called a lifting line. The lifting line is considered in $I(-b/2, b/2)$. Consider a rectangular coordinate system t, y, z taken paralled to spanwise, flow direction and lift respectively.

When an aerofoil of cordlength c is placed in a plane uniform flow v_0 at an angle α, the circulation x is given as

$$x = 1/2 a v_0 c \sin \alpha \simeq 1/2 a v_0 c \alpha$$

where a is the inclination of lift curve.

Since vortex can not disappear in an ideal (non viscous) flow a trailing vortex of $x^{(1)}(\xi)$ stems from the lifting line from the point $t = \xi$. Consequently, in the flow behind the wing, a sheet of trailing vortex is extended. By the trailing vortex of $x^{(1)}(\xi)$,an induced velocity $w(t)$ is introduced and we have

$$w(t) = -\, 1/2\pi \int_{-b/2}^{b/2} \frac{x^{(1)}(\xi)}{\xi - t}\, d\xi.$$

Consequently the effective angle of attack becomes

$$\alpha = \alpha_0 + w/v_0$$

where $\alpha_0(t)$ is determined by the geometrical configulation of the wing.

Finally we have

$$x(t) = 1/2 a c(\alpha_0 v_0 + w).$$

13. Solve the integral equation

$$x(t) - \lambda \int_{-b/2}^{b/2} \frac{x^{(1)}(\xi)}{t - \xi}\, d\xi = f(t)$$

Putting $x_0(t) = f(t)$, the equation can be solved successively.

References

Churchill, R.V. (1941) *Fourier Series and Boundary Value Problems* (McGraw, New York) pp.37–40.

Giraud, G. (1934) Equations à intégrales principes, étude suivie d'une application. *Ann. d l' Ecole Normale Superieur* **51**, 251–372.

Hille, E. and Tamarkin, J.D. (1930) On the theory of integral equations. *Annals of Mathematics* **31**.

Picard, E. (1911) Sur les équations intégrales de troisième espèce. *Ann. École supér.* 3rd series, **28**.

Poincaré, E. (1910) Remarque diverses sur l'équation de Fredholm. *Acta Mathematica* **33**, 57–89.

Whittaker, E.T. and Watson G.N. (1953) *A Course of Modern Analysis*, Cambridge, p. 341.

9 Nonlinear integral equations

In general, to solve an integral equation means to take the functional $F[x(\xi)_a^b, t] = f(t)$ of an unknown function $x(t)$ and to find its reverse functional $x(t) = F^{-1}[f(\xi_a^b), t]$. From Chapter 1 to the previous chapter we have treated F as a linear functional, while in this chapter we shall treat the case when it is nonlinear. In this case we may encounter many difficulties in studying the existence and uniqueness of the solution, and the method of solution is not easily handled. Although studies have been carried out concerning functionals of special forms, we still do not have a complete theory on these, such as we have in the case of the linear functional. In some special cases, the theories have been worked out, however the method is beyond the scope of this text, and therefore we shall present the results without going into detail of the theory. We might need another volume to explain the theory of nonlinear integral equations (See Muskhelishvill).

In § 9.5 we shall consider linear integro-differential equations.

9.1. Classification of the types of nonlinear integral equations

Nonlinear integral equations are classified into the cases where only the regular part is nonlinear:

$$x(t) - \int_a^b H[t, \xi, x(\xi)]d\xi = f(t), \qquad [F_2^{NR}]$$

where only the irregular part is nonlinear:

$$G[x(t)] - \int_a^b K(t, \xi)x(\xi)d\xi = f(t), \qquad [F_2^{NE}]$$

and where both the regular and the irregular parts are non-linear:

$$G[x(t)] - \int_a^b H[t, \xi, x(\xi)]d\xi = f(t). \qquad [F_2^{NG}]$$

Here G denotes a function of x; H a function of t, ξ, x; and $f(t)$

a known function. We call these three *regular*, *exceptional,* and *general nonlinear integral equations.* They are all of Fredholm type, have an exceptional part, are called the second kind, and are denoted by [F_2^{NR}], [F_2^{NE}], and [F_2^{NG}], since the nonlinearity of [F_2] stems from the regular or the exceptional part or both of them respectively.

As compared with these, we have the integral equation of the first kind:

$$\int_a^b H[t, \xi, x(\xi)]d\xi = f(t), \qquad [F_1^{NR}]$$

and the homogeneous integral equations:

$$x(t) = \int_a^b H[t, \xi, x(\xi)]d\xi, \qquad [F_0^{NR}]$$

$$G[x(t)] = \int_a^b K(t, \xi)x(\xi)d\xi, \qquad [F_0^{NE}]$$

$$G[x(t)] = \int_a^b H[t, \xi, x(\xi)]d\xi. \qquad [F_0^{NG}]$$

Besides these there are integral equations of the third kind, nonlinear integro-differential equations, associate integral equations and so on, which we will omit here as they are the same as those explained in § 1.3.

In addition to these, when $H \equiv 0$ or $K \equiv 0$ in $\xi > t$, we have the Volterra-type equation:

$$x(t) - \int_a^t H[t, \xi, x(\xi)]d\xi = f(t). \qquad [V_2^{NR}]$$

Also, we have [V_2^{NE}], [V_2^{NG}], [V_1^{NR}], [V_0^{NR}], [V_0^{NE}], and [V_0^{NG}] associated integral equations, and so on.

9.2. Method of solution of regular nonlinear integral equations

We can make use of the method of successive approximation. We have the following theorem corresponding to [Theorem 2.1].

THEOREM 9.1 [F_2^{NR}]

In the Fredholm-type regular nonlinear integral equation of the second kind:

$$x(t) - \int_a^b H[t, \xi, x(\xi)]d\xi = f(t), \qquad [F_2^{NR}]$$

when

(i) *the function* $H[t, \xi, x(\xi)]$ *is a single-valued function defined by* t, ξ *in the domain* S *and by* $x(\xi)$, *which satisfies the condition*:

$$A - m \leqq x(\xi) \leqq A + m \qquad (9.1)$$

(*where* A, m *are constants*) *and is continuous in* t *in* I;

(ii) *there exist constants* M, N *which hold for*

$$|H(t, \xi, x_i)| < M, \qquad (9.2)$$

$$|H(t, \xi, x_i) - H(t, \xi, x_j)| < N|x_i - x_j| \qquad (9.3)$$

for the arbitrary functions $x_i(\xi)$, $x_j(\xi)$ *that satisfy condition* (9.1);

(iii) $f(t)$ *is continuous in* I *and there exists* ε *which holds for the following expressions*:

$$A - \varepsilon \leqq f(t) \leqq A + \varepsilon, \qquad (9.4)$$

$$\varepsilon + M(b - a) < m; \qquad (9.5)$$

the sufficient condition for the existence of the unique continuous solution is

$$N(b - a) < 1. \qquad (9.6)$$

Then the solution is the limit of the successive approximate solutions

$$x_0(t) = f(t),$$

$$x_1(t) = f(t) + \int_a^b H(t, \xi, x_0(\xi))d\xi,$$

$$x_2(t) = f(t) + \int_a^b H(t, \xi, x_1(\xi))d\xi,$$

$$\dots,$$

$$x_n(t) = f(t) + \int_a^b H(t, \xi, x_{n-1}(\xi))d\xi, \qquad (9.7)$$

$$\dots.$$

Proof. (i) *Convergence.* If we have

$$x(t) = x_0 + (x_1 - x_0) + (x_2 - x_1) + \cdots + (x_n - x_{n-1}) + \cdots, \qquad (9.8)$$

from (9.4), (9.5) all the approximate solutions satisfy the condition (9.1) in I, then from (9.2), (9.3) we get

$$|x_1 - x_0| < M(b-a),$$
$$|x_2 - x_1| < MN(b-a)^2,$$
$$\ldots,$$
$$|x_n - x_{n-1}| < MN^{n-1}(b-a)^n,$$
$$\ldots.$$

Therefore, (9.8) is absolutely uniformly convergent with the condition (9.6). Hence (9.8) is continuous in I. It is obvious that this kind of function is the solution of [F_2^{NR}].

(ii) *Uniqueness.* Suppose that there exists a continuous solution $x_1(t)$ which satisfies the condition (9.1); then we have

$$\begin{aligned}|x - x_1| &= \int_a^b \{H[t, \xi, x(\xi)] - H[t, \xi, x_1(\xi)]\}d\xi \\ &< N\int_a^b |x - x_1|d\xi.\end{aligned} \tag{9.9}$$

If

$$|x - x_1| < 2m$$

in I, then we get

$$|x - x_1| < 2mN(b-a).$$

Putting this again into the right-hand side of (9.9):

$$|x - x_1| < 2mN^2(b-a)^2.$$

Repeating this process n times, we get

$$|x - x_1| < 2mN^n(b-a)^n.$$

Hence, when condition (9.6) applies we can prove that

$$x \equiv x_1.$$

THEOREM 9.1a [F_2^{NR}]

In the Fredholm-type regular nonlinear integral equation of the second kind:

$$x(t) - \lambda\int_a^b H(t, \xi, x(\xi))d\xi = f(t), \qquad [F_2^{NR}]$$

when all three conditions of Theorem 9.1 hold, the sufficient condition for the existence of the unique continuous solution is that λ satisfies

$$|\lambda| < \frac{m-\varepsilon}{M(b-a)} \quad \text{and} \quad |\lambda| < \frac{1}{N(b-a)} \tag{9.10}$$

at the same time. The solution is the limit of the successive approximate solutions:

$$x_n = f + \lambda \int_a^b H(t, \xi, x_{n-1}) d\xi.$$

The following theorem, which is given without proof, is employed for Volterra-type integral equations. This time, condition (9.6) is not necessary.

THEOREM 9.2 [V_2^{NR}]

In the Volterra-type regular nonlinear integral equation of the second kind:

$$x(t) - \int_a^t H[t, \xi, x(\xi)] d\xi = f(t), \qquad [V_2^{NR}]$$

when

(i) *the function $H[t, \xi, x(\xi)]$ is a single-valued function defined by t, ξ in the domain T and by $x(\xi)$, which satisfies the condition*

$$A - m \leqq x(\xi) \leqq A + m \tag{9.1}$$

(where A, m are constants) and is continuous in I;

(ii) *there exist constants M, N which hold for*

$$|H(t, \xi, x_i)| < M, \tag{9.2}$$

$$|H(t, \xi, x_i) - H(t, \xi, x_j)| < N|x_i - x_j| \tag{9.3}$$

for arbitrary functions $x_i(\xi)$, $x_j(\xi)$ that satisfy the condition (1);

(iii) *$f(t)$ is continuous in $t(a, t_0)$, $(t_0 < b)$ and there exists $\varepsilon(t_0)$ which holds for*

$$A - \varepsilon(t_0) \leqq f(t) \leqq A + \varepsilon(t_0), \tag{9.4a}$$

$$\varepsilon(t_0) + M(t_0 - a) < m; \tag{9.5a}$$

there exists a unique continuous solution, which is the limit of the successive approximate solutions:

$$x_0(t) = f(t),$$

$$x_1(t) = f(t) + \int_a^t H(t, \xi, x_0(\xi)) d\xi,$$

$$x_2(t) = f(t) + \int_a^t H(t, \xi, x_1(\xi))d\xi,$$

$$\ldots,$$

$$x_n(t) = f(t) + \int_a^t H(t, \xi, x_{n-1}(\xi))d\xi$$

$$\ldots. \tag{9.7a}$$

Example 9.1 [F_2^{NR}]

The approximate solution of [F_2^{NR}]:

$$x(t) - \int_0^{0.1} t\xi[1 + x^2(\xi)]d\xi = t + 1,$$

is

$$x(t) = 1.13t + 1.$$

We examine the conditions of Theorem 9.1.

(i) In $I(0, 0.1)$ we have

$$1.056 - 0.057 \leqq x(\xi) \leqq 1.056 + 0.057.$$

Then

$$A = 0.056, \ m = 0.057.$$

(ii) For the functions which satisfy the above condition we get

$$|t\xi(1 + x^2)| < \frac{2}{100},$$

$$t|\xi(1 + x_1^2) - t\xi(1 + x_2^2)|$$
$$= |t\xi|\,|(x_1 + x_2)(x_1 - x_2)| < \frac{3}{100}|x_1 - x_2|.$$

which give

$$M = \frac{2}{100}, \quad N = \frac{3}{100}.$$

(iii) $t + 1$ is continuous in I and we have

$$1.056 - 0.056 \leqq f(t) \leqq 1.056 + 0.056,$$

$$0.056 + \frac{2}{100} \times \frac{1}{10} < 0.057.$$

Then we may take $\varepsilon = 0.056$.

These results surely satisfy the condition $N(b - a) < 1$. Hence, the unique continuous solution, which is obtained by the method of successive approximations, exists.

That is, we have

$$x_0 = t + 1,$$
$$x_1 = 1.01t + 1$$
$$\ldots,$$
$$\ldots.$$

The exact solution is

$$x(t) = \frac{9920 - \sqrt{98272000}}{6}\, t + 1 = 1.13t + 1.$$

Thus Theorems 9.1 and 9.2 ensure that when $f(t)$ is within a certain interval $(A - \varepsilon, A + \varepsilon)$ in I, if we take a breadth m a little wider than ε, for a solution in the interval $(A - m, A + m)$, successive approximate solutions starting from $x_0 = f(t)$ converge, and the solution is unique. But, for the original integral equation, it is not clear whether there exist any solutions which do not satisfy the restriction (9.1).

For example,

In Example 9.1,there exists another continuous solution:

$$x(t) = \frac{9920 + \sqrt{28272000}}{6}\, t + 1.$$

Since this does not satisfy the condition

$$1.056 - 0.057 < x(t) < 1.056 + 0.057,$$

it is a solution that cannot be found by the above method of successive approximations.

When we treat nonlinear integral equations, we find more difficulties in the integral equations of the first kind than in those of the second kind. For example, in

$$\int_a^t H[t, \xi, x(\xi)]d\xi = f(t), \qquad [V_1^{NR}]$$

if we differentiate both sides with respect to t this becomes

$$H[t, t, x(t)] + \int_a^t H_t[t, \xi, x(\xi)]d\xi = f^{(1)}(t),$$

which results in $[V_2^{NG}]$.

In the case of exceptional nonlinear integral equations and general equations we may make use of the method of successive approximations if they take the form of the second kind. For example in $[F_2^{NG}]$, the asymptotic formula to determine the approximate solution of the nth order is read as

$$G[x_n(t)] = f(t) + \int_a^b H[t, \xi, x_{n-1}(\xi)]d\xi,$$

where the left-hand side is not linear in $x_n(t)$; then $x_n(t)$ is in general not unique. In addition, different non-unique solutions are obtained by the method of approximations, and this makes the problem become complex. This is the same with $[V_2^{NG}]$.

The method explained above applies to the cases containing multiple integrals and simultaneous equations which we studied in Chapter 7, among others, but we omit them as they are too complicated.

9.2.1. *Nonlinear integral equations and nonlinear differential equations*

In § 5.1 we found that the Volterra-type integral equation is similar to a linear differential equation, and in § 5.2 it became clear that the Fredholm-type integral equation is similar to the boundary-value problem of some sort of linear partial differential equation. We shall find that the same rules apply in the nonlinear case.

For example, when the nonlinear differential equation

$$dx/dt = H(x, t)$$

is given under the initial condition $x(a) = x_a$, the solution $x(t)$ is

$$x(t) = x_a + \int_a^t H(x(\xi), \xi)d\xi.$$

This is a special case of the Volterra-type nonlinear integral equation $[V_2^{NR}]$.

In the nonlinear partial differential equation, for example,

$$\frac{\partial^2 u}{\partial x \partial y} = H(u, x, y),$$

given the boundary condition

$$u(x, y) = h(y),\ u(x, b) = k(x) \quad \text{with} \quad h(b) = k(a),$$

the solution becomes

$$u(x, y) = \int_a^x d\xi \int_b^y H(u(\xi, \eta), \xi, \eta)d\eta + h(y) + k(x) - k(a).$$

This makes a special case when the regular part of the Volterra-type nonlinear integral equation is expressed in double integrals. The boundary-value problem shown above is called Riemann's problem.

9.3. Special method of solution of nonlinear integral equations

In Chapter 3 we gave special analytic methods of solution for linear integral equations. Many of those can be applied to nonlinear integral equations. Here we only show a few examples in which the calculation is carried out with ease. We may learn through the examples the particularities of nonlinear integral equations.

Example 9.2 [F_2^{NR}]

For

$$x(t) - 60\int_0^1 t\xi x(\xi)^2 d\xi = 1 + 20t - t^2, \qquad [F_2^{NR}]$$

let

$$x(t) = a + bt + ct^2$$

be the solution; then it is obvious that $a = 1$, $c = -1$. Substitute $x(t) = 1 + bt - t^2$ into the equation, and as the equation to determine b we obtain

$$b - 60\left[\frac{1}{2} + \frac{2}{3}b + \frac{1}{4}b - \frac{1}{2} - \frac{2}{5}b + \frac{1}{6}\right] = 20.$$

From this we get $b = -1$. Hence the solution is

$$x(t) = 1 - t - t^2.$$

This is continuous in $t(0, 1)$.

Example 9.3 [F_2^{NR}]

For

$$x(t) - \lambda\int_0^1 t\xi[1 + x^2(\xi)]d\xi = t + 1, \qquad [F_2^{NR}]$$

let

$$x(t) = at + 1.$$

Substitute this into the equation and compare the coefficients on both sides; then we get a bilinear equation in a:

$$\frac{\lambda}{4}a^2 + \left(\frac{2}{3}\lambda - 1\right)a + (\lambda + 1) = 0.$$

Now a has real roots when

$$\lambda(\lambda + 1) - \left(\frac{2}{3}\lambda - 1\right)^2 \leqq 0,$$

and it becomes approximately

$$-4.6 \leqq \lambda \leqq 0.4.$$

Generally there exist two different solutions. For example when $\lambda = -1$, we have the solutions

$$x(t) = 1 \quad \text{and} \quad x(t) = -\frac{20}{3}t + 1.$$

Example 9.4 $[F_2^{NE}]$

For

$$x^2(t) - \int_0^1 t\xi\, x(\xi)d\xi = 4 + 10t + 9t^2, \qquad [F_2^{NE}]$$

let

$$x(t) = at + b$$

be the solution; then the equation becomes

$$b^2 + 2abt + a^2t^2 - \left(\frac{b}{2} + \frac{a}{3}\right)t = 4 + 10t + 9t^2.$$

Comparison of the coefficients gives

$$b^2 = 4, \quad 2ab - \left(\frac{b}{2} + \frac{a}{3}\right) = 10, \quad a^2 = 9,$$

but among $b = \pm 2$, $a = \pm 3$ it is only $a = 3$, $b = 2$ that satisfy the second expression; then the solution is

$$x(t) = 3t + 2.$$

This is continuous in $t(0, 1)$.

Example 9.5 $[F_2^{NE}]$

For

$$x^2(t) - \lambda \int_0^1 t\xi\, x(\xi)d\xi = t^2 + 1, \qquad [F_2^{NE}]$$

let

$$x(t) = at + b$$

be the solution, and compare the coefficients; then we get

$$a^2 = 1, \qquad b^2 = 1,$$

$$\lambda = 2ab \Big/ \left(\frac{a}{3} + \frac{b}{2}\right),$$

which makes

$$\begin{aligned} x(t) &= t - 1 && \text{when } \lambda = 12, \\ x(t) &= t + 1 && \text{when } \lambda = 12/5, \\ x(t) &= -t - 1 && \text{when } \lambda = -12/5, \\ x(t) &= -t + 1 && \text{when } \lambda = -12. \end{aligned}$$

Example 9.6 $[F_2^{NG}]$

$$x^2(t) + \int_0^2 x^3(\xi)d\xi = t^2 - 2t - 1 \qquad [F_2^{NG}]$$

has the solution:

$$x(t) = t - 1.$$

Example 9.7 $[F_0^{NE}]$

$$x^2(t) = \frac{5}{2} \int_0^1 t\xi\, x(\xi)d\xi \qquad [F_0^{NE}]$$

has the continuous solution:

$$x(t) = \sqrt{t}.$$

Example 9.8 $[F_1^{NR}]$

$$\int_0^\pi t\xi\, x^2(\xi)d\xi = \frac{\pi^2}{4} t \qquad [F_1^{NR}]$$

has the continuous solution:

$$x(t) = \pm \sin t.$$

Example 9.9 [V_2^{NR}]

$$x(t) - \int_0^t (t\xi + 1)x^2(\xi)d\xi = 1 - t - \frac{t^2}{2} \qquad [V_2^{NR}]$$

has the continuous solution:

$$x(t) = 1.$$

When $f(t) = 0$, there exists no solution but

$$x(t) = 0.$$

Example 9.10 [V_2^{NE}]

$$x^2(t) - \int_0^t (t - \xi)x(\xi)d\xi = -\frac{t^2}{2} + \frac{t^1}{2} + 3t + 1 \qquad [V_2^{NE}]$$

has the continuous solution:

$$x(t) = t + 1.$$

9.3.1. Numerical method of solution of nonlinear integral equations

We may apply various kinds of numerical methods of solution that were covered in Chapter 4, but they cause trouble in treatment when a solution is not unique.

Example 9.11 [V_2^{NR}]

$$x(t) - \int_0^t (t\xi + 1)x^2(\xi)d\xi = -\frac{1}{4}t^5 - \frac{2}{3}t^4 - \frac{5}{6}t^3 - t^2 + 1 \qquad [V_2^{NR}]$$

has the exact solution:

$$x(t) = t + 1.$$

Let $\Delta t = 0.1$ be the length of a subdivision and find the numerical solution successively using Newton's integral formula. Write this as

$$x(0) = x_0,\ x(0.1) = x_1,\ x(0.2) = x_2,\ x(0.3) = x_3,\ \ldots,$$

and from the given equation we obtain

$$x_0 = 1.0000,$$

provided $t = 0$.

Next, if we put $t = 0.1$, we get

$$x_1 - \left(\frac{1}{2} \times 1.00 \times x_0^2 + \frac{1}{2} \times 1.01 \times x_1^2\right) \times 0.1 = 0.9891.$$

Then solving the quadratic equation in x_1 we find that the solution is either

$$x_1 = 1.1010 \quad \text{or} \quad 18.4020.$$

If $t = 0.2$, we have

$$x_2 - \left(\frac{1}{6} \times 1.00 \times x_0^2 + \frac{4}{6} \times 1.02 \times x_1^2 + \frac{1}{6} \times 1.04 \times x_2^2\right)$$
$$\times 0.2 = 0.9521.$$

Then when $x_1 - 1.1010$, we get either

$$x_2 = 1.2019 \quad \text{or} \quad 27.6442;$$

and when $x_1 = 18.4020$, we get $x_2 - 15 \pm 11i$, then we drop $x_1 =$ 18.4020.

When $t = 0.3$, we have

$$x_3 - \frac{1}{8}(1 \times 1.00 \times 1^2 + 3 \times 1.03 \times x_1^2 + 3 \times 1.06 \times x_2^2$$
$$+ 1.09 \times x_3^2) = 0.8815,$$

and when $x = 1.2019$, we get either

$$x_2 = 1.2722 \quad \text{or} \quad 23.1926.$$

When $x_2 = 27.6442$, x_3 becomes a complex number; therefore we drop it. Proceeding in a similar way, we can find the approximate solution successively, as Table 9.1.

TABLE 9.1. Precision of the numerical solution of a nonlinear integral equation

t	$x(t)$ Approximate solution	Exact solution	Error	Relative error
0.0	1.0000	1.0000	0.0000	0.00%
0.1	1.1010	1.1000	0.0010	0.09
0.2	1.2019	1.2000	0.0019	0.16
0.3	1.2722	1.3000	− 0.0278	2.14
0.4	1.3875	1.4000	− 0.0125	0.89
0.5	1.4844	1.5000	− 0.0156	0.10

9.3.2. Method of solution by Laplace transformation

Operational calculus, which employs the Laplace transform, has linear characteristics and we cannot apply this method to find the exact solution of a nonlinear problem. Using operational calculus in the approximation method, however, allows an easy calculation. Moreover, we can calculate each step of the successive approximations separately, and as a result we can avoid mistakes in calculation. This method is only used when a function has a kernel of the form $K(t-\xi)$.

For example, in the regular nonlinear integral equation

$$x(t) - \int_0^t K(t-\xi)[x(\xi) + \alpha x^2(\xi)]d\xi = f(t), \tag{9.11}$$

when we express the solution in the power series of α,

$$x(t) = x_0(t) + \alpha x_1(t) + \alpha^2 x_2(t) + \cdots, \tag{9.12}$$

in the image space, (9.11) becomes

$$\begin{aligned}(X_0 + \alpha X_1 + \alpha^2 X_2 + \cdots) - \boldsymbol{K}(X_0 + \alpha X_1 + \alpha^2 X_2 + \cdots \\ + \alpha \mathscr{L}[x_0 + \alpha x_1 + \alpha^2 x_2{}^2 + \cdots]^2) = F\end{aligned} \tag{9.13}$$

(see § 3.5 above). If $\mathscr{L}[\ \ ]$ denotes the Laplace transform, comparing the coefficients of α, we have

$$\begin{aligned}
&(E - \boldsymbol{K})X_0 = F,\\
&(E - \boldsymbol{K})X_1 = \boldsymbol{K}\mathscr{L}[x_0{}^2],\\
&(E - \boldsymbol{K})X_2 = \boldsymbol{K}\mathscr{L}[3x_0{}^2 x_1],\\
&(E - \boldsymbol{K})X_3 = \boldsymbol{K}\mathscr{L}[3x_0{}^2 x_2 + 3x_0 x_1{}^2],\\
&\cdots\cdot
\end{aligned}$$

Suppose next that the reciprocal function G of K exists, then we get

$$\begin{aligned}
&X_0 = (E - \boldsymbol{G})F,\\
&X_1 = (E - \boldsymbol{G})\boldsymbol{K}\mathscr{L}[x_0{}^3] = -\boldsymbol{G}\cdot\mathscr{L}[x_0{}^3],\\
&X_2 = -\boldsymbol{G}\cdot\mathscr{L}[3x_0{}^2 x_1],\\
&X_3 = -\boldsymbol{G}\cdot\mathscr{L}[3x_0{}^2 x_2 + 3x_0 x_1{}^2],\\
&\cdots\cdot
\end{aligned}$$

Therefore, when $G(t)$ is found we may calculate $x_0(t)$, $x_1(t)$, $x_2(t)$, ... successively. That is, we get

$$\left.\begin{aligned} x_0(t) &= f(t) - \int_0^t G(t-\xi)f(\xi)d\xi, \\ x_1(t) &= -\int_0^t G(t-\xi)x_0^3(\xi)d\xi, \\ x_2(t) &= -3\int_0^t G(t-\xi)x_0^2(\xi)x_1(\xi)d\xi, \\ x_3(t) &= -3\int_0^t G(t-\xi)x_0^2(\xi)x_2(\xi)d\xi \\ &\quad -3\int_0^t G(t-\xi)x_0(\xi)x_1^2(\xi)d\xi, \\ &\cdots\cdot \end{aligned}\right\} \tag{9.14}$$

Hence we can calculate the solution of successive approximations.

In addition, for example in the exceptional nonlinear integral equation

$$x(t) + \alpha x^2(t) - \int_0^t K(t-\xi)x(\xi)d\xi = f(t), \tag{9.15}$$

under the assumption that the solution takes the form of (9.12), in the image space, (9.15) becomes

$$\begin{aligned} X_0 + \alpha X_1 + \alpha^2 X_2 + \cdots + \alpha\mathscr{L}[(x_0 + \alpha x_1 + \alpha^2 x_2 + \cdots)^2] \\ - \boldsymbol{K}(X_0 + \alpha X_1 + \alpha^2 X_2 + \cdots) = F. \end{aligned}$$

Then, comparing the coefficients of α and assuming the existence of the reciprocal function G, we have

$$\begin{aligned} X_0 &= (E - \boldsymbol{G})F, \\ X_1 &= (E - \boldsymbol{G})\mathscr{L}[x_0^2], \\ X_2 &= (E - \boldsymbol{G})\mathscr{L}[3x_0^2 x_1], \\ X_3 &= (E - \boldsymbol{G})\mathscr{L}[3x_0^2 x_2 + 3x_0 x_1^2], \\ &\cdots\cdot \end{aligned}$$

Therefore, we get

$$\begin{aligned} x_0(t) &= f(t) - \int_0^t G(t-\xi)f(\xi)d\xi, \\ x_1(t) &= x_0^3(t) - \int_0^t G(t-\xi)x_0^3(\xi)d\xi, \end{aligned}$$

$$x_2(t) = 3x_0^2(t)x_1(t) - 3\int_0^t G(t-\xi)x_0^2(\xi)x_1(\xi)d\xi,$$

$$x_3(t) = 3x_0^2(t)x_2(t) + 3x_0(t)x_1^2(t) - 3\int_0^t G(t-\xi)x_0^2(\xi)x_2(\xi)d\xi$$
$$- 3\int_0^t G(t-\xi)x_0(\xi)x_1^2(\xi)d\xi,$$

....

A similar calculation is applied to integral equations of higher order and to the nonlinear integral equation of the first kind. In the case of the Fredholm type, we may use operational calculus with the Fourier transform when the kernel takes the form $K(t-\xi)$, but in using it we have to deal with considerably complex calculations.

9.4. Locally unique solutions

The solution of a nonlinear integral equation is generally not unique, but under certain conditions we can prove the existence of a locally unique solution.

In the n-dimensional spaces D and R, let the points belonging to D be $(x_1, x_2, ..., x_n)$ and the points belonging to R be $(y_1, y_2, ..., y_n)$; then the correspondence between D and R is expressed as

$$y_i = f_i(x_1, x_2, ..., x_n) \qquad (i = 1, 2, ..., n). \tag{9.16}$$

This is taken as the mapping of the point $((x_i))$ in D to the point $((y_i))$ in R. Then if $x^0(x_1^0, x_2^0, ..., x_n^0)$; $y^0(y_1^0, y_2^0, ..., y_n^0)$ are the corresponding two points, the condition for this mapping to be regular is that the Jacobian

$$J \equiv \frac{\partial(y_1, y_2, ..., y_n)}{\partial(x_1, x_2, ..., x_n)}$$

does not vanish in D. In this case one point in R corresponds to one point in D, and, conversely, to one point in R corresponds one point in D. If we express the fact that the points in D correspond to the points in R by

$$x_i = g_i(y_1, y_2, ..., y_n) \qquad (i = 1, 2, ..., n), \tag{9.17}$$

(9.17) is taken as the solutions of the simultaneous equation (9.16). Then if $J \neq 0$ in D, there exists the solution of (9.16) which is unique.

If J vanishes somewhere in D, yet the mapping is regular in the neighbourhood of the point x^0, provided J is not zero at x^0, we can say that the unique solution of the simultaneous equations (9.16) in the neighbourhood of x^0 exists.

Now in the case of the limit when the number of dimensions becomes infinitely large, the correspondence between the functional spaces $\mathscr{D}$ and $\mathscr{R}$ is expressed in the extended form of (9.16):

$$y(t) = F[x(\overset{b}{\underset{a}{\xi}}), t]. \tag{9.18}$$

The right-hand side is a functional of $x(t)$, and if we take $x(t)$ as an unknown function, then (9.18) is a general nonlinear integral equation. We can express the solution of the integral equation in general as

$$x(t) = G[y(\overset{b}{\underset{a}{\xi}}), t], \tag{9.19}$$

and this we may take as the inverse transformation of (9.18) that maps the elements of $\mathscr{R}$ into $\mathscr{D}$. Then examining the regularity of this mapping verifies the existence of the unique solution of the integral equation (9.18).

Now, the functional determinant J appears to be the coefficient determinant of linear simultaneous equations in dx_j when both sides of (9.16) are differentiated. That is, when (9.16) is differentiated to give

$$dy_i = \sum_{1}^{n}{}_j \frac{\partial f_i}{\partial x_j} dx_j, \tag{9.20}$$

J becomes

$$J \equiv \left[\frac{\partial f_i}{\partial x_j}\right] \qquad (i, j = 1, 2, \ldots, n). \tag{9.20a}$$

When $\delta y(t)$, $\delta x(t)$ denote the variation of $y(t)$, $x(t)$, for the functional transformation (9.18), (9.20) becomes

$$\delta y(t) = F_1[x(\overset{b}{\underset{a}{\xi}}), \delta x(\overset{b}{\underset{a}{\xi}}); t], \tag{9.21}$$

where F_1 is the derivative of F and the functional of $x(t)$ and $\delta x(t)$, and is linear for $\delta x(t)$. Then let $\delta x(t)$ be an unknown function, and (9.21) is a linear integral equation in $\delta x(t)$, where Fredholm's determinant is seen to be the extension of the coefficient determinant (9.20a) (see § 2.2). Therefore the following result is expected.

When the nonlinear integral equation (9.18) is satisfied by the pair of functions $x^0(t)$, $y^0(t)$, if Fredholm's determinant of the linear integral equation (9.21) is not zero for x^0, y^0, (9.18) has the unique solution (9.19) in the neighbourhood of x^0, and the solution it satisfied by x^0, y^0.

Instead of (9.21), take the implicit form

$$F[\underset{a}{\overset{b}{x(\xi)}}, \underset{a}{\overset{b}{y(\xi)}};\ t] = 0, \tag{9.22}$$

where $x(t)$ is an unknown function and $y(t)$ is a known function. This applies to the case to solve the n-dimensional simultaneous equation

$$f_i(x_1, x_2, \ldots, x_n; y_1, y_2, \ldots, y_n) = 0. \tag{9.23}$$

From (9.22) we have $\delta F = 0$; letting F_1 denote the differentiation of F, for (9.21) we have

$$F_1[\underset{a}{\overset{b}{x(\xi)}}, \underset{a}{\overset{b}{y(\xi)}}, \underset{a}{\overset{b}{\delta x(\xi)}}, \underset{a}{\overset{b}{\delta y(\xi)}};\ t] = 0. \tag{9.24}$$

This is linear in $\delta x(t)$, $\delta y(t)$.

Suppose that (9.22) is satisfied by $x_0(t)$, $y_0(t)$, then substituting $x(t) - x_0(t)$, $y(t) - y_0(t)$ for $x(t)$, $y(t)$, we find that (9.24) is satisfied by $x(t) \equiv 0$, $y(t) \equiv 0$; therefore, (9.24) becomes

$$\delta x(t) + \int_a^b K(t, \xi)\delta x(\xi)d\xi = G_1[\underset{a}{\overset{b}{\delta y(\xi)}}, t]. \tag{9.25}$$

When $\delta y(t)$ is known, the right-hand side of this equation is a known function and (9.25) is a linear integral equation in $\delta x(t)$. Hence we have proved that Fredholm's determinant of this expression is the coefficient determinant of (9.22) and that there exists a locally unique solution when the coefficient determinant of (9.22) is non-zero.

9.4.1. Special case

In the case when the functional of (9.16) or (9.23) takes a certain special form in a nonlinear integral equation, it is possible to treat it more precisely. V. Volterra treated minutely the case which is similar in form to the Taylor series.

For example, suppose that there is the integral equation

$$x(t) + \lambda\int_0^1 K_1(t, \xi)x(\xi)d\xi + \cdots$$

$$+\frac{\lambda^n}{n!}\int_0^1\cdots\int_0^1 K_n(t,\xi_1,\xi_2,\ldots,\xi_n)x(\xi_1)x(\xi_2)\ldots x(\xi_n)d\xi_1 d\xi_2\ldots d\xi_n$$
$$=\mu f(t), \tag{9.26}$$

where $x(t)$ is an unknown function, λ, μ are taken for the convenience of calculation, and $K_n(t, \xi_1, \xi_2, \ldots, \xi_n)$ is a symmetric expression to $\xi_1, \xi_2, \ldots, \xi_n$. We are going to find the solution of (9.26) so as to make $x(t) = 0$ when $\mu f(t) = 0$. This solution will be expressed in the power series of μ. Differentiate (9.26) by μ and put $\mu = 0$, then all the terms on the left except the first vanish since they contain $x(t)$, and the result is

$$f(t)=\left\{\frac{dx}{d\mu}\right\}_{\mu=0}+\lambda\int_0^1 K_1(t,\xi)\left(\frac{dx(\xi)}{d\mu}\right)_{\mu=0}d\xi. \tag{9.27}$$

Equation (9.27) if a Fredholm-type integral equation, and we can find the solution $[dx/d\mu]_{\mu=0}$ if the determinant is not zero. Further, differentiating (9.26) twice putting $\mu = 0$, and we get

$$0=\left[\frac{d^2x}{d\mu^2}\right]_{\mu=0}+\lambda\int_0^1 K_1(t,\xi)\left[\frac{d^2x(\xi)}{d\mu^2}\right]_{\mu=0}d\xi$$
$$+\lambda^2\int_0^1\int_0^1 K_2(t,\xi_1,\xi_2)\left[\frac{dx(\xi_1)}{d\mu}\right]_{\mu=0}\left[\frac{dx(\xi_2)}{d\mu}\right]_{\mu=0}d\xi_1 d\xi_2, \tag{9.28}$$

which is a Fredholm-type equation, has the same determinant as the previous one, and determines $[d^2x/d\mu^2]_{\mu=0}$. Proceeding in the same way, we can determine $[d^nx/d\mu^n]_{\mu=0}$ and find that $x(t)$ becomes

$$x(t)=\mu\left[\frac{dx}{d\mu}\right]_{\mu=0}+\frac{\mu^2}{2!}\left[\frac{d^2x}{d\mu^2}\right]_{\mu=0}+\cdots. \tag{9.29}$$

Now (9.29) is the solution of (9.26), which we may easily prove to be convergent when $|\mu f(t)| < \varepsilon$. Since (9.29) converges to zero when $\mu f(t)$ converges to zero, it follows that (9.29) is the unique solution in the neighbourhood of $x_0(t) \equiv 0$ of the integral equation (9.26).

9.5. Nonlinear integral equation of Hammerstein type

The nonlinear integral equation

$$x(t)-\int_0^1 K(t,\xi)F[\xi,x(\xi)]d\xi=0 \qquad [H] \tag{9.30}$$

is called to be of Hammerstein type when the following three conditions hold for the kernel.

(i) The kernel $K(t, \xi)$ satisfies the Fredholm's condition and the iterated kernel $\mathring{K}_2(t, \xi)$ is continuous;

(ii) the kernel $K(t, \xi)$ is symmetrical;

(iii) the kernel $K(t, \xi)$ is positive; that is, every characteristic value of the kernel is positive.

The equation is named for A. Hammerstein who investigated it.

The nonlinear integral equation of Hammerstein type can be solved applying the Hilbert–Schmidt theory for the integral equation of a symmetrical kernel. The series of characteristic values and the corresponding eigenfunctions are denoted as follows:

$$\lambda_1, \quad \lambda_2, \ldots, \quad \lambda_\nu, \ldots,$$
$$\varphi_1(t), \; \varphi_2(t), \; \ldots, \; \varphi_\nu(t), \; \ldots$$

Equation (9.30) can be expressed as

$$x(t) = \int_0^1 K(t, \xi)g(\xi)d\xi, \qquad g(\xi) = F[\,\xi, x(\xi)]. \tag{9.31}$$

This means that $x(t)$ can be expressed as

$$x(t) = \sum_{\nu=1}^{\infty} C_\nu \varphi_\nu(t) \tag{9.32}$$

by Hilbert–Schmidt theory. The series is uniformly convergent and the coefficient becomes

$$\begin{aligned} C_\nu &= \int_0^1 \varphi_\nu(t)x(t)dt = \int_0^1 \varphi_\nu(t)dt \int_0^1 K(t, \xi)F[\xi, x(\xi)]d\xi \\ &= \int_0^1 F[\xi, x(\xi)]d\xi \int_0^1 K(t, \xi)\varphi_\nu(t)dt \\ &= \frac{1}{\lambda_\nu} \int_0^1 F[\xi, x(\xi)]\varphi_\nu(\xi)d\xi. \end{aligned} \tag{9.33}$$

If expression (9.32) is introduced in the last expression, we have

$$C_\nu = \frac{1}{\lambda_\nu} \int_0^1 F\left[\xi, \sum_{h=1}^{\infty} C_h\varphi_h(\xi)\right] \varphi_\nu(\xi)d\xi \qquad (\nu = 1, 2, \ldots); \tag{9.34}$$

C_ν can be determined.

As an approximate solution of the integral equation (9.30), putting

$$x_n(t) = \sum_{n=1}^{n} C_{nm}\psi_m(t) \tag{9.35}$$

We can introduce simultaneous algebraic equations for C_{n1}, C_{n2}, ..., C_{nn}:

$$C_{nm} = \frac{1}{\lambda_m}\int_0^1 F\left[\xi, \sum_{h=1}^{n} C_{nh}\varphi(\xi)\right]\varphi_m(\xi)d\xi \qquad (m = 1, 2, \ldots, n) \tag{9.36}$$

from equation (9.34).

The uniqueness theorem can be proved by contradiction. We shall indicate a proof. Suppose that there are two solutions $x_1(t)$ and $x_2(t)$ for equation (9.30). We have

$$\left.\begin{aligned} x_1(t) - \int_0^1 K(t, \xi)F[\xi, x_1(\xi)]d\xi = 0,\\ x_2(t) - \int_0^1 K(t, \xi)F[\xi, x_2(\xi)]d\xi = 0.\end{aligned}\right\} \tag{9.37}$$

Then

$$x_1(t) - x_2(t) - \int_0^1 K(t, \xi)\{F[\xi, x_1(\xi)] - F[\xi, x_2(\xi)]\}d\xi = 0. \tag{9.38}$$

Hence, if we put

$$\begin{gathered} x_1(t) - x_2(t) = y(t),\\ G(\xi, y) = F[\xi, y(\xi) + x_2(\xi)] - F[\xi, x_2\,(\xi)], \end{gathered} \tag{9.39}$$

$y(t)$ satisfies an integral equation of Hammerstein type:

$$y(t) - \int_0^1 K(t, \xi)G[\xi, y(\xi)]d\xi = 0. \tag{9.40}$$

When the equation is integrated after multiplying by $G(t, y(t))$, we have

$$\int_0^1 y(t)G[t, y(t)]dt = \int_0^1\int_0^1 K(t, \xi)G[t, \xi y(t)]G[\xi, y(\xi)]d\xi dt. \tag{9.41}$$

When $G(t, y)$ satisfies

$$\left.\begin{aligned} G[t, y] \geqq 0 \quad \text{for} \quad y > 0,\\ G[t, y] \leqq 0 \quad \text{for} \quad y < 0,\end{aligned}\right\} \tag{9.42}$$

then

$$\int_0^1 y(t)G[t, y(t)]dt \geqq 0. \tag{9.43}$$

By condition (iii), we have

$$\int_0^1 \int_0^1 K(t, \xi) G[t_1, y(t)] G[\xi, y(\xi)] dt\, d\xi \leqq 0. \tag{9.44}$$

Therefore, we have

$$G[t, y(t)] \equiv 0 \tag{9.45}$$

as a condition for (9.41). Consequently, we have

$$y(t) \equiv 0, \tag{9.46}$$

which means that

$$x_1(t) \equiv x_2(t). \tag{9.47}$$

This shows that the uniqueness of the solution can be proved when the conditions (9.42) hold.

TABLE 9.2. Existence conditions for the integral equations of Hammerstein type

Conditions for $K(t, \xi)$	Conditions for $F[\xi, u]$		example
	$u \leqq 0$	$u \geqq 0$	
	$F[\xi, u]$ bounded or $\lvert F[\xi, u]\rvert \leqq C_1 \lvert u \rvert + C_2$ (where $0 < C_1 < \lambda_1,\ 0 < C_2$)		$\sin u$
Continuous	$\int_0^u F(\xi, u) du \geqq -\left(\frac{1}{2} k u^2 + C_2\right)$ (where $0 < C_1 < k < \gamma_1,\ 0 < C_3$) or $\int_0^u F(\xi, u) du$ is lower bounded		e^u or $\xi + u^2$
	$0 \leqq F \leqq C_1 \lvert u \rvert + C_2$	$F \geqq 0$	$1 + e^u \sin^2 \xi$
	$F \leqq 0$	$-(C_1 u + C_2) \leqq F \leqq 0$	$-(1 + e^{-u} \sin^2 \xi)$
$K(t, \xi) \geqq 0$	$-C_3 \leqq F \leqq C_1 \lvert u \rvert + C_2$	$F \geqq -C_3$	$p(\xi) e^u + q(\xi)$ $p \geqq 0$, p, q continuous
	$F \leqq C_3$	$-(C_1 u + C_2) \leqq F \leqq C_3$	$-p(\xi) e^{-u} + q(\xi)$ $p \geqq 0$, p, q continuous

Details of exsistence theorem are omitted.

9.5.1. The existence of the solution for the Hammerstein-type integral equation

The conditions for existence of a solution for the nonlinear integral equations of Hammerstein type are given in Table 9.2:

$$x(t) - \int_0^1 K(t, \xi)F[\xi, x(\xi)]d\xi = 0,$$

where the kernel $K(t, \xi)$ satisfies conditions (i), (ii) and (iii) while $F(\xi, x)$ is continuous.

9.5.2. *The uniqueness conditions for integral equations of Hammerstein type*

When an integral equation is nonlinear, the uniqueness of a solution is another difficult problem even if a solution exsists. The conditions on $F[\xi, x(t)]$ for the nonlinear integral equations of Hammerstein type are given below where the kernel $K(t, \xi)$ satisfies the three conditions (i), (ii), (iii), while $F[\xi, x(t)]$ is continuous and either (1), (2), (3) or (4) holds.

Conditions for $F[\xi, x(\xi)]$

(1) $F(\xi, u)$, $u = x(\xi)$ are nondecreating functions for any value of ξ in (0.1)

(2) If we put $G(\xi, u) = F(\xi, u + x(\xi) - f[\xi, x(\xi)]$ then we have $G(\xi, u) \leqq \alpha|u|$, $(0 < \alpha < \lambda_1)$.

(3) The Lipshitz condition
$|F(\xi, u_1) - F(\xi, u_2)| < \alpha|u_1 - u_2|$, $(0 < \alpha < \lambda_1)$
holds

(4) If $G(\xi, u) = F(\xi, u + x(\xi)] - f[\xi, x(\xi)]$ then we have $G(\xi, u) \geqq 0$ $(u < 0)$ and $G(\xi, u) < 0$ $(u < 0)$.

9.6. Integro-differential equations

A functional equation containing the derivative and the integral, or the integral of the derivative, of an unknown function is called an integro-differential equation. In § 1.1 we briefly deseribed various kinds of equations while here we are going to give a more detailed explanation for the linear case. The classification of integro-differential equations is made corresponding to that of differential equations. When the exceptional part contains a derivative, the power of the derivative determines the order of the integro-differential equation. An integro-differential equation with one variable in its unknown function is called an *ordinary integro-differential equation*, and one with a partial derivative is called a *partial integro-differential equation.*

The latter are subdivided into *elliptic type, hyperbolic type*, and *parabolic type.*

We may sometimes convert an integro-differential equation into a general integral equation. For example, in the Fredholm-type integro-differential equation

$$x^{(1)}(t) - \lambda \int_a^b K(t, \xi)x(\xi)d\xi = f(t), \tag{9.48}$$

with the given initial value x_a, by integrating both sides with respect to t, we get

$$x(t) - \lambda \int_a^b K^t(t, \xi)x(\xi)d\xi = F(t) + x_a. \tag{9.49}$$

Here we have

$$K^t(t, \xi) = \int_a^t K(\tau, \xi)d\tau, \qquad F(t) = \int_a^t f(\tau)d\tau,$$

all of which are known functions; therefore, the equation (9.49) is $[F_2]$.

While for the Volterra type, for example, in

$$x^{(2)}(t) + ax(t) - \int_a^t K(t, \xi)x(\xi)d\xi = f(t), \tag{9.50}$$

with the given initial values x_a, $x_a{}^{(1)}$, we get

$$(\alpha + 1)x(t) - \int_a^t K^{tt}(t, \xi)x(\xi)d\xi = F(t) + x_a + x_a{}^{(1)}t, \tag{9.51}$$

where

$$K^{tt}(t, \xi) = \int_\xi^t \left[-\alpha + \int_\xi^\tau K(\eta, \xi)d\eta \right] d\tau,$$

$$F(t) = \int_a^t d\tau \int_a^\tau f(\xi)d\xi,$$

which, therefore, becomes $[V_2]$.

We may convert a partial integro-differential equation into an ordinary integro-differential equation by separating variables. For example, in the hyperbolic-type integral equation

$$\frac{\partial^2 u(x, t)}{\partial t^2} - \frac{\partial^2 u(x, t)}{\partial x^2} - \int_0^t K(t, \xi)u_{xx}(x, \xi)d\xi = 0, \tag{9.52}$$

let

$$u(x, t) = \sin m(x + \alpha)v(t)$$

be the solution, and we obtain an ordinary integro-differential equation in $v(t)$:

$$v^{(2)}(t) + m^2\left[v(t) + \int_0^t K(t, \xi)x(\xi)d\xi\right] = 0. \tag{9.53}$$

Thus, [V_2] is obtained from this equation by the previous method.

Knowledge of the theory of differential equations is utilized as has been shown above.

Lastly, when a derivative is found in the regular part as

$$x(t) - \lambda \int_a^b K(t, \xi)x^{(1)}(\xi)d\xi = f(t), \tag{9.54}$$

we differentiate this in t, which results in

$$x^{(1)}(t) - \lambda \int_a^b K_t(t, \xi)x^{(1)}(\xi)d\xi = f^{(1)}(t), \tag{9.54a}$$

and this is taken as [F_2] for $x^{(1)}(t)$. We can determine the initial value with

$$x(a) = \lambda \int_a^b K(a, \xi)x^{(1)}(\xi)d\xi + f(a).$$

Hence the solution of (9.54)′ is determined.

In this case, by successive approximations in the symbolic expression, (9.54a) turns out to be

$$\begin{aligned} x &= f + \lambda K \circ x^{(1)} \\ &= f + \lambda K \circ f^{(1)} + \lambda^2 \mathring{K}^2 \circ x^{(2)} \\ &= \ldots, \\ &= f + \sum_{i=1}^{\infty} \lambda^i \mathring{K}^i \circ f^{(i)}. \end{aligned} \tag{9.55}$$

We may easily examine the convergence of such a solution. Yet, although an integro-differential equation like (9.48) or (9.54) may be regarded as the extension of the n-dimensional simultaneous linear differential equation, Fredholm's theory cannot be applied to it without modification. As has been noticed already, speciality is not necessarily found in an integro-differential equation of this kind of extension. We may use the various kinds of special methods of solution which we covered in Chapter 3.

Especially, we can solve the Volterra-type or convolution-type integro-differential equation, by operational calculus, making use of the Laplace transform

$$x^{(2)}(t) + \alpha x(t) - \int_0^t K(t - \xi)x(\xi)d\xi = f(t). \tag{9.56}$$

Here if $x_0 = x_0^{(1)} = 0$, in the image space, (9.56) becomes

$$(s^2 + \alpha)X - \boldsymbol{K}X = F;$$

then we have

$$X = \frac{F}{s^2 + \alpha - \boldsymbol{K}}.$$

Therefore, substituting

$$\frac{1}{s^2 + \alpha - \boldsymbol{K}} = 1 - \boldsymbol{G}, \quad \text{or} \quad \boldsymbol{G} = 1 - \frac{1}{s^2 + \alpha - K}, \tag{9.57}$$

we obtain

$$X = (1 - \boldsymbol{G})F;$$

hence we get

$$x(t) = f(t) - \int_0^t G(t - \xi)f(\xi)d\xi. \tag{9.58}$$

When the original function G of $\boldsymbol{G}$ is found, (9.58) is the solution. However, G is not the reciprocal function of K.

Exercises

1. Show that the nonlinear integro-differential equation

$$x(t)^2 - \lambda \int_0^1 t\xi x(\xi)d\xi = 4 + 10t + 9t^2$$

has the solutions $2 \pm 3t$ when $\lambda = \pm 1$.

2. Solve the nonlinear integral equation

$$x(t) - \lambda \int_0^1 t\xi x(\xi)^3 d\xi = 2 - 3t.$$

3. Solve the following nonlinear integral equations using operatioal calculus.

$$\text{(a) } x^2(t) + \int_0^t \sin(t-\xi)x(\xi)d\xi = e^t,$$

$$\text{(b) } x(t) - \int_0^t (t-\xi)^2 x^2(\xi)d\xi = t,$$

$$\text{(c) } x^2(t) - \int_0^t (t-\xi)^3 x^3(\xi)d\xi = t^2 + 1.$$

4. Show that the Laplace transform of the solution of the nonlinear integral equation

$$2x(t) - \int_0^t K(t-\xi)x(\xi)d\xi = f(t),$$

is

$$X(s) = \frac{F(s)}{1 + (1 - F(s))^{1/2}}$$

and prove that

$$x(t) = J_1(t)$$

when $f(t) = \sin t$.

5. Show that the solution of the integro-differential equation:

$$x(t) - \frac{2}{\sqrt{\pi}} \int_0^t (t-\xi)^{1/2} x^{(1)}(\xi)d\xi = \frac{2}{\sqrt{\pi}} t$$

at $x_0 = 0$, is

$$x(t) = e^t[\text{erf}\sqrt{t} + 1] - 1.$$

6. Numerically solve the nonlinear integral equations in question 3.

7. Show that the nonline ar equation

$$x(t) - \lambda \int_0^1 t\xi\,[\xi + x^2(\xi)]\,d\xi = 0$$

is Hammerstein type.

$$[K(t, \xi) = t\xi \text{ and } \mathring{K}^2(t, \xi) = \frac{1}{3}\,t\xi \text{ are continuous}$$

in S. $K(t, \xi) = t\xi$ is symmetrical.
Eigenvalue is 3 which is positive. Consequently the equation satisfies three conditions.]

8. By successive approximation solve the nonlinear equation

$$x(t) - \int_0^t (t - \xi)^2\, x^2\,(\xi)d\xi = t.$$

9. Nonlinear differential equation

$$x^{(2)} + n^2x + \alpha g(x, x^{(1)}) = f(t)$$

with the initial conditions

$$x_0,\ x_0{}^{(1)}$$

furnishes the fundamental equation of nonlinear oscillation. Putting $x^{(2)} = X$, deduce a nonlinear integral equation of Volterra type.

$$\Big[X + n^2\Big\{\int_0^t (t - \xi)X(\xi)d\xi + x_0{}^{(1)}t + x_0\Big\}$$
$$+ \alpha g\Big\{\int_0^t (t - \xi)X(\xi)d\xi + x_0{}^{(1)}t + x_0, \int_0^t X(\xi)d\xi + x_0{}^{(1)}\Big\}$$
$$= f(t)\Big]$$

10 Applications of integral equations (II)

In Chapter 5 we discussed various kinds of integral equations that were introduced in the first part, 'Fundamental Theories', and applications associated with these theories. In this chapter we are going to show the applications of the linear equations that have been treated in the second part, 'Special Theories'. Since the Sturm–Liouville boundary-value problem can be reduced to an integral equation of the symmetric kernel by means of the Green function, we may precisely treat the eigensolution of vibration problems, heat-conduction problems, and so on by Hilbert–Schmidt theory, and thus indicate the fundamental characteristics of these problems. For example, it is possible to express a variation principle in an integral equation as it is usually reduced to a differential equation.

Hysteresis is widely recognized not only as a natural phenomenon but also as a social phenomenon, and much attention has recently been paid to its application. In addition to these examples, we shall also consider double boundary problems and collision problems as applications of simultaneous integral equations and nonlinear integral equations. It is almost impossible to cover the whole field of applications of integral equations, since this field is ever expanding; however, we shall discuss mathematical properties of various applications and add some exercises involving the singular kernel in the last section.

10.1. The Sturm–Liouville boundary-value problem

(1) *The Sturm–Liouville boundary-value problem.* We define the linear functional $L[x(t)]$ of the argument function $x(t)$ as follows:

$$L[x(\overset{b}{\underset{a}{t}})] = px^{(2)} + p^{(1)}x^{(1)} - qx = \frac{d}{dt}\left(p\frac{dx}{dt}\right) - qx, \qquad (10.1)$$

where $p(t)$, $p^{(1)}(t)$, $q(t)$ are continuous functions of t in $I(a, b)$ and $p(t) > 0$. This equation is called a self-adjoint linear differential expression of the second order.

Let there be the differential equation

$$L[x(t)] = -\lambda f(t), \tag{10.2}$$

and the homogeneous boundary conditions

$$\left.\begin{aligned} R_a(x) &\equiv Ax(a) + Bx^{(1)}(a) = 0, \\ R_b(x) &\equiv Cx(b) + Dx^{(1)}(b) = 0. \end{aligned}\right\} \tag{10.3}$$

Here $f(t)$ is either continuous or piecewise continuous, and A and B, and C and D do not become zero at the same time. Boundary-value problems of this sort of ordinary differential equations are called Sturm–Liouville boundary-value problems or the Sturm–Liouville system.†

A Sturm–Liouville boundary-value problem has two linearly independent solutions $x_1(t)$, $x_2(t)$, which are continuous in I as far as the second derivatives when $f(t) = 0$, and an arbitrary solution of (10.2), continuous up to the second derivative, is expressed as a linear combination of these, viz.

$$x(t) = c_1 x_1(t) + c_2 x_2(t).$$

We find that, when $f(t) = 0$, (10.2) has a meaningless solution $x(t) \equiv 0$; if we consider the boundary condition (10.4), then the value of λ which allows other solutions than this to exist is called an eigenvalue, and the solution for this eigenvalue is called an eigensolution. We shall proceed for a while with the premise that $\lambda = 0$ is not an eigenvalue.

Since L is linear, it is possible to superpose the solutions. Then instead of (10.2) put

$$L[x(t)] = -d_\xi(t), \tag{10.4}$$

$$d_\xi(t) = \begin{cases} 1, & t = \xi, \\ 0, & t \neq \xi, \end{cases}$$

provided $a < \xi < b$, and let

$$G(t, \xi)$$

be the solution which satisfies (10.4) and the homogeneous boundary condition (10.3). If we take $\delta_\xi(t)$ such that it is zero in $|t - \xi| > \varepsilon$ and

† The general and systematic treatment of this sort of boundary value problem was first developed by J.C.F. Sturm and J. Liouville in *Journal de mathématique* 1–3 (1836–1838).

$$\int_a^b \delta_\xi(t)dt = \int_{\xi-\varepsilon}^{\xi+\varepsilon} \delta_\xi(t)dt = d_\xi(t)$$

hold, then the right-hand side of (10.2) becomes

$$f(t) = \lambda \int_a^b f(\xi)\delta_\xi(\xi)d\xi,$$

and the solution follows by superposition:

$$x(t) = \lambda \int_a^b G(t, \xi)f(\xi)d\xi, \tag{10.5}$$

which is the solution of (10.2), satisfying the boundary condition (10.3). Precise proof of this will be given in the following theorem.

Definition. The Green Function

Let $G(t, \xi)$ be a function with the following properties:

(i) $G(t, \xi)$ satisfies the homogeneous differential equation $L[G(t, \xi)] = 0$ for $t \neq \xi$.

(ii) $G(t, \xi)$ is continuous in t in I and satisfies the homogeneous boundary conditions $R_a[G(t, \xi)] = 0$, $R_b[G(t, \xi)] = 0$.

(iii) dG/dt, d^2G/dt^2 is continuous except at $t = \xi$, where it has a jump,

$$\left[\frac{dG}{dt}\right]_{\xi-0}^{\xi+0} \equiv \left[\frac{dG}{dt}\right]_{t=\xi+0} - \left[\frac{dG}{dt}\right]_{t=\xi-0} = -\frac{1}{p(\xi)}. \tag{10.6}$$

A function which satisfies the above three conditions is called a *Green function* of the differential form L (or the linear functional L).

LEMMA 10.1 (*Hilbert's fundamental equation*) *The function*

$$x(t) = \lambda \int_a^b G(t, \xi)f(\xi)d\xi$$

defined the by Green function $G(t, \xi)$ *satisfies the differential equation* $L[x(t)] = -\lambda f(t)$ *and the homogeneous boundary condition* (10.3).

Conversely, the solution which satisfies the differential equation (10.2) *and the homogeneous boundary condition* (10.3) *is expressed in the form of* (10.5) *by means of a Green function.*

Proof. (i) First we begin with the proof that (10.5) is the solution of the Sturm–Liouville system.

Since G is continuous in t, x is also continuous in t from (10.5). We differentiate both sides of (10.5) in t, and we have

$$\frac{dx}{dt} = \lambda \int_a^b G_t(t, \xi) f(\xi) d\xi;$$

$G_t(t, \xi)$ is discontinuous only at $\xi = t$, then we differentiate it again in t and

$$\frac{d^2x}{dt^2} = \lambda \frac{d}{dt}\left\{\int_a^t G_t(t, \xi) f(\xi) d\xi + \int_t^b G_t(t, \xi) f(\xi) d\xi\right\}$$
$$= \lambda \int_a^b G_{tt}(t, \xi) f(\xi) d\xi + \lambda \Big[G_t(t, \xi) f(\xi)\Big]_{\xi=t+0}^{\xi=t-0}.$$

Then if $f(\xi)$ is continuous at $\xi = t$, we have

$$\frac{d^2x}{dt^2} = \lambda \int_a^b G_{tt}(t, \xi) f(\xi) d\xi + \lambda \Big[G_t(t, \xi)\Big]_{\xi=t+0}^{\xi=t-0} f(t).$$

Therefore from (10.6) and the following Lemma 10.2 we obtain

$$\frac{d^2x}{dt^2} = \lambda \int_a^b G_{tt}(t, \xi) f(\xi) d\xi - \lambda f(t)/p(t).$$

Using all these we arrive at

$$L[x] = \lambda \int_a^b (pG_{tt} + p^{(1)} G_t - qG) f(\xi) d\xi - \lambda f$$
$$= \lambda \int_a^b L[G] f(\xi) d\xi - \lambda f = -\lambda \int_a^b d_\xi(t) f(\xi) d\xi - \lambda f$$
$$= -\lambda f.$$

If $f(t)$ is discontinuous at $\xi = t$, let $\alpha(t)$ be the jump; that is, $f(t + 0) = f(t - 0) + \alpha(t)$; then $\lim_{t \to t-0} \alpha(t) = 0$. Since we have

$$\frac{d^2x}{dt^2} = \lambda \int_a^b G_{tt}(t, \xi) f(\xi) d\xi - \lambda f(t-0)/p(t) - \lambda \alpha G_t(t, t+0),$$

we get $\lim_{t \to t-0} L[x] = -\lambda f(t-0)$. Similarly $\lim_{t \to t+0} L[x] = -\lambda f(t+0)$ may be proved.

(ii) All the solutions of the Sturm–Liouville system take the form of (10.5). In

$$\frac{d}{dt}\left(p\frac{dx}{dt}\right) - qx = -\lambda f(t), \tag{10.2}$$

$$\frac{d}{dt}\left(p\frac{dG}{dt}\right) - qG = -d_\xi(t), \tag{10.4}$$

multiply (10.2) by $-G$, (10.4) by x and add them, then take the integral in t from a to $\xi - 0$, which yields

$$\int_a^{\xi-0}\left[x\frac{d}{dt}\left(p\frac{dG}{dt}\right) - G\frac{d}{dt}\left(p\frac{dx}{dt}\right)\right]dt = \lambda\int_a^{\xi-0} G(t, \xi)f(t)dt;$$

that is,

$$\left[p\left(x\frac{dG}{dt} - G\frac{dx}{dt}\right)\right]_a^{\xi-0} = \lambda\int_a^{\xi-0} G(t, \xi)f(t)dt.$$

Similarly, take the integral from $\xi + 0$ to b and we get

$$\left[p\left(x\frac{dG}{dt} - G\frac{dx}{dt}\right)\right]_{\xi+0}^{b} = \lambda\int_{\xi+0}^{b} G(t, \xi)f(t)dt.$$

Since we get $(x\, dG/dt - G\, dx/dt) = 0$ from the homogeneous boundary condition, add the above two expressions to yield

$$\left[p\left(x\frac{dG}{dt} - G\frac{dx}{dt}\right)\right]_{\xi+0}^{\xi-0} = \lambda\int_a^{b} G(t, \xi)f(t)dt.$$

While pGx_t is a coninuous function, we have

$$\left[px\frac{dG}{dt}\right]_{\xi+0}^{\xi-0} = \lambda\int_a^{b} G(t, \xi)f(t)dt,$$

and from (10.6) we get

$$x(\xi) = \lambda\int_a^{b} G(t, \xi)f(t)dt.$$

From Lemma 10.2 it will be obvious that $G(t, \xi) = G(\xi, t)$, and then we have

$$x(t) = \lambda\int_a^{b} G(t, \xi)f(\xi)d\xi.$$

Hence all the arbitrary solutions take the form of (10.5).

LEMMA 10.2 Symmetricity
The Green function of $L[x]$ *is symmetric. That is,*

$$G(t, \xi) = G(\xi, t).$$

Proof. Put $a < \xi < \eta < b$ and from (10.4) take

$$\frac{d}{dt}[p(t)G_t(t, \xi)] - q(t)G(t, \xi) = 0 \qquad (t \neq \xi);$$

$$\frac{d}{dt}[p(t)G_t(t, \eta)] - q(t)G(t, \eta) = 0 \qquad (t \neq \eta);$$

then after multiplying the former by $G(t, \eta)$ and the latter by $G(t, \xi)$, add the results, to get

$$\frac{d}{dt}[p(t)G_t(t, \xi)]G(t, \eta) - \frac{d}{dt}[p(t)G_t(t, \eta)]G(t, \xi) = 0.$$

Now substitute $p(t)\{ G(t, \eta)G_t(t, \xi) - G(t, \xi)G_t(t, \xi) \} \equiv D$. Then, since $d[D] = 0$ in the above expression, take the integral from a to b, paying attention to discontinuous points and we obtain

$$[D]_a^{\xi-0} + [D]_{\xi+0}^{\eta-0} + [D]_{\eta+0}^{b} = 0.$$

Taking the boundary condition (10.3) into consideration, and noticing that

$$-[D]_{\xi-0}^{\xi+0} - [D]_{\eta-0}^{\eta+0} = 0,$$

then taking account of the fact that $G_t(t, \xi)$ is continuous at $t = \eta$, we finally obtain

$$-p(\xi)G(\xi, \eta)[G_t(t, \xi)]_{\xi-0}^{\xi+0} + p(\eta)G(\eta, \xi)[G_t(t, \eta)]_{\eta-0}^{\eta+0} = 0.$$

Hence

$$G(\xi, \eta) = G(\eta, \xi).$$

Therefore, Green's function is symmetric.

(2) *Constructicn of the Green function and the modified Green function.* A Green function shows its singularity in the solution to the Sturm–Liouville boundary-value problem when $f(t)$ becomes a special function $d_\xi(t)$ at $t = \xi$. It cannot be expected that $G(t, \xi)$ covers the whole region of I in a simple expression. But when we find the differential equation and the boundary condition: the solution $x_a(t)$ which satisfies $L[x_a(t)] = 0$, $R_a(x_a) = 0$, the solution $x_b(t)$ which satisfies $L[x_b(t)] = 0$, $R_b(x_b) = 0$, then we can form the Green function from these solutions $x_a(t)$, $x_b(t)$. Since $L[x]$ is linear, letting C_a, C_b be arbitrary constants, we have only to determine these properly so that G_t may jump as much as $-1/p(\xi)$ at $t = \xi$.

That is, if $x_a(t)$, $x_b(t)$ are linearly independent, by Wronski's theorem (Whittaker and Watson 1953) we have $x_a(t)x_b{}^{(1)}(t) - x_a{}^{(1)}(t)x_b(t) \neq 0$, and then $G(t, \xi)$ becomes as follows:

$$\left.\begin{aligned} &\text{in } t \leqq \xi \quad G(t, \xi) = -\frac{1}{c}x_b(\xi)x_a(t), \\ &\text{in } t \geqq \xi \quad G(t, \xi) = -\frac{1}{c}x_a(\xi)x_b(t) \end{aligned}\right\} c = p(\xi)\begin{vmatrix} x_a(\xi) & x_b(\xi) \\ x_a{}^{(1)}(\xi) & x_b{}^{(1)}(\xi) \end{vmatrix}.$$

Hence the Green function is determined to be uniform and so is the solution from (10.5).

Next, when $x_a(t)$, $x_b(t)$ are linearly dependent, $x_a(t)$ becomes the continuous solution of $L[x] = 0$, which satisfies both of the homogeneous boundary conditions and is not identically zero. Therefore $\lambda = 0$ becomes the eigenvalue.

If there exists a solution $x(t)$ not identically zero which satisfies $L[x] = 0$, $R_a(x) = R_b(x) = 0$, at the same time, x_t becomes continuour at $t = \xi$, and, therefore, we cannot construct a Green function. Then for the continuous or piecewise-continuous distrubance function $f(t)$, either the homogeneous boundary condition of the differential equation $L[x(t)] = -\lambda f(t)$ has a unique solution, or $L[x(t)] = 0$ has a solution not identically zero. We shall treat the case when there exists a solution $x_c(t)$, not uniformly zero, which makes $L[x_c(t)] = 0$ in Lemma 10.3 below.

Definition. Modified Green function

Let $G_1(t, \xi)$ have the following properties;

(i) $G_1(t, \xi)$ satisfies the inhomogeneous differential equation $L[G_1(t, \xi)] = \lambda p(t)x_c(t)x_c(\xi)$ for $t = \xi$, where

$$\int_a^b p(t)x_c(t)dt = 1.$$

(ii) $G_1(t, \xi)$ is continuous in t in I, and satisfies the homogeneous boundary conditions.

(iii) $dG/d\xi$, $d^2G/d\xi^2$ are continuous except at $t = \xi$, and have the following equation:

$$\left[\frac{dG_1}{dt}\right]_{\xi-0}^{\xi+0} = -\frac{1}{p(\xi)}.$$

(iv) $G_1(t, \xi)$ satisfies

$$\int_a^b G_1(t, \xi)p(t)x_c(t)dt = 0.$$

We call a function which satisfies the above four conditions a *modified Green function* of the differential expression L (or the linear functional L).

LEMMA 10.3 Modified Green's function
A function which is determined by a modified Green's function

$$x(t) = \lambda \int_a^b G_1(t, \xi)f(\xi)d\xi + cx_c(t), \tag{10.7}$$

satisfies the differential equation $L[x(t)] = -\lambda f(t)$ *and the homogeneous boundary conditions* (10.3), *where*

$$\int_a^b x_c(t)f(t)dt = 0 \tag{10.8}$$

Conversely, a solution which satisfies the differential equation (10.2) *and the homogeneous boundary conditions* (10.3) *can be expressed by a modified Green function in the form of* (10.7).

We omit the proof, but see Question 2. We know from the following process that the condition (10.8) is necessary:

$$\frac{d}{dt}\left(p\frac{dx_c}{dt}\right) - qx_c = -\lambda d_\xi(t), \qquad \frac{d}{dt}\left(p\frac{dx}{dt}\right) - qx = -\lambda f(t).$$

Multiply each by $x(t)$, $-x_c(t)$ and add them together and take the integral with respect to t from a to b, which makes

$$\left[p\left(x\frac{dx_c}{dt} - x_c\frac{dx}{dt}\right)\right]_a^b = \int_c^b x_c(t)f(t)dt,$$

The left-hand side becomes zero from the boundary conditions.

LEMMA 10.4 Symmetricity
The modified Green's function of $L[x]$ *is symmetric. That is,*

$$G_1(t, \xi) = G_1(\xi, t).$$

Again, see Question 2.

(3) *Integral equations and boundary-value problems.*

In the differential equation

$$L[x(t)] + \lambda \rho(t)x(t) = -\lambda f(t) \tag{10.9}$$

and the homogeneous boundary conditions

$$\left.\begin{aligned} R_a(x) &\equiv Ax(a) + Bx^{(1)}(a) = 0, \\ R_b(x) &\equiv Cx(b) + Dx^{(1)}(b) = 0, \end{aligned}\right\} \tag{10.3}$$

put $\rho(t)x(t) + f(t) = f_1(t)$, and (10.9) becomes

$$L[x(t)] = -\lambda f_1(t),$$

which is called the Sturm–Liouville boundary-value problem. Therefore, using the Green function $G(t, \xi)$ of the functional L, we find that the solution becomes

$$x(t) = \lambda \int_a^b G(t, \xi)\rho(\xi)x(\xi)d\xi + g(t). \tag{10.10}$$

Where

$$g(t) = \lambda \int_a^b G(t, \xi)f(\xi)d\xi. \tag{10.11}$$

Hence, the boundary-value problem comes to the integral equation (10.11). While conversely from Lemma 10.1 we know that the solution of the integral equation (10.10) agrees with that of the boundary-value problem.

In addition, making use of the modified Green function $G_1(t, \xi)$ of L, we find the solution to be

$$x(t) = \lambda \int_a^b G_1(t, \xi)\rho(\xi)x(\xi)d\xi + g(t), \tag{10.10a}$$

where

$$g(t) = \lambda \int_a^b G_1(t, \xi)f(\xi)d\xi + cx_c(t); \tag{10.11a}$$

and here again the boundary-value problem and the integral equation are equivalent.

In particular, in (10.9), when $f(t) \equiv 0$, we get $g(t) \equiv 0$, then putting

$$\sqrt{(\rho(x))}x(t) \equiv X(t),$$

the integral equation becomes [F_0]:

$$X(t) = \lambda \int_a^b K(t, \xi)X(\xi)d\xi, \tag{10.12}$$

where we put

$$K(t, \xi) = \sqrt{(\rho(t)\rho(\xi))}G(t, \xi). \tag{10.13}$$

Equation (10.12) is a homogeneous integral equation with a symmetric kernel.

Hence the eigenvalue of the integral equation (10.12) agrees with that of the differential equation $L[x(t)] + \lambda\rho(t)x(t) = 0$, and also the eigenfunction of the integral equation agrees with the eigenfunction of this differential equation.

The following properties are to be disclosed by the theory of symmetric kernels.

(i) It has at least one eigenvalue (Theorem 6.3).

(ii) Every eigenvalue is real (Theorem 6.4).

(iii) The index p and the multiplicity m agree (Theorem 6.5).

(iv) There exists a complete orthogonal eigenfunction system $\{\psi_i(t)\}$ for the eigenvalue $\{\lambda_i\}$ (Theorem 6.6).

(v) If $\sum_i \psi_i(t)\psi_i(\xi)/\lambda_i$ is uniformly convergent, we have

$$\sum_i \frac{\psi_i(t)\psi_i(\xi)}{\lambda_i} = K(t, \xi)$$

(Theorem 6 7).

(vi) If $f(t)$ can be expressed as

$$f(t) = \int_a^b G(t, \xi)g(\xi)d\xi, \tag{10.14}$$

from the continuous function (or piecewise-continuous function) $g(t)$, then

$$f(t) = \sum c_n\psi_n(t), \qquad c_n = f \cdot \psi_n \tag{10.15}$$

is absolutely and uniformly convergent in I (Theorem 6.8).

Further, from the mathematical properties of a Green function, we have the following theorem.

THEOREM 10.1

There exist an infinite number of eigenvalues.

Proof. Suppose that the number of eigenvalues is finite; we have

$$G(t, \xi) = \sum \frac{\psi_\nu(t)\psi_\nu(\xi)}{\lambda_\nu}, \tag{10.16}$$

where G is continuous up to the second derivative, but this is contradictory, since dG/dt is discontinuous at $t = \xi$.

THEOREM 10.2

If a function $x(t)$, which is continuous up to its second derivative, satisfies the homogeneous boundary condition (10.3), then

$$x(t) = \sum c_n\psi_n(t) \tag{10.17}$$

is absolutely and uniformly convergent in I.

Proof. We get $x(t) = \int_a^b G(t, \xi)(-L[x])dt$, and since $L[x]$ is continuous this theorem follows directly from (vi).

THEOREM 10.3

The index of an eigenvalue is always 1.

Proof. If the index is more than 3, then more than three different eigenfunctions correspond to one eigenvalue, but since the differential equation of the Sturm–Liouville boundary-value problem is of the second order, it cannot have more than three linearly independent solutions. Hence the index must be less than two.

If there exist two eigenfunctions $\varphi_1(t)$, $\varphi_2(t)$ we have

$$\varphi_1 L[\varphi_2] - \varphi_2 L[\varphi_1] = \frac{d}{dt} p(\varphi_1 \varphi_2{}^{(1)} - \varphi_1{}^{(1)} \varphi_2),$$

but if φ_1, φ_2 are linearly independent

$$\varphi_1 \varphi_2{}^{(1)} - \varphi_1{}^{(1)} \varphi_2 = \text{constant}(\neq 0).$$

Then from the boundary conditions (10.3) we have

$$\left.\begin{aligned} A\varphi_1(a) + B\varphi_1{}^{(1)}(a) &= 0, \\ A\varphi_2(a) + B\varphi_2{}^{(1)}(a) &= 0 \end{aligned}\right\},$$

where in order that A and B do not become zero at the same time we must have

$$\varphi_1 \varphi_2{}^{(1)} - \varphi_1{}^{(1)} \varphi_2 = 0.$$

This contradicts the proposition.

THEOREM 10.4

If

$$[pxx^{(1)}]_a^b = 0 \tag{10.18}$$

and $p(x) > 0$, *then the minimum eigenvalue exists, and if* $p(x) < 0$ *the maximum eigenvalue exists.*

Proof We have

$$L[\psi_n] + \lambda_n \psi_n = 0, \qquad R_a(\psi_n) = 0, \quad R_b(\psi_n) = 0.$$

Multiply both sides by ψ_n and take the integral, then we get

$$\lambda_n = -\int_a^b \psi_n L[\psi_n]dt = -\int_a^b \left[\psi_n \frac{d}{dt}(p\psi_n{}^{(1)}) - q\psi_n{}^2\right]dt. \tag{10.19}$$

The above expression further takes the form

$$\lambda_n = -[p\psi_n\psi_n^{(1)}]_a^b + \int_a^b p\psi_n^{(1)2}dt + \int_a^b q\psi_n^2 dt. \qquad (10.20)$$

Now q is a continuous function in I; then let M be the maximum, and m be the minimum, and from $\psi_n^2\,dt = 1$ we derive

$$\lambda_n \leqq \int_a^b p\psi_n^{(1)2}dt + M,\ \lambda_n \geqq \int_a^b p\psi_n^{(1)2}dt + m.$$

Therefore, when $p(t) > 0$ we have $\lambda_n > m$, and when $p(t) < 0$ we have $\lambda_n < M$.

THEOREM 10.4a

If $[px\,x^{(1)}{}_a^b] = 0$ *then when* $p > 0$, $q \geqq 0$ *in* I, *every eigenvalue is positive; when* $p < 0$, $q \leqq 0$, *every eigenvalue is negative.*

10.1.1. Boundary-value problem of the ordinary high-order-differential equation

So far we have treated the boundary-value problem of the ordinary second-order differential equation, and this problem can be extended without any amendments in the case of the ordinary higher-order differential equation. We shall explain a linear functional:

$$\underset{a}{\overset{b}{H}}[x(t)] = x^{(4)} - \lambda x \quad \text{or} \quad \underset{a}{\overset{b}{H_1}}[x(t)] = x^{(4)} - \lambda p(t)x, \qquad (10.21)$$

as an example.

Consider the differential equation

$$H[x(t)] = -\lambda f(t) \quad \text{or} \quad H_1[x(t)] = -\lambda f(t), \qquad (10.22)$$

with the boundary value at $t = a$, b as the homogeneous boundary condition.

Since H, H_1 are linear, it follows that the superposition of solutions is possible. Then, as in the previous process, let $G(t, \xi)$ be the solution satisfying the homogeneous boundary condition of

$$H[x(t)] = -d_\xi(t), \qquad (10.23)$$

and the solution is

$$x(t) = \lambda \int_a^b G(t, \xi)f(\xi)d\xi. \qquad (10.24)$$

Here $G(t, \xi)$ is a Green function, defined as follows:

(i) $G(t, \xi)$ satisfies the homogeneous differential equation $H[x(t)] = 0$ for $t \neq \xi$;

(ii) $G(t, \xi)$ is continuous together with $G_t(t, \xi)$, $G_\xi(t, \xi)$, and satisfies the homogeneous boundary conditions;

(iii) d^2G/dt^2, d^3G/dt^3, d^4G/dt^4 are continuous except at $t = \xi$, and at $t = \xi$, d^3G/dt^3 has the following jump:

$$\left[\frac{d^3G}{dt^3}\right]_{t=\xi-0}^{t=\xi+0} = -1.$$

Moreover, two lemmas hold that are quite similar to the previous one. We give them without proof.

LEMMA 10.5
Equation (10.24) satisfies the differential equation (10.22) and the boundary conditions. Conversely, the solution of (10.22) which satisfies the boundary conditions can be expressed in the form of (10.24).

LEMMA 10.6

$$G(t, \xi) = G(\xi, t).$$

In this case, too, we can convert the boundary-value problem of a differential equation into the integral equation $[F_1]$, putting

$$K(t, \xi) = G(t, \xi)\sqrt{(\rho(t)\rho(\xi))}.$$

10.1.2. Boundary-value problem of a partial differential equation

We can also associate a partial differential equation with a Green function corresponding to the homogeneous boundary value of a linear functional, and it is possible to reduce the problem to an integral equation by expressing the solution as a Green function. Of course, this operation can be extended for general use. For example, in a harmonic function, let us give an explanation about a linear functional:

$$\nabla^2 u = \frac{\partial^2 u}{\partial x^2} + \frac{\partial^2 u}{\partial y^2}. \tag{10.25}$$

Consider the problem of solving the partial differential equation

$$\nabla^2 u = -\lambda\varphi(x, y) \tag{10.26}$$

with the homogeneous boundary value on the boundary R. For example, on R

$$u = 0 \quad \text{or} \quad \frac{\partial u}{\partial n} = 0, \tag{10.26a}$$

where n means a normal to R, and with these we solve (10.26) in the domain D.

Now the Green function is introduced in the following way. Take a small circle of radius ε with the point (ξ, η) as its centre in the domain D, that is, $(x-\xi)^2+(y-\eta)^2=\varepsilon^2$, and a function $d(x, y)$ which becomes zero outside the domain. Where

$$\nabla^2[u(x, y)] = -\, d(\xi, \eta), \tag{10.27}$$

provided $d(\xi, \eta) = \iint \delta(\xi, \eta)dxdy$, where

$$\delta(\xi, \eta) = \begin{cases} \text{not } 0, & (x-\xi)^2+(y-\eta)^2 \leqq \varepsilon^2, \\ 0, & (x-\xi)^2+(y-\eta)^2 > \varepsilon^2, \end{cases}$$

and

$$\iint \delta(x, y)dxdy = 1.$$

Going back and using the solution of (10.3), we can express the solution of (10.2) as

$$u(x, y) = \lambda \iint_D G(x, y; \xi, \eta)\varphi(\xi, \eta)d\xi d\eta, \tag{10.28}$$

where $G(x, y; \xi, \eta)$ is a Green function, which is defined as follows:

(i) $G(x, y, \xi; \eta)$ satisfies the homogeneous differential equation $\Delta[u(x, y)] = 0$ for $x \neq \xi$, $y \neq \eta$;

(ii) $G(x, y; \xi, \eta)$ is continuous in x, y and satisfies the homogeneous boundary condition;

(iii) G is continuous together with $\partial G/\partial x$, $\partial G/\partial y$, $\partial^2 G/\partial x^2$, $\partial^2 G/\partial y^2$, $\partial^2 G/\partial x \partial y$, and

$$G(x, y; \xi, \eta) = -\frac{1}{2\pi}\log r + \gamma(x, y; \xi, \eta), \tag{10.29}$$

where $r = ((x-\xi)^2+(y-\eta)^2)^{1/2}$, and γ is continuous up to the second differential coefficients. (And of course $\nabla^2\gamma = 0$ from (i).)

Furthermore, the following lemma holds, though we shall not prove this.

LEMMA 10.7

When $\varphi(x, y)$ and its first differential coefficient† *are continuous in*

† Thus in the case of a partial differential equation more strict conditions are required for the function on the right-hand side than in the case of an ordinary differential equation. But this condition is not necessary in the converse of the theorem.

D, then (10.28) *is continuous and satisfies the boundary condition, and first and second derivatives exist that satisfy the differential equation* (10.26). *Conversely, the solution of* (10.26), *which satisfies the homogeneous boundary condition, and is continuous in D, and has continuous first derivatives and piecewise continuous second derivatives, can be expressed in the form of* (10.28).

LEMMA 10.8 Symmetricity
A Green function is symmetric:

$$G(x, y; \xi, \eta) = G(\xi, \eta; x, y).$$

10.2. Problem of vibration

We have previously introduced an integral equation with symmetric kernel with regard to the uniform vibration of a string. When the tension of the string, elastic coefficient of the stick, and density of the string and stick are not uniform, then the fundamental equation, with x as the point and $y(x, t)$ as the sidewise displacement at time t, is

$$\rho(x)\frac{\partial^2 y}{\partial t^2} = \frac{\partial}{\partial x}\left((p(x)\frac{\partial y}{\partial x}\right) \tag{10.30}$$

(see § 5.3.3).

Given

$$y(x, 0) = f(x), \qquad y_t(x, 0) = g(x) \tag{10.31}$$

as the initial conditions, and, as the boundary conditions,
(a) when both ends are fixed,

$$y(a, t) = y(b, t) = 0, \tag{10.32a}$$

(b) when both ends are free,

$$y_x(a, t) = y_x(b, t) = 0, \tag{10.32b}$$

(c) when one end is fixed and the other is elastically sustained,

$$y(a, t) = 0, \quad y_x(b, t) + hy(b, t) = 0 \qquad (h > 0). \tag{10.32c}$$

Separate the variables to solve the equation and put

$$y(x, t) = u(x)v(t); \tag{10.33}$$

then we have

$$\frac{d}{dx}\left(p\frac{du}{dx}\right) + \lambda\rho u = 0, \qquad v^{(2)} + \lambda v = 0. \tag{10.34}$$

The boundary conditions become

$$\left.\begin{aligned} &u(a) = u(b) = 0, \\ &u^{(1)}(a) = u^{(1)}(b) = 0, \\ &u(a) = 0, \quad u^{(1)}(b) + hu(b) = 0, \qquad (h > 0), \end{aligned}\right\} \tag{10.35}$$

respectively. Then (10.34) and (10.35) furnish the Sturm–Liouville boundary-value problem.

Let $\varphi_0(x)$ be the eigenfunction corresponding to an eigenvalue λ_0, and from (10.34) we obtain

$$\lambda_0\rho\varphi_0 = -\frac{d}{dx}\left(p\frac{d\varphi_0}{dx}\right);$$

then multiply this expression by φ_0 and take the integral from a to b, which gives

$$\lambda_0\int_a^b \rho(x)\varphi_0(x)dx = -\int_a^b \frac{d}{dx}\left(p\frac{d\varphi_0}{dx}\right)\varphi_0 dx.$$

Taking the partial integral of the right-hand side we get

$$\lambda_0\int_a^b \rho(x)\varphi_0^2(x)dx = p(a)\varphi_0(a)\varphi_0^{(1)}(a) - p(b)\varphi_0(b)\varphi_0^{(1)}(b)$$
$$+\int_a^b p[\varphi_0^{(1)}]^2 dx,$$

where the first two terms are zero in the case of (10.35), they are $h\varphi_0^2$ and positive, and $\rho(x)$, $p(x)$ are positive; then finally $\lambda_0 > 0$. Therefore, every eigenvalue is positive.

If $G(t, \xi)$ denotes the Green function corresponding to (10.34) and (10.35), then we get a homogeneous integral equation:

$$u(x) = \lambda\int_a^b G(t, \xi)x(\xi)d\xi.$$

This is an integral equation with a symmetric kernel. From the results we have obtained so far, we find the following facts.

Put all the positive eigenvalues in order of magnitude as

$$0 < \lambda_1 < \lambda_2 < \lambda_3 < \cdots; \tag{10.36}$$

then their indices are all 1. When the corresponding normal orthogonal eigenfunction system is expressed in

$$\psi_1(x),\ \psi_2(x),\ \psi_3(x),\ \ldots, \tag{10.37}$$

the solution is read as

$$y(x, t) = \sum(a_n \cos\sqrt{\lambda_n}t + b_n \sin\sqrt{\lambda_n}t)\psi_n(x). \tag{10.38}$$

The initial condition becomes (10.31):

$$f(x) = \sum a_n\psi_n(x), \qquad g(x) = \sum\sqrt{\lambda_n}b_n\psi_n(x); \tag{10.39}$$

then the coefficients of (10.38) are determined as Fourier coefficients:

$$a_n = f \cdot \psi_n, \qquad b_n = \frac{1}{\sqrt{\lambda_n}} g \cdot \psi_n. \tag{10.38a}$$

In (10.38), $\sqrt{\lambda_n}$ is called an eigenfrequency (characteristic frequency) and

$$(a_n \cos\sqrt{\lambda_n}t + b_n \sin\sqrt{\lambda_n}t)\psi_n(t)$$

is the eigenvibration (characteristic vibration) for this frequency.

Example 10.1 Lateral vibration of a heavy cord (Lovitt, 1924, pp. 215–222; Kneser, 1924, pp. 43–47)

A heavy cord AB of length 1 is hung down perpendicularly from the fixed point A. It is assumed that the cut surface of the cord is uniform and the density is constant. Let the cord move back and forth with an infinitesimal deformation and a certain initial velocity in the perpendicular plane along AB, then examine the side vibration of the cord. Let B be the origin, put the x-axis in the BA-direction and let y be the side displacement of an arbitrary point x. Since the tension of this point is proportional to y, as 5.2.3, we get the equation of motion

$$\frac{\partial^2 y}{\partial t^2} = c^2 \frac{\partial}{\partial x}\left(x \frac{\partial y}{\partial x}\right), \tag{10.40}$$

the initial conditions

$$y(x, 0) = f(x), \qquad y_t(x, 0) = g(x), \tag{10.41}$$

and the boundary conditions

$$y(1, t) = 0, \qquad y(0, t): \quad \text{bounded.} \tag{10.42}$$

Let us put

$$y(x, t) = u(x)v(t), \tag{10.43}$$

and let it apply in (10.40), which takes the form

$$\frac{v^{(2)}(t)}{v(t)} = c^2 \frac{\dfrac{d}{dx}\left(x \dfrac{du}{dx}\right)}{u(x)}.$$

Here we see that since the left-hand side is a function of t only and the right-hand side is a function of x only, the value of this expression is constant so that the equation may hold. Then let $-\lambda a^2$ be the constant.

Thus we get

$$v^{(2)} + \lambda c^2 v = 0, \tag{10.44}$$

and

$$\frac{d}{dx}\left(x \frac{du}{dx}\right) + \lambda u = 0, \tag{10.45}$$

and (10.42) becomes

$$u(1) = 0, \qquad u(0): \qquad \text{bounded.} \tag{10.46}$$

Making use of the operational calculus, put

$$u(x) \supset U(s);$$

then in the image space, (10.45) becomes

$$-s^2 \frac{dU}{ds} - 2sU + u_0 + sU - u_0 + \lambda U = 0,$$

where $U(0) = u_0$, and the solution is

$$U = C \frac{1}{s} e^{-\lambda/s}.$$

While

$$\frac{1}{s} e^{-\lambda/s} \subset J_0(2\sqrt{(\lambda x)}).$$

Then we get

$$u(x) = CJ_0(2\sqrt{(\lambda x)}). \tag{10.47}$$

From (10.46) we obtain

$$J_0(2\sqrt{\lambda}) = 0, \tag{10.48}$$

and thus the eigenvalue is determined.

Now, put

$$2\sqrt{\lambda} = k \tag{10.49}$$

and put the zero point of Bessel's function in due order as

$$k_1 = 2.405,\ k_2 = 5.520,\ k_3 = 8.654,\ k_4 = 11.792,\ \ldots$$

then we have $(n - {}^1/_2)\pi < k_n < n\pi$, in general.

Let λ_n be the eigenvalue for k_n; then the solution of (10.34) for this eigenvalue becomes

$$v(t) = a_n \cos\frac{ck_n}{2}\, t + b_n \sin\frac{ck_n}{2} t, \tag{10.50}$$

and, from (10.43), the solution of the problem is

$$y(x, t) = \sum C \left(a_n \cos\frac{ck_n}{2}\, t + b_n \sin\frac{ck_n}{2}\, t\right) J_0(k_n\sqrt{x}). \tag{10.51}$$

The eigenfunction for λ_n is

$$\varphi_n(x) = CJ_0(2(\sqrt{\lambda_n x})) = CJ_0(k_n\sqrt{x}), \tag{10.52}$$

which we rearrange into

$$C = \frac{1}{\left(\int_0^1 J_0^2(k_n\sqrt{x})dx\right)^{\frac{1}{2}}}$$

to normalize it.

While from the formula of the Bessel function (Churchill (1941), formula (75)) we have

$$\int_0^1 J_0^2(k_n\sqrt{x})dx = \frac{1}{[J_0'(k_n)]^2},$$

then normally the orthogonal eigenfunction becomes

$$\psi_n(x) = \frac{J_0(k_n\sqrt{x})}{J_0'(k_n)}. \tag{10.53}$$

Thus the solution finally becomes

$$y(x, t) = \sum_1^\infty \left(A_n \cos\frac{ck_n}{2}\, t + B_n \sin\frac{ck_n}{2} t\right)\psi_n(x). \tag{10.54}$$

The coefficients A_n, B_n are determined by the remaining condition (10.50), which appears to be

$$\sum A_n\psi_n = f(x), \qquad \sum \frac{ak_n}{2} B_n\psi_n = g(x).$$

Therefore, A_n, B_n become Fourier coefficients in $\{\psi_n\}$, and are expressed as

$$A_n = f \cdot \psi_n = \int_0^1 f(x)\psi_n(x)dx,$$

$$\frac{Ck_n}{2} B_n = g \cdot \psi_n = \int_0^1 g(x)\psi_n(x)dx. \qquad (10.55)$$

Now, since (10.45) and (10.46) are the boundary-value problems of the Sturm–Liouville system, they must be reduced to integral equations with symmetric kernels. Form a Green function $G(x, \xi)$ for this purpose. This satisfies the following conditions;

(i) $G(x,\xi)$ satisfies the homogeneous differential equation

$$\frac{d}{dx}\left(x\frac{dK}{dx}\right) = 0$$

for $x \neq \xi$;

(ii) $G(x, \xi)$ is continuous in x at (0, 1) and satisfies the homogeneous boundary condition $G(0,\xi)$ bounded, $G(1,\xi) = 0$.

(iii) dG/dx, d^2G/dx^2 are continuous except at $t = \xi$, and at $t = \xi$ they have the following jump:

$$\left[\frac{dG}{dx}\right]_{x=\xi+0} - \left[\frac{dG}{dx}\right]_{x=\xi-0} = -\frac{1}{\xi}.$$

From (i) we get

$$G(x, \xi) = \begin{cases} \alpha_0 \log x + \beta_0, & (0, \xi) \\ \alpha_1 \log x + \beta_1, & (\xi, 1), \end{cases}$$

and from (ii), $\alpha_0 = 0$, $\beta_1 = 0$, but in order for this to be continuous in $x(0, 1)$, we only need $\beta_0 = a_1 \log\xi$. From (iii),

$$\frac{\alpha_1}{\xi} - 0 = -\frac{1}{\xi},$$

then $\alpha_1 = -1$, and consequently we get

$$G(x, \xi) = \begin{cases} -\log \xi & (0, \xi), \\ -\log t & (\xi, t). \end{cases} \qquad (10.56)$$

Hence, the problem becomes a homogeneous integral equation:

$$u(x) = \lambda \int_0^1 G(x, \xi) u(\xi) d\xi. \tag{10.57}$$

It is evident from (10.46) that this integral equation has a symmetric kernel, but it is discontinuous at $t = \xi = 0$. Schmidt (1906) proved that Hilbert's various theorems on the symmetric kernel also hold in the case of a discontinuous kernel if, whenever $f(x)$ is continuous in I,

$$\int_a^b K(t, \xi) f(\xi) d\xi$$

is continuous in I, and the iterated kernel $\mathring{K}^2$ is continuous and not uniformly zero. Therefore, in our case there exist enumerable infinite number of eigenvalues:

$$0 < \lambda_1 < \lambda_2 < \lambda_3 < \cdots,$$

and the corresponding eigenfunction system forms a normally orthogonal function system. This agrees with the result we have already obtained.

10.3. Problem of heat conduction

In § 5.1, we treated the problem of the one-dimensional heat conduction of a conductor having a uniform section whose surface was insulated, and we showed that the problem can be reduced to an integral equation with symmetric kernels. In this section, we treat the case where heat radiates from the surface of a fine bar or a string.

Take the x-axis along the longitudinal direction and let $u(x, t)$ be the temperature at time t. Let $s(x)$ be specific heat, $\sigma(x)$ be the sectional area, $k(x)$ be the heat conductivity, $\rho(x)$ be the line density, $l(x)$ be the length of the perimeter, and 0° the temperature in the environment of the conductor. Put hlu as the radiating calories, and neglect the sectional area for the length of the conductor and suppose the temperature to be isothermal in every section. Then the fundamental equation of heat conduction is

$$s\rho\sigma \frac{\partial u}{\partial t} = \frac{\partial}{\partial x}\left(k\sigma \frac{\partial u}{\partial x}\right) - hlu. \tag{10.58}$$

Here, put

$$s\rho\sigma = g(x), \qquad k\sigma = p(x), \qquad hl = q(x),$$

and we have $g(x) > 0$, $p(x) > 0$, $q(x) \geqq 0$, and when the surface is insulated we have $q(x) = 0$.

From the condition of heat current at both ends of the conductor $x = a$, $x = b$, we have

$$k(a)u_x(a, t) - h(a)u(a, t) = 0,$$
$$k(b)u_x(b, t) + h(b)u(b, t) = 0.$$

Put $h(a)/k(a) = H_0(\geqq 0)$, $h(b)/k(b) = H_1(\geqq 0)$, and for the boundary conditions we obtain

$$\left.\begin{aligned} u_x(a, t) - H_0 u(a, t) &= 0, \\ u_x(b, t) - H_1 u(b, t) &= 0. \end{aligned}\right\} \tag{10.59}$$

We use a temperature distribution as the initial condition, which gives

$$u(x, 0) = f(x). \tag{10.60}$$

From (10.59) it is necessary that $f(t)$ on the boundary satisfies the conditions

$$\left.\begin{aligned} f^{(1)}(a) - H_0 f(a) &= 0, \\ f^{(1)}(b) + H_1 f(b) &= 0. \end{aligned}\right\} \tag{10.61}$$

Use separation of variables:

$$u(x, t) = U(x)v(t), \tag{10.62}$$

then for $U(x)$ we get the Sturm–Liouville boundary–value problem:

$$\frac{d}{dx}\left(p\frac{dU}{dx}\right) - q(x)U + \lambda g U = 0, \tag{10.63}$$

$$\left.\begin{aligned} U^{(1)}(a) - H_0 U(a) &= 0, \\ U^{(1)}(b) + H_1 U(b) &= 0.\dagger \end{aligned}\right\} \tag{10.64}$$

For $v(t)$ we obtain

$$\frac{dv}{dt} + \lambda v = 0. \tag{10.65}$$

Here, if φ_0 denotes the eigenfunction corresponding to an arbitrary eigenvalue λ_0, from (10.63) and (10.64) we get

† Using the Green function we can see that the boundary-value problem (10.63) (10.64) are equal to the homogeneous integral equation

$$U(x) = \lambda \int_a^b G(t, \xi) g(\xi) U(\xi) d\xi.$$

$$L[\varphi_0] + \lambda_0\varphi_1 g = 0, \tag{10.66}$$

$$\varphi_0^{(1)}(a) - H_0\varphi_0(a) = 0, \qquad \varphi_1^{(1)}(b) + H_0\varphi_0(b) = 0. \tag{10.67}$$

From (10.66) we get

$$\lambda_0\int_a^b g\varphi_0^2 dx = -\int_a^b \varphi_0 L[\varphi_0]dx.$$

Take the partial integral, which comes to

$$\lambda_0\int_a^b g\varphi_0^2 dx = -p(b)\varphi_0(b)\varphi_0^{(1)}(b) + p(a)\varphi_0(a)\varphi_0^{(1)}(a)$$
$$+\int_a^b (p\varphi_0^{(1)2} + q\varphi_0^2)dx.$$

From (10.67) we get

$$\lambda_0\int_a^b g\varphi_0^2 dx = p(b)\varphi_0^2(b)H_1 + p(a)\varphi_0^2(a)H_0$$
$$+\int_a^b (p\varphi_0^{1)2} + q\varphi_0^2)dx. \tag{10.68}$$

Since $p(b)$, $\varphi_0^2(b)$, $p(a)$, $\varphi_0^2(a)$, p are positive and H_1, H_0, $\varphi_0^{(1)2}$, φ_0^2, q are either positive or zero, we get

$$\lambda_0 \geqq 0.$$

Therefore, the eigenvalues are

$$0 \leqq \lambda_1 < \lambda_2 < \lambda_3 < \cdots,$$

and each of their indexes is 1, and the normally orthogonal eigenfunction system

$$\psi_1(x),\ \psi_2(x),\ \psi_3(x),\ \ldots$$

correspond to them, and when the solution of (10.65) for λ_n, is expressed in $C_n\exp[-\lambda_n t]$ the solution (10.62) becomes

$$u(x, t) = \sum_1^\infty C_n\exp[-\lambda_n t]\psi_n(x), \tag{10.69}$$

and thus it is known that C_n is a Fourier coefficient.
The condition (10.60) becomes

$$f(x) = \sum_1^\infty C_n\psi_n(x), \tag{10.70}$$

$$C_n = f\cdot\psi_n. \tag{10.70a}$$

Example 10.2 *One-dimensional heat conduction with radiation* (Churchill (1944), pp. 243–256)

Take the particular case when

$$g(x) = 1, \qquad p(x) = 1$$

in (10.58). This is the one-dimensional heat-conduction problem of a uniform homogeneous bar where only the heat conductivity is variable. The standard equation for the problem is

$$\frac{\partial u}{\partial t} = \frac{\partial^2 u}{\partial x^2} - q(x)u. \tag{10.71}$$

Let one be the length of the bar, while the boundary condition and the initial condition are

$$u(0, t) = 0, \qquad u(1, t) = 0; \tag{10.72}$$

$$u(x, 0) = f(x); \tag{10.73}$$

where $f(0) = f(1) = 0$.

By operational calculus, changing t into s, put

$$u(x, t) \supset U(x, s);$$

then taking (10.73) into consideration, (10.71) becomes

$$\frac{d^2U}{dx^2} - q(x)U - sU = -f(x) \tag{10.74}$$

in the image space. The conditions (10.72) become

$$U(0, s) = 0, \qquad U(1, s) = 0. \tag{10.75}$$

Take the homogeneous equation

$$\frac{d^2U}{dx^2} - qU - sU = 0. \tag{10.76}$$

for (10.61). This equation has two solutions $U_1(x, s)$, $U_2(x, s)$, which can be put as

$$U_1(0, s) = 0, \ U_1^{(1)}(0, s) = 1,$$

$$U_2(1, s) = 0, \ U_2^{(1)}(1, s) = 1. \tag{10.77}$$

Here, we can directly indicate that the solution which satisfies the condition (10.77) of (10.76) is

$$U(x, s) = \int_0^1 G(x, \xi; s) f(\xi) d\xi, \tag{10.78}$$

and that $G(x, \xi; s)$ is expressed as

$$G(x, \xi; s) = \begin{cases} -\dfrac{U_2(x, s)U_1(\xi, s)}{U_1(1, s),}, & \xi \leqq x, \\ -\dfrac{U_1(x, s)U_2(\xi, s)}{U_1(1, s),}, & \xi > x. \end{cases} \tag{10.79}$$

Now put

$$D = U_1(x, s)U_2{}^{(1)}(x, s) - U_1{}^{(1)}(x, s)U_2(x, s) \tag{10.80}$$

and differentiate this over x, which results in

$$D^{(1)} = U_1(x, s)U_2{}^{(2)}(x, s) - U_1{}^{(2)}(x, s)U_2(x, s),$$

where the right-hand side disappears since both U_1 and U_2 are solutions of (10.76), and this shows that D has no dependence on x; then, in particular, put $x = 1$, and noting (10.77), we get

$$D = U_1(1, s). \tag{10.81}$$

When $\rho_n(x, t)$ denotes the residue for the pole $s = s_n$ of $e^{st}U(x, s)$, the original function $u(x, t)$ of $U(x, s)$ becomes

$$u(x, t) = \sum \rho_n(x, t), \tag{10.82}$$

and the initial condition (10.73) becomes

$$f(x) - \sum_1^\infty \rho_n(x, 0). \tag{10.83}$$

From (10.79), the pole of Ue^{st} is

$$U_1(1, s_n) = 0; \tag{10.84}$$

then, from $D = 0$ we have $U_1U_2{}^{(1)} - U_1{}^{(1)}U_2 = 0$ and $d(U_1/U_2) = 0$, and therefore we get

$$U_2(x, s_n) = CU_1(x, s_n). \tag{10.85}$$

In addition, since $U_1(0, s) = 0$, $U_1(1, s_n) = 0$, it follows that $U_1(x, s_n)$ satisfies

$$U_1{}^{(2)} - qU_1 = s_nU_1, \; U_1(0, s_n) = U_1(1, s_n) = 0. \tag{10.86}$$

Next we shall show that the poles are all simple. First, as $U_1(x, s)$ is the solution of (10.76), we have

$$U_1^{(2)}(x, s) - (q + s)U_1(x, s) = 0$$

which, being partially differentiated in s on both sides, becomes

$$\frac{\partial}{\partial s} U_1^{(2)}(x, s) - (q + s)\frac{\partial}{\partial s} U_1(x, s) = U_1(x, s).$$

If we put

$$\left[\frac{\partial}{\partial s} U_1(x, s)\right]_{s=s_n} = A(x, s_n), \tag{10.87}$$

it can be considered from the nature of the problem that U_1 is analytic in s, except in $s = s_n$, and then we can change the order of differentiation, which gives

$$A^{(2)}(x, s_n) - (q + s_n)A(x, s_n) = U_1(x, s_n).$$

At the same time we have

$$U_1^{(2)}(x, s_n) - (q + s_n)U_1(x, s_n) = 0;$$

then eliminating $q + s_n$ from the above two expressions we get $A^{(2)} U_1 - AU_1^{(2)} = U_1^2$; , viz.

$$\frac{d}{dx}[A^{(1)}(x, s_n)U_1(x, s_n) - U_1^{(1)}(x, s_n)A(x, s_n)] = [U_1(x, s_n)]^2.$$

Noting that $A(0, s_n) = 0$, $U_1(0, s_n) = 0$, $U_1(1, s_n) = 0$ from (10.64), then integrating both sides in x from 0 to 1 gives

$$- A(1, s_n)U_1^{(1)}(1, s_n) = \int_0^1 [U_1(x, s_n)]^2 dx. \tag{10.88}$$

As will be shown later, the eigenfunction $U_1(x, s_n)$ is a real function; then the right-hand side is positive, and therefore we have

$$A(1, s_n) \neq 0, \quad \text{viz.} \quad \left[\frac{\partial}{\partial s} U_1(1, s)\right]_{s=s_n} \neq 0.$$

This expression show that $s = s_n$ is the simple root of $U_1(1, s) = 0$.

Next, let g be the original function of G and, putting

$$G(x, \xi;\ s) \subset g(x, \xi;\ t),$$

from (10.66) we obtain

$$g(x, \xi;\ t) = \begin{cases} -\sum \dfrac{U_2(x, s_n)U_1(\xi, s_n)}{A(1, s_n)} \exp[s_n t], & \xi \leqq x; \\ -\sum \dfrac{U_1(x, s_n)U_2(\xi, s_n)}{A(1, s_n)} \exp[s_n t], & \xi > x; \end{cases}$$

then noting (10.85), we have

$$g(x, \xi;\ t) = -\sum \frac{CU_1(x, s_n)\exp[s_n t]}{A(1, s_n)} U_1(\xi, s_n). \qquad (10.89)$$

Again from (10.77) and (10.58) we get

$$CU_1^{(1)}(1, s_n) = U_2^{(1)}(1, s_n) = 1,$$

and, from (10.88),

$$-\frac{1}{A(1, s_n)} = \frac{U_1^{(1)}(1, s_n)}{\int_0^1 [U_1(x, s_n)]^2 dx}.$$

Finally, (10.89) becomes

$$g(x, \xi;\ t) = \sum \frac{U_1(x, s_n)\exp[s_n t]}{\int_0^1 [U_1(x, s_n)]^2 dx} U_1(\xi, s_n). \qquad (10.90)$$

Therefore, from (10.78) the solution is

$$u(x, t) = \sum \frac{U_1(x, s_n)\exp[s_n t]}{\int_0^1 [U_1(x, s_n)]^2 dx} \int_0^1 U_1(\xi, s_n) f(\xi) d\xi. \qquad (10.91)$$

Here, if we put

$$\psi_n(x) = U_1(x, s_n) \Big/ \left\{ \int_0^1 [U_1(x, s_n)]^2 dx \right\}^{\frac{1}{2}}, \qquad (10.92)$$

we have $\int_0^1 [\psi_n(x)]^2 dx = 1$, and then $\{\psi_n(x)\}$ is a normalized eigen-function system.

In the following presentation we shall show that every eigenvalue is real. Suppose that an eigenfunction is expressed as

$$s = \alpha + i\beta,$$

and its corresponding eigenfunction is

$$U(x, s) = X(x, s) + iY(x, s);$$

then since $U(x, s)$ must satisfy (10.76) and (10.77), separating it into real and imaginary parts we get

$$X^{(2)} - (\alpha + q)X + \beta Y = 0,$$
$$Y^{(2)} - (\alpha + q)Y - \beta X = 0,$$
$$X(0, s) = X(1, s) = 0, \; Y(0, s) = Y(1, s) = 0. \tag{10.93}$$

Therefore we have

$$-\beta(X^2 + Y^2) = X^{(2)}Y - Y^{(2)}X. \tag{10.94}$$

Taking the integral on both sides from 0 to 1, thus making

$$-\beta\int_0^1 (X^2 + Y^2)dx = [X^{(1)}Y - Y^{(1)}X]_0^1 = 0,$$

we obtain

$$\beta = 0.$$

Further, from (10.94)

$$X^{(1)}Y - Y^{(1)}X = \text{constant},$$

and we find that this value is zero from (10.93). Consequently we get

$$Y = kX,$$

(where k is constant). Therefore the eigenfunction corresponding to $s = \alpha$ is

$$U(x, s) = (1 + ik)X(x, s),$$

where

$$(1 - ik)U(x, s)$$

satisfies (10.75) and (10.76). Thus without loss of generality we may suppose that the eigenfunction corresponding to a real eigenvalue is a real function.

Next, let $U(x, s_m)$, $U(x, s_n)$ be the eigenfunctions corresponding to two different eigenvalues s_m, s_n; then from (10.76) we have

$$U^{(2)}(x, s_m) = (s_m + q)U(x, s_m); \qquad U^{(2)}(x, s_n) = (s_n + q)U(x, s_n).$$

Therefore

$$U^{(2)}(x, s_m)U(x, s_n) - U^{(2)}(x, s_n)U(x, s_m)$$
$$= (s_m - s_n)U(x, s_m)U(x, s_n)$$

Integrating both sides over x from 0 to 1 and taking (10.75) into consideration, we get

$$(s_m - s_n)\int_0^1 U(x, s_m)U(x, s_n)dx = 0.$$

This equation shows that the eigenfunctions corresponding to the different eigenvalues intersect orthogonally. Therefore (10.92) forms a normalized orthogonal eigenfunction system.

The solution (10.91) becomes

$$u(x, t) = \sum c_n\psi_n(x)\exp[s_n t],\ c_n = \int_0^1 f(\xi)\psi_n(\xi)d\xi. \tag{10.95}$$

To show that (10.95) actually converges absolutely and uniformly we have to calculate the Green function further and calculate the eigenvalue and the eigenfunction precisely.

Thus, by using a differential equation, although it is troublesome, we can reach an accurate result without difficulty if we introduce an integral equation with a symmetric kernel making use of the Green function.

Separate the variables of the solution and put

$$u(x, t) = U(x)v(t), \tag{10.96}$$

then as for $U(x)$ from (10.71) we have

$$U^{(2)} - qU = -\lambda U, \tag{10.97}$$

and for the boundary conditions from (10.72) we have

$$U(0) = 0, \qquad U(1) = 0. \tag{10.98}$$

Comparing (10.76) and (10.97) we can see that the operator s is equivalent to $-\lambda$.

Then let $U_1(x)$, $U_2(x)$ be the two solutions of (10.97), and let each of them satisfy

$$\begin{aligned} U_1(0) &= 0, \qquad U_1^{(1)}(0) = 1,\\ U_2(1) &= 0, \qquad U_2^{(1)}(1) = 1. \end{aligned} \tag{10.99}$$

Now if

$$G(x, \xi) = \begin{cases} cU_1(x)U_2(\xi), & \xi \leqq x,\\ cU_1(\xi)U_2(x), & \xi > x, \end{cases} \tag{10.100}$$

it follows that

(i) $G(x, \xi)$ satisfies the homogeneous integral equation for $x \neq \xi$;
(ii) $G(x, \xi)$ is continuous in x, and satisfies the homogeneous boundary condition (10.98);
(iii) dG/dx, d^2G/dx^2 are continuous except at $x = \xi$, and at $x = \xi$ we have

$$\left[\frac{dG}{dx}\right]_{\xi-0}^{\xi+0} = cU_1(\xi)U_2{}^{(1)}(\xi) - cU_1{}^{(1)}(\xi)U_2(\xi) = -1. \quad (10.101)$$

Now put

$$D(x) = cU_1(x)U_2{}^{(1)}(x) - cU_1{}^{(1)}(x)U_2(x),$$

and differentiate this in x and note that $U_1(x)$, $U_2(x)$ are the solutions of (40); then we have

$$\frac{d}{dx}D(x) = 0.$$

Therefore, $D(x)$ is independent of x.

Then in (10.101), put $\xi = 1$, and noting (10.99), we get

$$c = -1/U_1(1).$$

Therefore the, Green function becomes

$$G(x, \xi) = \begin{cases} -\dfrac{U_1(x)U_2(\xi),}{U_1(1)}, & \xi \leqq x; \\ -\dfrac{U_1(\xi)U_2(x)}{U_1(1)}, & \xi > x. \end{cases} \quad (10.102)$$

In this equation $[pUU^{(1)}]_0^1 = 0$ and $p = 1 > 0$, $q > 0$; therefore every eigenfunction is positive and the solution of the problem for the normalized orthogonal eigenfunction system $\{\psi_i\}$ may be expressed by (10.95), which is absolutely and uniformly convergent.

10.4. Integral equations and variation problems

A variation problem is to find a function $x(t)$, continuous up to its first derivative, which makes the value of the functional of the argument function $x(t)$

$$I[x(t)] = \int_a^b f(t, x(t), x^{(1)}(t))dt \quad (10.103)$$

a minimum, and satisfies the condition

$$x(a) = \alpha,\ x(b) = \beta \quad (\alpha,\ \beta = \text{constant}). \tag{10.104}$$

The function $f(t, x(t), x^{(1)}(t))$ is supposed to be continuous in all of its variables t, $x(t)$, $x^{(1)}(t)$, up to its third partial derivative. If we express its partial derivatives in $x(t)$, $x^{(1)}(t)$ by f_x, $f_{x'}$, and so on, it is required that the solution of the variation problem satisfies the differential equation

$$f_x - \frac{d}{dt} f_{x'} = 0.\dagger \tag{10.105}$$

This is called Euler's differential equation.

When given the condition

$$\int_a^b g(t, x(t), x^{(1)}(t))dt = l \quad (l = \text{constant}), \tag{10.106}$$

is added to the condition (10.104), we call the problem an isoperimetric variational problem. The function g, as well as f, is supposed to be continuous up to its third partial derivatives. Then take an arbitrary constant λ, put

$$h \equiv f - \lambda g,$$

and we have only to select that which minimizes

$$\int_a^b h(t, x, x^{(1)})dt$$

and satisfies the conditions (10.104) and (10.106). Then (10.105) becomes

$$h_x - \frac{d}{dx} h_{x'} = 0,$$

viz.

$$f_x - \lambda g_x + \frac{d}{dt}(f_{x'} - \lambda g_{x'}) = 0. \tag{10.107}$$

(1) *Dirichlet's problem.* D. Hilbert called the isoperimetric variational problem *Dirichlet's problem.* With it we find a function $x(t)$ that minimizes the functional

† We give here only the result. For a detailed explanation consult, for example, (Bolza, 1904).

$$D[x(t)] = \int_a^b \left[p\left(\frac{dx}{dt}\right)^2 + qx^2 \right] dt, \tag{10.108}$$

and satisfies the conditions

$$x(a) = 0, \qquad x(b) = 0 \tag{10.109}$$

and

$$\int_a^b x^2(t)dt = 1. \tag{10.110}$$

Here, $x(t)$ is continuous up to its second derivative, $p(t)(>0)$ is continuous up to its first derivative, and $q(t)$ is also continuous.

Then we have

$$f = p(x^{(1)})^2 + qx^2, \qquad g = x^2.$$

Therefore, (10.107) becomes

$$\frac{d}{dt}\left(p\frac{dx}{dt}\right) - (q - \lambda)x = 0. \tag{10.111a}$$

This is expressed in

$$L[x] + \lambda x = 0, \tag{10.111}$$

which being combined with the boundary conditions (10.104) becomes the Sturm–Liouville system. Therefore, the results obtained in § 10.1 apply without exception.

It is known that the solution of the variation problem exists only when λ takes eigenvalues, that the eigenvalues are all real and countable, and that each of them has index 1. Further, we have $p > 0$; then from Theorem 10.4 there is the minimum eigenvalue and all the eigenvalues are taken as

$$\lambda_1 < \lambda_2 < \lambda_3 < \cdots,$$

and to these correspond the normalized orthogonal eigenfunction system

$$\psi_1, \psi_2, \psi_3, \ldots.$$

Then, for $n = 1, 2, \ldots,$

$$L(\psi_n) + \lambda_n \psi_n = 0,$$

$$\psi_n(a) = 0, \qquad \psi_n(b) = 0,$$

$$\int_a^b \psi_n^2(t)dt = 1.$$

While if Dirichlet's problem has a solution it must take the form

$$x = c\psi_n$$

for $\lambda = \lambda_n$. But from the condition (10.110), we have

$$c^2 \int_a^b \psi_n{}^2 dt = 1;$$

then $c = \pm 1$ and hence we have

$$x = \pm\psi_n(t).$$

Yet from (10.20) we have

$$\lambda_n = \int_a^b \left[p\left(\frac{d\psi_n}{dt}\right)^2 + q\psi_n{}^2 \right] dt;$$

then, that which minimizes $D[x(t)]$ is, finally,

$$x(t) = \pm\psi_1(t),$$

where the value of λ is λ_1.

Then we have the following lemma.

LEMMA 10.9 *Dirichlet's problem*

The function, which is continuous up to its second derivative, which minimizes

$$D[x(t)] = \int_a^b \left[p\left(\frac{dx}{dt}\right)^2 + qx^2 \right] dt,$$

and which satisfies the conditions $x(a) = 0$, $x(b) = 0$ *and*

$$\int_a^b x^2(t) dt = 1,$$

is $x(t) = \pm\, \psi_1(t)$.

Proof. (i) By taking the partial integral we get

$$\int_a^b (px^{(1)})x^{(1)} dt = [xpx^{(1)}]_a^b - \int_a^b x \frac{d}{dt}\left(p \frac{dx}{dt}\right) dt,$$

the first term of which disappears from the boundary condition (10.109). Then we have

$$D[x] = -\int_a^b x \left[\frac{d}{dt}\left(p \frac{dx}{dt}\right) - qx \right] dt.$$

Therefore,

$$D[x] = -\int_a^b xL[x]dt.$$

If we put

$$-L[x] = g(t),$$

$g(t)$ is continuous in I and yields

$$D[x] = +\int_a^b x(t)g(t)dt.$$

While, since we have $L[x] + g = 0$, $x(a) = 0$, $x(b) = 0$, we get

$$x(t) = \int_a^b K(t, \xi)g(\xi)d\xi$$

from Lemma 10.1. Integrating the equation over t and multiplying both sides by $g(t)$, we obtain

$$D[x] = \int_a^b \int_a^b K(t, \xi)g(t)g(\xi)dt\, d\xi. \tag{10.112}$$

(ii) From question (6) in Chapter 6 we have

$$x(t) = \int_a^b K(t, \xi)g(\xi)d\xi = \sum_1^\infty \frac{g \cdot \psi_i}{\lambda_i}\psi_i(t),$$

where the series on the right-hand side is absolutely and uniformly convergent. Put

$$g \cdot \varphi_i = \int_a^b g(t)\psi_i(t)dt \equiv c_i,$$

and then

$$x(t) = \sum_1^\infty \frac{c_i}{\lambda_i}\psi_i(t); \tag{10.113}$$

then if we put

$$x^2(t) = \sum_{ij} \frac{c_i c_j}{\lambda_i \lambda_j}\psi_i(t)\psi_j(t),$$

it is also absolutely and uniformly convergent. Therefore, we have

$$\int_a^b x^2(t)dt = \sum_{ij} \frac{c_i c_j}{\lambda_i \lambda_j}\int_a^b \psi_i(t)\psi_j(t)dt = 1,$$

while $\{\psi_i\}(t)$ is a normalized orthogonal eigenfunction system. Finally, we obtain

$$\sum \frac{c_i^2}{\lambda_i^2} = 1. \tag{10.114}$$

From Question 6 in Chapter 6 we have

$$D[x] = \sum_1^\infty \frac{(g \cdot \varphi_i)^2}{\lambda_i} = \sum_1^\infty \frac{c_i^2}{\lambda_i}. \tag{10.115}$$

(iii) Next, we calculate

$$D[x] - D[\pm \psi_1] = D[x] - \lambda_1.$$

From (10.114) we get

$$\lambda_1 = \sum_{i=1}^\infty \frac{c_i^2 \lambda_1}{\lambda_i^2}.$$

Then from (10.115) we have

$$D[x] - \lambda_1 = \sum \frac{c_i^2}{\lambda_i^2} (\lambda_i - \lambda_1) \geqq 0.$$

Therefore,

$$D[x] - D[\pm \psi_1] \geqq 0.$$

When $c_i = 0$ stands for all of $i = 2, 3, \ldots$, the equality sign holds. Now, we have $c_1 = \pm \lambda_1$ from (10.114), but by (10.113) we have $x = \pm \psi_1$ in this case. Therefore, as long as $x \not\equiv \pm \psi_1$, we have

$$D[x] - D[\pm \psi_1] > 0.\dagger$$

In addition to these, for the function $\eta(t)$ continuous up to its second derivative, if we rewrite the second variation

$$\delta^2 I = \int_a^b (f_{xx}\eta^2 + 2f_{xx^{(1)}}\eta\eta^{(1)} + f_{x^{(1)}x^{(1)}}\eta^{(1)2})dt \geqq 0,$$

$$\eta(a) = 0, \qquad \eta(b) = 0 \tag{10.116}$$

in the form

$$\delta^2 I = \int_a^b [P\eta^2 + 2Q\eta\eta^{(1)} + R\eta^{(1)2}]dt \geqq 0,$$

we have the following lemma.

LEMMA 10.10

In the second variation $\delta^2 I$, if in $R > 0$, λ_1 is the minimum eigenvalue in the boundary-value problem

† In this proof we began by supposing that $\gamma = 0$ is not an eigenvalue. For the extension to this case, see (Lovitt, 1924, pp. 201–240).

$$\frac{d}{dt}\left(R\frac{dx}{dt}\right) - (P - Q)x = 0, \quad x(a) = 0, \quad x(b) = 0,$$

the necessary and sufficient condition for

$$\delta^2 I \geqq 0$$

is that $\lambda_1 \geqq 0$.

Proof. See (Lovitt 1924, pp. 204–206).

(2) *Prediction of time series* (Kondo 1954). Suppose that the time series $f(t)$ is a continuous function in t, is bounded and has the autocorrelation function

$$\varphi(\tau) = \lim_{T\to\infty} \frac{1}{2T}\int_{-T}^{T} f(t+\tau)f(t)dt. \tag{10.117}$$

Then suppose also that $\varphi(\tau)$ is continuous and there exists the Fourier transformation

$$\int_{-\infty}^{\infty} e^{iu\tau}\varphi(\tau)d\tau. \tag{10.118}$$

The problem of prediction of time series is to preestimate the future value $f(t+\alpha)$ $(\alpha < 0)$ of $f(t)$ knowing it value in $(-\infty, t)$. N. Wiener (1949) assumed that the effect of the value $f(t-\tau)$ at one moment in the past, $t-\tau$ $(0 \leqq \tau < \infty)$, on the estimated value, depends on the time elapsed τ only (uniform for time), and is cumulative. Then taking the influence function $K(\tau)$, he tried to determine the form of $K(\tau)$ so as to best approximate future value by means of

$$\int_{0}^{\infty} f(t-\tau)K(\tau)d\tau. \tag{10.119}$$

We use the word 'best' in the sense of least squares (root mean square) and, choosing the proper form of $K(\tau)$, this problem reduces to the variational problem to minimize the functional

$$I[K(\tau)] \equiv \lim_{T\to\infty} \frac{1}{2T}\int_{-T}^{T}\left\{f(t+\alpha) - \int_{0}^{\infty} f(t-\tau)K(\tau)d\tau\right\}^2 dt. \tag{10.120}$$

Expand the right-hand side of (10.120) and change the order of integral and limit; then we have

$$I[K] = \lim_{T\to\infty} \frac{1}{2T}\int_{-T}^{T} f(t+\alpha)f(t+\alpha)dt$$

$$-2\int_0^\infty K(\tau)d\tau \lim_{T\to\infty}\frac{1}{2T}\int_{-T}^{T} f(t+\alpha)f(t-\tau)dt$$

$$+\int_0^\infty K(\tau)d\tau \int_0^\infty K(\sigma)d\sigma \lim_{T\to\infty}\frac{1}{2T}\int_{-T}^{T} f(t-\tau)f(t-\sigma)dt,$$

which is rearranged as

$$I[K]=\varphi(0)-2\int_0^\infty \varphi(\alpha+\tau)K(\tau)d\tau+\int_0^\infty K(\tau)d\tau\int_0^\infty \varphi(\tau-\sigma)K(\sigma)d\sigma.$$

Now, in order to find the solution of this variation problem, take a function of bounded variation $M(\tau)$ in $t(0, \infty)$ and calculate the function group $K(\tau)+\varepsilon M(\tau)$, with an arbitrary real number ε. Then the above expression becomes

$$I[K+\varepsilon M]=I[K]-2\varepsilon\int_0^\infty \varphi(\tau+\alpha)M(\tau)d\tau+2\varepsilon\int_0^\infty M(\tau)d\tau$$

$$\times\int_0^\infty \varphi(\tau-\sigma)K(\sigma)d\sigma+\varepsilon^2\int_0^\infty M(\tau)d\tau\int_0^\infty \varphi(\tau-\sigma)M(\sigma)d\sigma. \qquad (10.121)$$

If $K(\tau)$ is the minimal solution,

$$\left(\frac{\partial I[K+\varepsilon M]}{\partial\varepsilon}\right)_{\varepsilon=0}=0$$

is necessary, therefore

$$\int_0^\infty\left[\varphi(\tau+\alpha)-\int_0^\infty \varphi(\tau-\sigma)K(\sigma)d\sigma\right]M(\sigma)d\sigma=0,$$

where M is an arbitrary function, then finally

$$\varphi(\tau+\alpha)=\int_0^\infty \varphi(\tau-\sigma)K(\sigma)d\sigma, \qquad \tau\geqq 0 \qquad (10.122)$$

is obtained as the necessary condition.

Further, it is made clear as follows that (10.122) serves also as the sufficient condition for $K(\tau)$ to be minimal. That is, when (10.122) holds, (10.121) becomes

$$I[K+\varepsilon M]-I[K]=\varepsilon^2\int_0^\infty M(\tau)d\tau\int_0^\infty \varphi(\tau-\sigma)M(\sigma)d\sigma,$$

while we have

$$\lim_{T\to\infty}\frac{1}{2T}\int_{-T}^{T}\left[\int_0^\infty f(t-\tau)M(\tau)d\tau\right]^2 dt$$

$$= \int_0^\infty M(\tau)d\tau \int_0^\infty \varphi(\tau - \sigma)M(\sigma)d\sigma.$$

Then we get

$$I[K + \varepsilon M] - I[K] \geqq 0,$$

and from $I[K + \varepsilon M]$ we see that this takes a minimum value when $M \equiv 0$.

Thus, to determine $K(\tau)$ in order to get the best estimate, we finally solve the integral equation

$$\int_0^\infty \varphi(\tau - \sigma)K(\sigma)d\sigma = \varphi(\tau + \alpha), \qquad \tau \geqq 0,$$

where $\varphi(\tau)$ is the autocorrelation function of the time series $f(t)$ and is supposed to be known by observational data. We consider the problem in the following way. Suppose that $\varphi(\tau)$ is expressible with $\psi_1(\tau)$, $\psi_2(\tau)$ in

$$\varphi(\tau) = \int_{-\infty}^\infty \psi_2(\sigma)\psi_1(\tau - \sigma)d\sigma, \tag{10.123}$$

provided

$$\psi_1(\tau) = 0, \qquad \tau < 0;$$
$$\psi_2(\tau) = 0, \qquad \tau > 0.$$

Then, putting (10.123) into the integral equation (10.122) we get

$$\varphi(\tau + \alpha) = \int_{-\infty}^0 \psi_2(\nu)d\nu \int_0^\infty \psi_1(\tau - \nu - \sigma)K(\sigma)d\sigma. \tag{10.124}$$

Here, if we write $\varphi(\tau)$ as

$$\varphi(\tau) = \int_{-\infty}^0 \psi_1(\tau - \nu)\psi_2(\nu)d\nu, \tag{10.125}$$

(10.124) becomes

$$\int_{-\infty}^0 \psi_1(\tau + \alpha - \nu)\psi_2(\nu)d\nu = \int_{-\infty}^0 \psi_2(\nu)d\nu \int_0^\infty \psi_1(\tau - \nu - \sigma)K(\sigma)d\sigma.$$

Then we have

$$\int_{-\infty}^0 \psi_2(\nu)\left[\psi_1(\tau + \alpha - \nu) - \int_0^\infty \psi_1(\tau - \nu - \sigma)K(\sigma)d\sigma\right]d\nu = 0.$$

Therefore, the integral equation (10.121) finally becomes

$$\psi_1(\tau + \alpha - \nu) - \int_0^\infty \psi_1(\tau - \nu - \sigma)K(\sigma)d\sigma = 0, \qquad \tau > 0, \nu < 0.$$

Here, if we replace $\tau - \nu$ with τ this expression becomes

$$\psi_1(\tau + \alpha) - \int_0^\infty \psi_1(\tau - \sigma)K(\sigma)d\sigma = 0, \quad \tau > 0.$$

While since

$$\psi_1(\tau - \sigma) = 0, \qquad \tau < \sigma,$$

the above expression becomes of Volterra type, as we observed in § 7.1, and takes the form

$$\psi_1(\tau + \alpha) - \int_0^\tau \psi_1(\tau - \sigma)K(\sigma)d\sigma = 0.$$

Then we use the Laplace transformation with

$$\psi_1(\tau) \supset \Psi_1(s), \qquad K(\tau) \supset K(s)$$

as the image function, and (10.121) becomes

$$e^{\alpha s}[\Psi_1(s) - \Psi_1(s, \alpha)] - \Psi_1(s)K(s) = 0.$$

Therefore

$$K(s) = e^{\alpha s}[\Psi_1(s) - \Psi_1(s, \alpha)]/\Psi_1(s). \tag{10.126}$$

Hence, finally we have only to find $\psi_1(t)$ or $\Psi_1(s)$ which satisfies the condition (10.123). Wiener called this the factorization problem. By the theorem of Fourier transforms (Titchmarsh (1937)), when we put

$$\int_{-\infty}^\infty e^{-s\tau}\varphi(\tau)d\tau = \Phi(s),$$

if we factorize $\Phi(s)$ into

$$\Phi(s) = \Psi_1(s)\overline{\Psi_1(s)}, \tag{10.127}$$

and suppose that $\Psi_1(s)$ is analytic on the right-hand side of the s-plane, we can express $\Psi_1(s)$ by a function $\Psi_1(\tau)$ identically equal to zero in $\tau < 0$ as

$$\Psi_1(s) = \int_0^\infty e^{-s\tau}\psi_1(\tau)d\tau.$$

Therefore, we have only to factorize $\Phi(s)$ in the form of (10.127) and take one factor.

For example, when $\varphi(\tau) = e^{-|\tau|}$, for $\Phi(s) = 2/(1 - s^2)$ we have $\psi_1(s) = \sqrt{2}/(1 + s)$; then

$$\psi_1(\tau) = \begin{cases} \sqrt{2}e^{-\tau}, & \tau \geqq 0; \\ 0, & \tau < 0. \end{cases}$$

Therefore, by (10.126) we get

$$K(s) = e^{-\alpha}.$$

Hence, we obtain

$$f(t + \alpha) = \int_0^\infty f(t - \tau)dK(\tau) = e^{-\alpha}f(t).$$

As $\Phi(s)$ generally becomes a rational function in s, like this, we may factorize it as shown above.

10.5. Hysteresis

When the condition of a certain system in time t is determined by the condition in that instant or in the neighbourhood of t we say that it exhibits non-hysteresis. On the other hand, in the case when the condition in t is determined by the condition at the instant $t - \tau$, the time τ before t, or the condition in the neighbourhood of that instant, we say it has time lag τ. Further, the phenomenon which depends on the whole condition of $t(-\infty, t)$ is called hysteresis.

Hysteresis is generally expressed by a functional in (Volterra, 1929)

$$z(t) = F[\underset{-\infty}{\overset{t}{f(\xi)}}, t]. \tag{10.128}$$

This expression shows that the value of z at the time t depends on all the values of $f(t)$ in the past. When a hysteresis effect is cumulative, it is expressed as

$$z(t) = \int_{-\infty}^{t} K(t, \xi)f(\xi)d\xi. \tag{10.129}$$

Here, the kernel $K(t, \xi)$ is a function indicating the hysteresis characteristics and is called a hysteresis function. This function, which also shows the effect of the increment at $t = t$ of the unit of f on the value z at the moment $t = \xi$, therefore, is also called an effect function. When hysteresis depends only on time-lapse, a hysteresis is a function of $t - \xi$. In such a case we say that hysteresis is uniform in time and is independent of the instant ξ. For the hereditary phenomenon uniform in time, its hereditary function is written as $K(t)$. Then when

$$K(t) = 0 \quad \text{for } t > T,$$

T is called the effective period of hysteresis. Where, if $y = f(t)$ changes periodically with period T, then $z = z(t) = F[f(\xi), t]$ also changes periodically, with period T. If we indicate this phenomenon by points on the (y, z) plane, they form a closed curve at every period. Therefore, when the hysteresis effect lasts for only a finite time, this hysteresis phenomenon is said to be of closed-cycle type. For any arbitrary positive number ε there exists t_0 which makes

$$|K(t)| < \varepsilon, \; t > t_0.$$

In the case even when the effect period is infinitely long, the hysteresis effect is considered in general to be infinitely small if the lapsed time is sufficiently long; then the form of $K(t)$ is supposed to be

$$K(t) = \sum_1^n A_i \exp(-\alpha_i t). \tag{10.130}$$

Here the number of terms is called the order of hysteresis.

Many natural phenomena as well as social ones have the characteristics of hysteresis and the fundamental equation without regard to hysteresis is to be considered as a formed first approximation. For example, C.E. Picard stated as follows. 'In the study of classical dynamics, a rule indicating the idea of motion is expressed in the form of a differential equation, and is given in the relative expression between the variables and the differential coefficients. In this treatment, under the supposition that the future of a certain system depends only on the present conditions, or more generally, taking that force depends on velocity, and depends on the present condition and only that past which is extremely close to the present, we have to keep it in mind that we are practically formulating non-hysteresis characteristics. We have many instances in which the future condition of a system is related not only to the present but to the past. Hysteresis problems arise from these instances. The integral which then appears contains elements from the remote past to the present and this integral serves as a past for the hysteresis as a matter of course. Those who developed classical dynamics seem to have had the conception that hysteresis ostensively exists and that it was because observed variables are too few that this sort of observation came about, yet this conception is right only when the case is simple under complex conditions.'

Making use of Lagrange's generalized coordinates, let $q_1, q_2, \ldots, q_n$ be the independent coordinate system of the system, T be the kinetic energy, and $-Q$ be the potential energy, then when they are expressed as

$$T = \frac{1}{2} \sum_i \sum_j a_{ij} q_i^{(1)} q_j^{(1)},$$

$$Q = \frac{1}{2} \sum_i \sum_j b_{ij} q_i q_j,$$

where a_{ij}, b_{ij} are constants, and if $Q_1, Q_2, \ldots, Q_n$ denote the external forces, the equation of motion in the case of non-hysteresis is expressed as

$$\sum_j a_{ij} q_j^{(2)} + \sum_j b_{ij} q_j = Q_i.$$

In the case of hysteresis, we have

$$\sum_j a_{ij} q_j^{(2)} + \sum_j b_{ij} q_j + \sum_j \int_{-\infty}^{t} K_{ij}(t, \xi) q_j(\xi) d\xi = Q_i, \quad (10.131)$$

which is uniform in time, and when the effect takes place for a period T, it takes the form

$$\sum_j a_{ij} q_j^{(2)} + \sum_j b_{ij} q_j + \sum_j \int_{t-T}^{t} K_{ij}(t-\xi) q_j(\xi) d\xi = Q_i \quad (10.132)$$

or

$$\sum_j a_{ij} q_j^{(2)} + \sum_j b_{ij} q_j + \sum_j \int_{0}^{T} K_{ij}(\xi) q_j(t-\xi) d\xi = Q_i. \quad (10.132)'$$

Here, take the external force $Q_i(t)$ to be unknown and the problem to seek $q_i(t)$ terminates in the simultanecus integro-differential equation (10.132).

In an elastic material, take a three-dimensional orthogonal coordinate system and let $[\varepsilon_{ij}]$, $[\sigma_{ij}]$ $(i, j = 1, 2, 3)$ be the matrices denoting strain and stress. We can calculate the problem of strain under the assumption of the relation between strain and stress. If we suppose that strain is a linear function of stress, we have

$$\sigma_{ij} = \sum_{vs} a_{ij.rs} \varepsilon_{rs}, \quad (10.133)$$

where a_{ij} is constant and this is just Hooke's law. While if we suppose hysteresis instead of (10.133), put

$$\sigma_{ij} = \sum_{rs} a_{ij.rs}\varepsilon_{rs} + F_{rs}\left[\overset{t}{\underset{-\infty}{\varepsilon_{11}}}, \overset{t}{\underset{-\infty}{\varepsilon_{12}}}, \ldots, \overset{t}{\underset{-\infty}{\varepsilon_{23}}}; t\right], \tag{10.134}$$

which we observe to have some relation to the stress condition from $-\infty$ to t. In a linear hysteresis phenomenon this becomes

$$a_{ij} = \sum_{rs} a_{ij.rs}\varepsilon_{rs} + \int_{-\infty}^{t} \sum K_{ij.rs}(t, \xi)\varepsilon_{rs}(\xi)d\xi. \tag{10.135}$$

In a case such as this, we may obtain an integro-partial differential equation of strain.

10.5.1. *Theory of large strain*

(1) *Fundamental equation.* In a material of plastic deformation, both strain and stress are functions of time t, and their relation is given by

$$\sigma = f(\varepsilon) + \int_0^t H\left[t, \tau, \frac{d\varepsilon(\tau)}{d\tau}\right]d\tau.$$

The first term indicates the relation between strain and stress in a static deformation, and the second term indicates their relation in a dynamic deformation.

Under the assumption that the relation between stress and strain is linear and uniform in time, we have

$$\sigma = E\varepsilon(t) + \int_0^t K(t - \xi)\,\frac{d\varepsilon(\xi)}{d\xi}\,d\xi, \tag{10.136}$$

where $K(t)$ is what we call a hysteresis function and is supposed to take the form

$$K(t) = \sum_1^n A_n e^{-\alpha_n t} \tag{10.137}$$

as a monotone decreasing function. Here n is the degree of hysteresis. Suppose that the initial strain is zero, and take the partial integral of (10.136), which becomes

$$\sigma = [E + K(0)]\varepsilon(t) + \int_0^t K_t(t - \xi)\varepsilon(\xi)d\xi. \tag{10.138}$$

Let M be the mass of the material of plastic deformation, A_0 be the section area and l_0 be the initial length, and use the expression (10.136) as the relation between strain and stress; then we obtain, with $x(l)$ denoting length, the integro-differential equation

$$Mx^{(2)} + \frac{A_0}{l_0}[E + K(0)]x + \frac{A_0}{l_0}\int_0^t K_t(t-\xi)x(\xi)d\xi = 0. \quad (10.139)$$

Suppose that we have a one-degree linear hysteresis:

$$K(t) = A_1 e^{-\alpha_1 t}, \quad (10.140)$$

and as the initial condition, take

$$x_0 = 0, \qquad x_0{}^{(1)} = v_0;$$

then (10.139) is seen to become

$$\left[Ms^2 + \frac{A_0}{l_0}(E + A_1) + \frac{A_0}{l_0}\left(\frac{A_1 s}{s + a_1} - A_1\right)\right]X = v_0.$$

by means of Laplace transformation. Therefore, let $-r$, $-p + iq$, $-p - iq$ be the three roots of the characteristic equation

$$z^3 + a_1 z^2 + \frac{(E + A_1)A_0}{l_0 M}z + \frac{E\alpha A_0}{l_0 M} = 0,$$

and we obtain the solution

$$\begin{aligned} x(s) = {} & \frac{2pv_0}{q^2 + (r-p)^2}[e^{-rt} - e^{-pt}\cos qt] \\ & + \frac{[q^2 + r^2 - p^2]v_0}{q[q^3 + (r-p)^3]}e^{-pt}\sin qt. \end{aligned} \quad (10.141)$$

High agreement of the results obtained by this sort of theoretical calculation and experiment have been noted (Volterra 1950, 1951).

When we give a forced vibration to a plastic material with hysteresis of the first order, giving it a periodical external force $F_0 \sin \omega t$, its equation of motion is

$$\begin{aligned} & Mx^{(2)}(t) + \frac{A_0}{l_0}[E + A_1]x(t) \\ & - \frac{A_0}{l_0}\int_0^t A_1\alpha_1 \exp[-\alpha_1(t-\xi)]x(\xi)d\xi = F_0 \sin \omega t, \end{aligned} \quad (10.142)$$

and the solution is

$$x(t) = C\sin(\omega t - \varphi). \quad (10.143)$$

Here we have

$$C = F_0 \frac{\{[\alpha_1{}^2(a^2 - \omega^2) - \omega^2(\omega^2 - a^2 + \omega^2)]^2 + (a_1\omega b^2)\}^{1/2}}{\alpha_1{}^2(a^2 - \omega^2) - \omega^2(\omega^2 - a^2 + \omega^2)} \quad (10.143a)$$

$$\varphi = \tan^{-1} \frac{\alpha_1 \omega b^2}{\alpha_1{}^2(a^2 - \omega^2) - \omega^2(\omega^2 - a^2 + \omega^2)} \quad (10.143b)$$

provided

$$a^2 = \frac{EA_0}{Ml_0}, \qquad b^2 = \frac{A_1 A_0}{Ml_0},$$

If we put $\omega = a$, especially in (10.141), we have

$$\tan\varphi = \frac{a_1}{\alpha}, \qquad C = \frac{F_0}{b^2 \cos\varphi}.$$

Then the parameters of the hysteresis function α_1, A_0 become

$$\alpha_1 = (a \tan\varphi)_{\omega=\alpha},$$

$$A_0{}^2 = \frac{Ml_0}{A_1}\left[\frac{F_0}{\cos\varphi}\right]_{\omega=\alpha}.$$

Since a is determined by the experiment of static deformation, we can determine the first-order hysteresis function by observing the phase difference φ of the forced variation with the frequency of the external force ω.

(2) Hang a mass M in a model with a spring of strength K and another spring of strength K_1, with a buffer equipment of capacity C attached to it, as is shown in Fig. 10.1. Take the coordinates, as is indicated in the figure, and let $x(t)$, $x_1(t)$ denote the elongation of the spring from

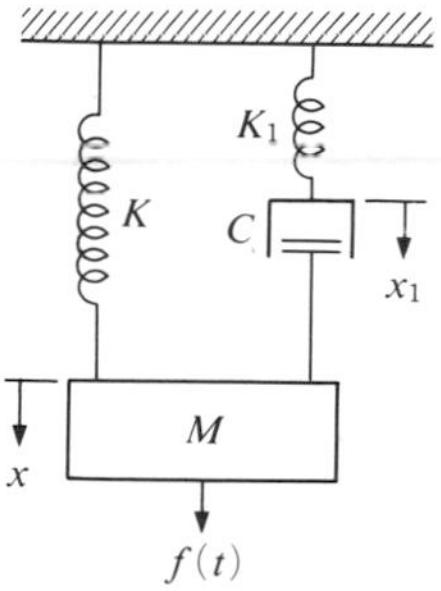

Fig. 10.1. Model of a machine with first-order hysteresis

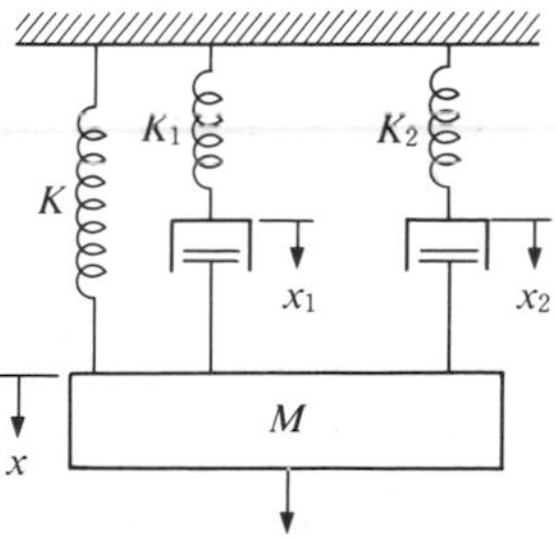

Fig. 10.2. Model of a machine with second-order hysteresis

the equilibrium position and $f(t)$ be the external force; then the equations of motion are

$$\left.\begin{aligned} Mx^{(2)}(t) &= -Kx(t) - K_1x_1(t) + f(t), \\ K_1x_1(t) &= C[x^{(1)}(t) - x_1{}^{(1)}(t). \end{aligned}\right\} \tag{10.144}$$

Since we have

$$\frac{d}{dt}\left[\exp\left(\frac{K_1}{C}t\right) \cdot x_1(t)\right] = \exp\left(\frac{K_1}{C}t\right) \cdot x^{(1)}(t),$$

from the second expression we get

$$x_1(t) = \exp\left(-\frac{K_1}{C}t\right)\int_0^t \exp\left(\frac{K_1}{C}\xi\right)x^{(1)}(\xi)d\xi.$$

Then the first expression becomes the integro-differential equation including only $x(t)$:

$$Mx^{(2)}(t) + Kx(t) + K_1\int_0^t \exp\left[-\frac{K_1}{C}(t-\xi)\right]x^{(1)}(\xi)d\xi = f(t).$$

This agrees with (10.142).

Figure 10.2 shows the mechanical model with second-order hysteresis. In this way we can make a model with hysteresis of an arbitrary order.

(3) When we take an effect period into consideration, making T be the effect period of hysteresis, we have

$$\begin{aligned} \sigma &= [E + K(0)]\varepsilon(t) + \int_{t-T}^t K_t(t-\xi)\varepsilon(\xi)d\xi \\ &= [E + K(0)]\varepsilon(t) + \int_0^T K_t(\xi)\varepsilon(t-\xi)d\xi. \end{aligned}$$

Then the equation of motion of the deformed material is, as in (10.132),

$$Mx^{(2)}(t) + \frac{A_0}{l_0}[E + K(0)]x(t) + \frac{A_0}{l_0}\int_0^T K_t(\xi)x(t-\xi)d\xi = f(t).$$

Now, if we put

$$t/T = t, \qquad \frac{A_0}{Ml_0}[E + K(0)] = a^2, \qquad \frac{A_0}{Ml_0} = b^2, \qquad \frac{1}{M}f(t) = f(t),$$

once again taking the effect period T as a unit of time, we obtain

$$x^{(2)}(t) + a^2x(t) + b^2 \int_0^1 K_t(\xi)x(t-\xi)d\xi = f(t). \qquad (10.145)$$

This time the expression makes a Fredholm-type integro-differential equation. This may be solved when $K_t(t)$ is given.

10.5.2. Beam with hysteresis

Let $u(x, t)$ be the deflection of a uniform beam at a point x on a central axis at time t, and l be the length of the beam; then taking $K(t - \xi)$ as the hysteresis function, we express the equation of free vibration of the beam as

$$m\frac{\partial^2 u}{\partial t^2} + EI\frac{\partial^4 u}{\partial x^4} + I\int_0^t K(t-\xi)\frac{\partial^4 u}{\partial x^4 \partial \xi}\, d\xi = 0, \qquad (10.146)$$

in which m is the mass for a unit length and the second-order moment of the section is I. Separate the variables and put

$$u(x, t) = w(x)\, z(t), \qquad (10.147)$$

then by making use of the parameter k we have

$$w(x)^{(4)} - k^4 w(t) = 0, \qquad (10.148)$$

$$z(t)^{(2)} + \frac{k^4 I}{m}\left[Ez(t) + \int_0^t K(t-\xi)z^{(1)}(\xi)d\xi\right] = 0. \qquad (10.149)$$

We can solve them by giving boundary conditions for (10.148) and initial conditions for (10.149). For example, when both ends are simply supported, the boundary conditions are

$$w(0) = w(l) = 0, \qquad (10.150)$$

$$w^{(2)}(0) = w^{(2)}(l) = 0;$$

then k becomes $k_j = j\pi/l$ $(j = 1, 2, 3, \ldots)$, which gives

$$w_j(x) = \sin j\,\frac{\pi x}{l}. \qquad (10.151)$$

In (10.149) if we put the solution

$$z_j(t),$$

when $k = k_j$, the free vibration of the beam becomes

$$u(x, t) = \sum_1^\infty {}_j\, w_j(x)z_j(t). \qquad (10.152)$$

10.5.3. Problem of control

There is a physical system which allows generation of the output $\theta_O(t)$ in accordance with the input $\theta_I(t)$. When this system is linear, the relation between the input and the output is given by

$$\theta_O(t) = \theta_I(0)K(t) + \int_0^t \theta_I{}^{(1)}(\xi)K(t-\xi)d\xi, \tag{10.153}$$

where $K(t)$ denotes the output for the unit input starting from $t = 0$, and $K(0) = 0$ due to the nature of the problem. Then if $dK(t)/dt = W(t)$, the above expression is rewritten as

$$\theta_O(t) = \int_0^t W(t-\xi)\theta_I(\xi)d\xi.† \tag{10.154}$$

We call $W(t)$ a weighting function (Fig. 10.3). Therefore, if either $K(t)$ or $W(t)$ is given, we can find the output for an arbitrary input. In most cases, $K(t)$ can be approximated by the sum of exponential functions:

$$K(t) = \sum a_i \exp(-k_i t). \tag{10.155}$$

Therefore, $W(t)$ becomes

$$W(t) = \sum b_i \exp(-k_i\, t). \tag{10.156}$$

Given a disturbance output $f(t)$ caused by an exterior input of the system, the problem of keeping $\varepsilon(t)$ to a desired value by giving an appropriate input $x(t)$ is called the problem of control. This is expressed as

$$f(t) + \int_0^t W(t-\xi)x(\xi)d\xi = \varepsilon(t). \tag{10.157}$$

This is a Volterra-type integral equation of the first kind. In particular if we put $\varepsilon(t) = 0$, we have

$$\int_0^t W(t-\xi)x(\xi)d\xi = -f(t). \tag{10.158}$$

For the dynamic-stability problem of an aeroplane, the textbooks of Bairstow (1939) and Goldstein (1932) in this way determined the steering $x(t)$ which is necessary to maintain horizontal flight after

† Alternatively we may define $W(t)$ first and then give the definition of $K(t)$ as the output for the unit input $E(t)$.

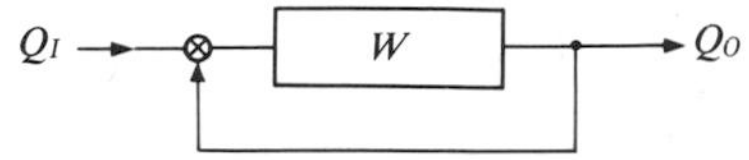

Fig. 10.3. Control of the physical system

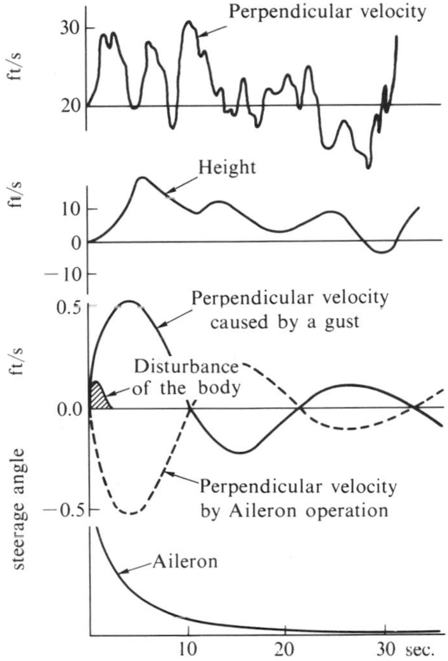

Fig. 10.4. Motion of a plane with steerage

the passing of a perpendicular gust. The output to be controlled here is the vertical velocity of the centre of gravity of the aeroplane, and the input is the aileron angle $x(t)$.

10.6. Problem of a double boundary

The problem of potential for a complicated connected domain can lead to simultaneous integral equations. For example, when the inner and outer boundary lines of a double boundary are marked by Jordan closed curves, let $x_0(\theta)$, $x_i(\theta)$ respectively be the source distributions on the above boundary lines: then the potential ψ_P at a point P in the domain is expressed by

$$-\psi_P = \frac{1}{2\pi}\int^o \chi_o d\theta + \frac{1}{2\pi}\int^i \chi_i d\theta, \tag{10.159}$$

as in § 5.3 (Example 2). Here, θ denotes the angle between the cross axle and the line connecting the source elements Q and P. The integral $\int^o$ relates to Q on the outer boundary and $\int^i$ relates to Q on the inner boundary.

In particular, when P is on the outer boundary we have

$$-\psi_o = \frac{1}{2\pi}\int_o^o \chi_o d\theta + \frac{1}{2\pi}\int_o^i \chi_i d\theta, \tag{10.160}$$

and when P is on the inner boundary we have

$$-\psi_i = \frac{1}{2\pi}\int_i^o \chi_o d\theta + \frac{1}{2\pi}\int_i^i \chi_i d\theta. \tag{10.161}$$

The symbol o at the lower end of the integral sign indicates that it is the integral going round the outer boundary, similarly, i indicates the integral going round the inner boundary. Potentials on the boundaries are indicated by ψ_o, ψ_i, and when these are known, (10.160) and (10.161) become simultaneous integral equations.

They can be solved by the method explained in § 7.3. That is, put

$$\chi_o = \chi_{1o} + \chi_{3o} + \chi_{5o} + \cdots, \tag{10.162}$$

$$\chi_i = \chi_{1i} + \chi_{3i} + \chi_{5i} + \cdots, \tag{10.163}$$

and solve approximately in the following way:

$$-\psi_o = \frac{1}{2\pi}\int_o^o \chi_{1o} d\theta, \tag{10.164}$$

$$-\psi_i - \frac{1}{2\pi}\int_i^o \chi_{1o} d\theta = \frac{1}{2\pi}\int_i^i \chi_{1i} d\theta, \tag{10.165}$$

$$-\frac{1}{2\pi}\int_o^i \chi_{1i} d\theta = \frac{1}{2\pi}\int_o^o \chi_{3o} d\theta, \tag{10.166}$$

$$-\frac{1}{2\pi}\int_i^o \chi_{3o} d\theta = \frac{1}{2\pi}\int_i^i \chi_{3i} d\theta, \tag{10.167}$$

This method can be explained physically in the following way.

(i) First suppose that the inner boundary does not exist, and find a source distribution x_{1o} such that ψ_o is the potential for the outer boundary. Expression (10.164).

(ii) Find the potential $(1/2\pi)\int_i^o x_{1o}d\theta$ on the inner boundary by the source distribution x_{1o} on the outer boundary, then supposing that the outer boundary does not exist, find a source distribution x_{1i} such that $\psi_i + (1/2\pi)\int_i^o x_{1o}d\theta$ is the potential on the inner boundary. Expression (10.165).

(iii) Find the potential on the inner boundary by the source distribution x_{1i} and assuming that the inner boundary does not exist, determine a source distribution such that the potential is zero on the outer boundary. Expression (10.166).

(iv) Find the potential on the inner boundary by the source distribution x_{3o}, and assuming that the outer boundary does not exist, find a source distribution such that the potential is zero on the inner boundary.

(v) Repeat these iterative steps, and, if the solution approaches zero, (10.162) and (10.163) are the result we are seeking. If the solution approaches a constant non-zero value c, we have only to put

$$x_o = (x_{1o} - c) + (x_{3o} - c) + (x_{5o} - c) + \cdots$$

as the solution.

Example 10.3 *Torsion of a hollow column*

L. Bairstow treated the problem on the torsion of a hollow wheel axle as shown in Fig. 10.5 by the above method, and obtained very good agreement with the precise solution obtained analytically by A. Berry. In this case, the stress function ψ satisfies the following differential equation and boundary conditions.:

$$\nabla^2\psi = \frac{\partial^2\psi}{\partial x^2} + \frac{\partial^2\psi}{\partial y^2} = 0,$$

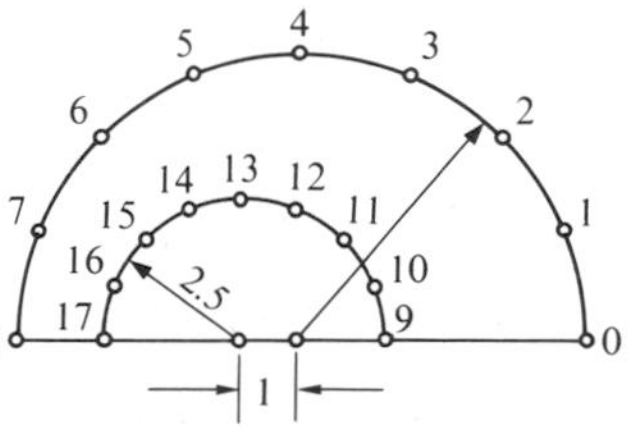

Fig. 10.5. Hollow column

$$\psi_o - \frac{1}{2}(x^2 + y^2) = \text{constant},$$

$$\psi_i - \frac{1}{2}(x^2 + y^2) = \text{constant}.$$

As for the inner circle, we can rewrite the above boundary condition as

$$\psi_i = \frac{1}{2}(r_i^2 + d^2 + 2r_i d \cos \beta) + \text{constant}.$$

Here, distance between centres is said to be d, and radius of smaller circle r_i, but Fig. 10.5 has specific numbers for these. For example, at the point 1 we have

$$x_{1o} = 0.00, \quad x_{3o} = -2.04, \quad x_{5o} = -0.50,$$

$$x_{7o} = -0.12, \quad x_{9o} = -0.02,$$

and x_o takes the value -2.68 as the sum of these, while the precise solution is -2.82. And as for the inner boundary it is close to zero in x_{7i}, and at the point 9, for instance, the approximate solution is -6.52 and the precise solution is -6.64.

1. *Theory of the biplane* (T. Moriya 1936). The problem of a potential with two boundaries, such as with biplanes and slotted wings, can lead to simultaneous integral equations under the supposition of the existence of vorticity distribution on the boundary as we discussed in § 5.5. Let C_1, C_2 be the boundary lines; $x_1(s)$, $x_2(s)$ be the rotation or the velocity distribution; s be the line element; and v_{1t}, v_{2t} be the tangent velocity of the main flow on C_1, C_2; then we obtain

$$x_1(s) - \int_{(1)}^{(1)} x_1(\xi)\frac{\cos\theta(s, \xi)}{\pi\, r(s, \xi)} d\xi - \int_{(1)}^{(2)} x_2(\xi)\frac{\cos\theta(s, \xi)}{\pi\, r(s, \xi)} d\xi = 2v_{1t},$$

$$x_2(s) - \int_{(2)}^{(1)} x_1(\xi)\frac{\cos(s, \xi)}{\pi\, r(s, \xi)} d\xi - \int_{(2)}^{(2)} x_2(\xi)\frac{\cos\theta(s, \xi)}{\pi\, r(s, \xi)} d\xi = 2v_{2t}.$$

Here, $r(s, \xi)$ denotes the length of the line connecting the vorticity and the marked point, $\theta(s, \xi)$ denotes the angle made by this line with the normal line of the boundary line, the number at the lower end of the integral sign shows the number of the curve over which the integral is taken, and the number at the upper end shows the number of the curve on which the assumed vorticity exists.

Use the method of successive approximation to solve the integral equations, put

$$x_1 = x_{11} + x_{31} + x_{51} + \dots,$$

$$x_2 = x_{12} + x_{32} + x_{52} + \ldots;$$

then x_{11} is determined as the solution of

$$x_{11}(s) - \int_{(1)}^{(1)} x_{11}(\xi) \frac{\cos \theta(s, \xi)}{\pi \, r(s, \xi)} d\xi = 2v_{1t},$$

but since $\lambda = 1$ is the eigenvalue of this integral equation, it is not uniquely determined. The solution is determined when we take the velocity at the trailing edge to be zero. (This corresponds to assuming a certain circulation around the aerofoil.)

In the next place, put

$$x_{21}(s) - \int_{(2)}^{(2)} x_{21}(\xi) \frac{\cos \theta(s, \xi)}{\pi \, r(s, \xi)} d\xi = 2v_{2t} + \int_{(2)}^{(1)} x_{11}(\xi) \frac{\cos \theta(s, \xi)}{\pi \, r(s, \xi)} d\xi$$

and find $x_{21}(s)$. We determine the solution when $x_{21} = 0$ at the trailing edge of C_2. Hereafter, we have only to solve in turn, putting

$$x_{31}(s) - \int_{(1)}^{(1)} x_{31}(\xi) \frac{\cos \theta(s, \xi)}{\pi \, r(s, \xi)} d\xi = \int_{(1)}^{(2)} x_{21}(\xi) \frac{\cos \theta(s, \xi)}{\pi \, r(s, \xi)} d\xi,$$

$$x_{32}(s) - \int_{(2)}^{(2)} x_{32}(\xi) \frac{\cos \theta(s, \xi)}{\pi \, r(s, \xi)} d\xi = \int_{(2)}^{(1)} x_{31}(\xi) \frac{\cos \theta(s, \xi)}{\pi \, r(s, \xi)} d\xi.$$

Similar methods are applied to cases with more boundaries.

10.7. Problem of collision

In this section, we are going to explain a historically famous problem of the elastic collision between a beam and a ball as an application of a nonlinear integral equation. S. Timoshenko (1913) expressed the fundamental equation of the problem in the form of a nonlinear integral equation applying Hertz's theory of collision. Since then, many results have been presented on the method of approximation, or on the comparison between theory and experiment.

When two elastic materials collide, they proceed from the geometrical position of point contact to that of surface contact by deformation, and reach elastic equilibrium, with a compression force operating on each through the surface. Thus, the distance between barycentres decreases more than in the case of point contact, and we call the decrease in the distance the *approach*. According to Hertz, between whole pressure P and approach α there exists a relation

$$\alpha = kP^{2/3}, \tag{10.168}$$

where k is a constant determined by the materials colliding.

Next, take a uniform beam of length l of which both ends are simply supported and cause a ball of mass m to collide with the centre of the beam at an initial velocity v_0. Take the instant of the collision as the origin of time t, put one end of the beam on the origin of length x. Put the x-axis in the direction of the beam and let $u(t, x)$ be the deflection of the beam at the position x at the time t. The equation of deflection of the beam with concentrated load is

$$\frac{\partial^2 u}{\partial t^2} + a^2 \frac{\partial^4 u}{\partial x^4} = \frac{a^2}{EI} p(t) d(x). \tag{10.169}$$

Here $a^2 = EI/A\rho$, EI is the sectional modulus, p is the line density; $p(t)$ is imposed on the centre of the beam by the collision, and $d(x)$ is defined by

$$d(x) = \begin{cases} 0, & x \neq \frac{1}{2}l, \\ \infty, & x = \frac{1}{2}l, \end{cases} \tag{10.170}$$

$$\int_0^l d(x)dx = 1. \tag{10.170a}$$

The boundary condition, under the condition that the both ends are supported, is

$$u(t, 0) = u(t, l) = 0, \qquad u_{xx}(t, 0) = u_{xx}(t, l) = 0; \tag{10.171}$$

and the initial condition, under the condition of a stationary state, is

$$u(0, x) = 0, \qquad u_t(0, x) = 0. \tag{10.172}$$

We solve this problem and find the amount of deflection $u(t, l/2)$ of the beam at the contact point $x = l/2$. Since the beam is of finite length, we have only to take a finite Fourier transform for x and a Laplace transform for t. Then put

$$u(t, x) \supset \supset U(s, n),$$

$$p(t) \supset P(s).$$

$$d(x) \supset D(n),$$

and with the conditions (10.171) and (10.172) being taken into consideration, (10.169) becomes

$$\left[s^2 + a^2\frac{n^4\pi^4}{l^4}\right]U(s, n) = \frac{a^2}{EI}P(s)D(n). \tag{10.173}$$

From (10.173) we have

$$D(n) = \begin{cases} 0, & n \text{ even}, \\ (-1)^{n+1}, & n \text{ odd}. \end{cases} \tag{10.174}$$

Put n back into x in (10.173), then putting

$$U(s, n) \subset U(s, x),$$

we have

$$U(s, x) = \frac{2}{l}\Sigma(-1)^{n+1}\frac{a^2}{EI}\,\frac{P(s)}{s^2 + (a\,n^2\pi^2/l^2)^2}\sin\frac{n\pi x}{l}.$$

The sum on the right-hand side is over only the positive odd numbers of n. Therefore, we get

$$U\left(s, \frac{l}{2}\right) - \frac{2}{l}\Sigma\frac{a^2}{EI}\,\frac{P(s)}{s^2 + (a\,n^2\pi^2 l/l^2)^2}$$

If we put $a\,n^2\pi^2/l^2 = \alpha_n$, when we transform s back to t, we get

$$u\left(t, \frac{l}{2}\right) = \frac{2}{l}\Sigma\frac{a^2}{EI}\int_0^t p(\xi)\frac{\sin\alpha_n(t-\xi)}{\alpha_n}d\xi$$

$$= \frac{2}{A\rho l}\Sigma\int_0^t p(\xi)\frac{\sin\alpha_n(t-\xi)}{\alpha_n}d\xi. \tag{10.175}$$

clearly Σ on the right-hand side is the sum over the odd numbers of n.

Now the velocity of the ball is

$$v(t) = v_0 - \frac{1}{m}\int_0^t p(\xi)d\xi;$$

then the deformation is

$$s(t) = v_0 t - \frac{1}{m}\int_0^t (t-\xi)p(\xi)d\xi. \tag{10.176}$$

Since the deformation of the ball is equivalent to the sum of the deflection of the beam u and the approach α, we obtain the following integral equation from (10.168), (10.175), and (10.176):

$$k\,p(t)^{\frac{2}{3}} + \frac{2}{A\rho l}\Sigma\int_0^t p(\xi)\frac{\sin\alpha_n(t-\xi)}{\alpha_n}d\xi + \int_0^t (t-\xi)p(\xi)d\xi$$

$$= v_0 t. \tag{10.177}$$

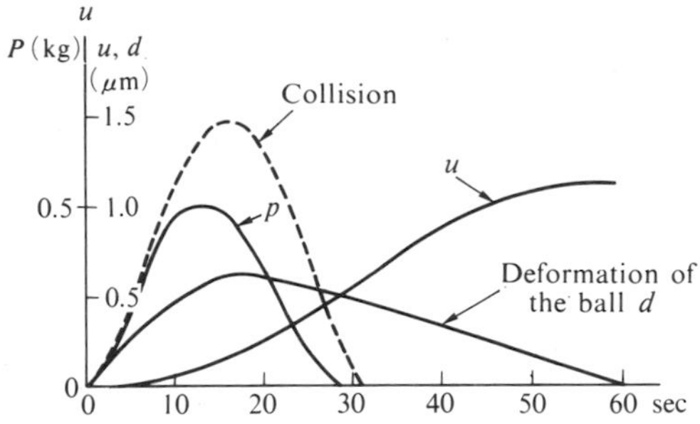

Fig. 10.6. Collision process of ball and beam

TABLE 10.1.

Beam	Ball
E: 2.2×10^6kg/m^2	r: 1 cm
ρg: 7.96 g/cm^3 (steel)	m: $\frac{4}{3}\pi\frac{0.00796}{981}$ kg sec^2/cm
A: 1×1 cm^2	v_0: 1 cm/sec
l: 15.35 cm	

This is the fundamental equation to determine the pressure which operates through the contact surface at the elastic collision of the ball and the beam. This integral equation is specially nonlinear. We may apply various sorts of approximation to this, as we did in Chap. 9 S. Timoshenko subdivided t and carried out the calculation under the supposition that $p(t)$ does not change during an infinitesimal time. Figure 10.6 is an example of the calculation when a ball collides with a beam of the same quality with the characteristics shown in Table 10.1. In the numerical calculation, we have divided the period

$$T\left(=\frac{2l^2}{\pi}\left(\frac{\rho}{EI}\right)^{\frac{1}{2}}=1\times10^{-3}\,\text{sec}\right)$$

of the free vibration of the beam intc 180 equal parts.

J. Lennertz (1937) supposed the solution to be

$$p(t) = A\sin\lambda t, \tag{10.178}$$

from the curve of $p(t)$ in Fig. 10.6, and he took only the fundamental vibration for the beam, neglecting the higher-order vibrations. If we use this method, we have

$$\frac{1}{\alpha_1}\int_0^t p(\xi)\sin\alpha_1(t-\xi)d\xi = \frac{A}{\alpha_1}\left[\frac{\lambda}{\lambda^2-\alpha_1{}^2}\sin\alpha_1 t - \frac{\alpha_1}{\lambda^2-\alpha_1{}^2}\sin\lambda t\right].$$

But here if we take α_1 to be sufficiently small and put

$$\sin\alpha_1 t \simeq \alpha_1 t, \qquad k\,A^{2/3}\sin^{2/3}\lambda t \simeq k\,A^{2/3}\sin\lambda t, \tag{10.179}$$

(10.177) becomes

$$v_0 t - \frac{A}{m}\left(\frac{t}{\lambda} - \frac{1}{\lambda^2}\sin\lambda t\right) - k\,A^{2/3}\sin\lambda t$$

$$= \frac{2}{A\rho l}A\left(\frac{\lambda}{\lambda^2-\alpha_1{}^2}t - \frac{1}{\lambda^2-\alpha_1{}^2}\sin\lambda t\right).$$

Comparing the coefficients of t and of $\sin\lambda t$, we get

$$v_0 - \frac{A}{m\lambda} = \frac{2}{A\rho l}\frac{\lambda}{\lambda^2-\alpha_1{}^2}A, \qquad \frac{A}{m\lambda^2} - k\,A^{2/3} = -\frac{2}{A\rho l}\frac{A}{\lambda^2-\alpha_1{}^2}.$$

Solving them in A, they become

$$A = v_0\frac{m\,M\lambda(\lambda^2-\alpha_1{}^2)}{(m+M)\lambda^2 - M\,\alpha_1{}^2}, \qquad A = \frac{k^3m^3M^3\lambda^6(\lambda^2-\alpha_1{}^2)^3}{[(m+M)\lambda^2-M\,\alpha_1{}^2]^3},$$

$$M = \frac{A\rho l}{2}. \tag{10.180}$$

For λ we obtain from these expressions

$$\lambda^5 = v_0\frac{[(m+M)\lambda^2 - M\,\alpha_1{}^2]^2}{k^2m^2M^2(\lambda^2-\alpha_1{}^2)^2}. \tag{10.181}$$

We have only to put $\lambda \gg p$ and approximate successively with

$$\lambda_1{}^5 = v_0(m+M)^2k^{-3}m^{-2}M^{-2}$$

as the first approximation, then determine A from (10.180) making use of this λ, and we find the solution (10.178).

E.H. Lee (1940) put $k\,A^{2/3}\sin^{2/3}\lambda t \simeq k\,A^{2/3}\,\eta\,\sin\lambda t$ in the approximation of (10.179), and carried out the same sort of calculation starting from

$$\int_0^\pi \sin\theta\,d\theta = \eta\int_0^\pi \sin^{3/2}\theta\,d\theta$$

with $\eta = 1.093$ and obtained the better result

$$\lambda^5 = v_0\frac{[(m+M)\lambda^2 - M\,\alpha_1{}^2]^2}{\eta^3k^3m^2M^2(\lambda^2-\alpha_1{}^2)^2}$$

instead of (10.181).

10.8. Characteristics of the applications of integral equations

It is only when we discuss microscopic details of momentary balance or local balance of phenomena and form the relative fundamental equations expressing the causes and effects that we use differential equations to explain natural or social phenomena by mathematics. On the other hand, if we consider the macroscopic changes that take place over a long period of time or the entire balance of a system and establish the fundamental rules as mathematical expressions we are introduced to integral equations. In the boundary-value problem, if we superpose the elements of a solution satisfying a differential equation so that it satisfies the boundary conditions, then we can introduce integral equations such as those used in treating Dirichlet's problem of potential theory, Neumann's problem, or the Sturm–Liouville problem. In many cases, it is more convenient for a problem, such as Hamilton's principle or the law of minimum energy expressible as a variation problem, to move toward an integral equation, even though such a problem hitherto has been treated by a differential equation. In addition, if we develop a fundamental expression discussing discrete (discontinuous) effects into a continuous case, we can sometimes obtain an integral equation.

If we use an integral equation in place of a differential equation, the characteristics of the eigenvalue or eigenfunction are clarified and it becomes far more convenient to solve using the numerical method. This is specially marked in the case of nonlinear problems. Due to the advancement of computers complicated numerical calculations are used extensively. Integral equations are more suitable for obtaining precise numerical solutions than differential equations when analytical solutions are hard to find.

In other words, the integral equation is an effective analytic measure regardless of whether a differential equation is available or not, and the field of its application is expected to expand much further in the future.

Miscellaneous applications of integral equations are given below. There we show some historically important examples.

(1) *Abel's problem.* Given in a perpendicular plane a smooth curve PQO under the gravitational field, on which a particle is allowed to

move gently from the point P, then consider the problem of determining such a curve that makes the given function $f(y)$ equal to the required period for the particle to reach the lowest point O, where y is the perpendicular distance OP.

Put the coordinates within the plane of the curve, make O the origin, have the x-axis horizontal and y-axis perpendicularly upwards. Let $x = x(y)$ be the expression of the curve, s being the length along the curve, and we have

$$ds = X(y)dy, \qquad X(y) = \left(1 + \left(\frac{dx}{dy}\right)^2\right)^{\frac{1}{2}}.$$

Let η be the y-coordinate of an arbitrary point Q, on the curve; then the tangent velocity at Q is $(2g(y-\eta))^{\frac{1}{2}}$, which leads to the integral equation

$$\sqrt{(2g)}f(y) = \int_0^y \frac{X(\eta)}{(y-\eta)^{\frac{1}{2}}}\,d\eta.$$

If we solve this we know the shape of the curve $x = x(y)$ when $X(y)$ is known. The solution is

$$X(y) = \frac{\sqrt{(2g)}}{\pi}\,\frac{d}{dy}\int_0^y \frac{f(\eta)}{(y-\eta)^{\frac{1}{2}}}\,d\eta$$

(see § 8.4.1).

In the particular case when $f(y)$ is a constant c, the curve is called a tautochrone. Here, we have

$$X(y) = \frac{\sqrt{(2g)}}{\pi}\,\frac{c}{\sqrt{y}},$$

from which we find that this curve is a cycloid.

(2) *Liouville's problem.* Take an orthogonal coordinate system and when a mass is uniformly distributed on the y-axis, we take the gravity on a point A$(x, 0)$ caused by a mass dy at an arbitrary point P on the y-axis to be $X(r)dy$, as it depends only on the distance PA$(= r)$. But the law of this gravity is unknown, so we take $X(r)$ to be an unknown function. The total gravity $F(x)$ at every point on the x-axis, when known, is presented as

$$F(x) = \int_{-\infty}^{\infty} F(r)\frac{x}{r}\,dy = 2\int_0^{\infty} F(r)\frac{x}{r}\,dy.$$

Then with the change of variables

$$r^2 = \xi, \qquad x^2 = t,$$

we get

$$y^2 = \xi - t,$$

and if we write

$$X(\sqrt{\xi})/\sqrt{\xi} = x(\xi), \qquad F(\sqrt{t})/\sqrt{t} = f(t),$$

then we get the integral equation

$$\int_t^\infty \frac{x(\xi)}{(\xi - t)^{\frac{1}{2}}} d\xi = f(t).$$

Put $\xi = 1/\eta$, $t = 1/\tau$ here, and it is clearly Abel's integral equation. Therefore, the solution is

$$x(t) = -\frac{1}{\pi}\frac{d}{dt}\int_t^\infty \frac{f(\xi)}{(\xi - t)^{\frac{1}{2}}} d\xi.$$

This result can also be obtained directly from the following calculation:

$$\int_t^\infty \frac{f(\xi)}{(\xi - t)^{\frac{1}{2}}} d\xi = \int_t^\infty \frac{d\xi}{(\xi - t)^{\frac{1}{2}}} \int_\xi^\infty \frac{x(\eta)}{(\eta - \xi)^{\frac{1}{2}}} d\eta$$
$$= \int_t^\infty x(\eta) d\eta \int_t^\eta \frac{d\xi}{(\xi - t)^{\frac{1}{2}} (\eta - \xi)^{\frac{1}{2}}} = \pi \int_t^\infty \varphi(\xi) d\xi.$$

(3) *The problem of geometrical optisc* (Riemann and Weber 1935, pp. 383–386). Put a luminescent source and a mapping screen on the lines of the ξ-axis and x-axis, which are perpendicular to the optical axis of the lens system, and arrange them so that the image of the luminescent source converges on this screen. In this case, the image of a unit-point luminescent source on ξ does not focus to a point, but has a certain breadth centring upon the corresponding point $x = \eta$ on the mapping screen. When η is the centre of the image, we describe the distribution of light of the unit luminescent source on the x-axis by $K(t,\eta)$. Then the distribution of light of the image of a continuously luminating body, the distribution of whose luminosity may be expressed by $\varphi(\xi)$, is given by

$$\int_a^b K(x, \eta)\, \varphi(\eta) d\eta,$$

by superposing the image. Here, $\eta = a$, $\eta = b$ are the positions of the central points of the image corresponding to the ends of the continuously illuminating source. In this case, the problem of finding the luminosity distribution $\varphi(\xi)$ by measuring the light distribution of the image becomes the integral equation

$$\int_a^b K(x, \eta)\varphi(\eta) d\eta = f(x), \qquad -\infty < x < \infty,$$

where $\varphi(\eta)$ is similar to $\varphi(\xi)$.

Now when the kernel $K(x, \xi)$ is related only to $x - \xi$, we say that the lens is uniform. For example, in the Fraunhofer refractive image of a rectangular luminescent source, we have

$$K(x - \xi) = C \frac{\sin^2 d(x - \xi)}{|x - \xi|},$$

where C a constant; d is a constant determined by the wavelength of light and the breadth of the rectangle. This kernel is not singular since it becomes zero when $x = \xi$.

(4) *Problem of the spectrum* (Doetch 1936). We cannot measure the spectrum radiating from an atom or a molecule unless we heat the material intensely. In such a case, the spectrum is not a mathematical line but has a certain breadth. This is due to the fact that the wavelength changes by Doppler's effect as the atom radiating light moves at a high velocity. According to Rayleigh's calculation, the light of frequency x_0 shows such an intensity distribution as

$$f(x) = C \frac{1}{(2\pi\kappa T)^{\frac{1}{2}}} \exp\left\{- \frac{(x - x_0)^2}{2\kappa T}\right\}, \qquad \kappa = \frac{k x_0^2}{mc^2},$$

C constant,

at absolute temperature T. Here T denotes the absolute temperature, m the mass of the atom, c the velocity of light, and k Boltzmann's constant.

Now, when an atom which has its original continuous spectrum $X(x_0)$ at a low temperature shows the distribution $F(x)$ at temperature T, then

$$C \frac{1}{(2\pi\kappa T)^{\frac{1}{2}}} \int_{-\infty}^{\infty} \exp\left\{-\left\{\frac{(x - x_0)^2}{2\kappa T}\right\}X(x_0) dx_0 = F(x).$$

Since $F(x)$ can be measured, we can solve $X(x_0)$ by solving this integral equation. This is a homogeneous integral equation with Gauss kernel. This problem can be extended to cover the problems of information theory, when signals are associated with noise.

(5) *Retrogradation problem of a stochastic process.* In a Gaussian process, let $f(x, t_0)$ be the distribution function at time t_0 and $F(x, t_0 + t)$ be the distribution function at time $t_0 + t$; then we have

$$F(x, t_0 + t) = \frac{1}{\sigma(2\pi t)^{\frac{1}{2}}} \int_{-\infty}^{\infty} \exp\left\{-\frac{(x-\xi)^2}{2\sigma^2 t}\right\} f(\xi, t_0)\, d\xi.$$

Here we find that F is a Gauss transform (see § 7.1 (d)) of the parameter $\sqrt{t}\sigma$. The retrogradation problem of a stochastic process is used to find the distribution function $f(x, t_0)$ at time t_0, t hours retrograded, with the distribution function $F(x, t_0 + t)$ given.

Generally, with regard to a stochastic process that is uniform in time and space, there exists the relative expression

$$F(x, t_0 + t) = \int_{-\infty}^{\infty} K(x - \xi;\ t, \sigma) f(\xi, t_0)\, d\xi$$

between the distribution functions at two instants. Here K is a function with parameters t, σ which gives diffusion. The problem of finding f when F is given is thus seen to be a Fredholm-type singular integral equation.

10.9. Method of solution of an integro-differential equation of wing theory

In Prandtl's wing theory, he replaces the main wing with a vortex filament $t(-a, a)$ and treats the nonlinear integro-differential equation in the circulation $\Gamma(t)$:

$$\frac{\Gamma(t)}{B(t)} + \frac{1}{\pi} \int_{-a}^{a} \frac{\Gamma^{(1)}(\xi)}{t - \xi}\, d\xi = f(t). \tag{10.182}$$

Now let $2a$ be the wing span, U be the velocity of the main stream, $\alpha(t)$ the geometrical attack angle, and $b(t)$ the chord length, then we have

$$B(t) = \frac{5.5b(t)}{8}, \qquad f(t) = 4U\alpha(t). \tag{10.183}$$

Since the plan form of the wing is symmetrical about the central line,

$\Gamma(t)$, $B(t)$, $f(t)$ and so on, are even functions. Suppose that there is no circulation at the wing edge and assume that

$$\Gamma(-a) = (a) = 0. \tag{10.184}$$

Equation (10.182) is an integro-differential equation with a singular kernel.

Practical methods of solution of this kind of equation have often been proposed in response to requests from the field of aeronautics (Glauert 1926, Chapter X; Trefftz 1921; Lotz 1933; Tani 1940) etc. Here we explain the method of I.N. Vekua and L.G. Magnaradze, which is introduced by Muskhelishvili (1946). They solve (10.182) by transforming it into an ordinary Fredholm-type integral equation.

Suppose that $B(t)$ does not become zero in $t(-a, a)$, and take the Hilbert transform of (10.182); then we have

$$\Gamma^{(1)}(t) = \frac{1}{\pi(a^2-t^2)^{\frac{1}{2}}}\int_{-a}^{a}\frac{(a^2-\xi^2)^{\frac{1}{2}}}{B(\xi)(t-\xi)}\,\Gamma(\xi)d\xi + F(t) + \frac{C}{(a^2-t^2)^{\frac{1}{2}}} \tag{10.185}$$

$$F(t) = \frac{-1}{\pi(a^2-t^2)^{\frac{1}{2}}}\int_{-a}^{a}\frac{(a^2-\xi^2)^{\frac{1}{2}}}{t-\xi}\,f(\xi)d\xi. \tag{10.186}$$

Since $\Gamma(t)$ is an even function, we get $\Gamma^{(1)}(t) = -\Gamma^{(1)}(-t)$; then $\Gamma^{(1)}(0) = 0$, and therefore $C = 0$.

Integrate both sides of (10.185) with respect to t from 0 to t, which makes

$$\Gamma(t) - \frac{1}{\pi}\int_{-a}^{a}\log\left|\frac{i(t-\xi)+(a^2-t^2)^{\frac{1}{2}}-(a^2-\xi^2)^{\frac{1}{2}}}{i(t-\xi)+(a^2-t^2)^{\frac{1}{2}}+(a^2-\xi^2)^{\frac{1}{2}}}\right|\frac{\Gamma(\xi)}{B(\xi)}\,d\xi$$
$$= \int_0^t F(\xi)d\xi + C_0 \tag{10.187}$$

Here C_0 is an as yet unknown integral constant. Put

$$t = -a\cos\tau, \qquad \xi = -a\cos\eta,$$

and (10.187) becomes

$$\Gamma(t) - \frac{a}{\pi}\int_0^{\pi}\frac{\sin\eta}{B(\eta)}\log\left|\frac{\sin\dfrac{\tau-\eta}{2}}{\sin\dfrac{\tau+\eta}{2}}\right|\Gamma(\eta)d\eta = G(\tau) + C_0, \tag{10.188}$$

which turns out to be a kind of $[F_2]$.

For ease of application, we have only to put

$$B(t)\Gamma^{(1)}(t) = \frac{1}{\pi}\int_{-a}^{a}\frac{\Gamma(\xi)}{t-\xi}\,d\xi + B(t)F(t)$$

$$-\frac{B(t)}{\pi(a^2-t^2)^{\frac{1}{2}}}\int_{-a}^{a}R(t,\xi)\,\Gamma(\xi)d\xi \tag{10.189}$$

from (10.185), and put

$$R(t,\xi) = \frac{1}{t-\xi}\left\{\frac{(a^2-t^2)^{\frac{1}{2}}}{B(t)} - \frac{(a^2-\xi^2)^{\frac{1}{2}}}{B(\xi)}\right\}. \tag{10.189a)'}$$

To faciliate the discussion, we suppose that

$$P(t) = \frac{(a^2-t^2)^{\frac{1}{2}}}{B(t)}, \qquad -a \leqq t \leqq a \tag{10.190}$$

has a continuous derivative.

From (10.189) we obtain

$$\frac{d}{dt}[B(t)\Gamma^{(1)}(t)] - \frac{1}{\pi}\int_{-a}^{a}\frac{\Gamma^{(1)}(\xi)}{t-\xi}\,d\xi = g(t), \tag{10.191}$$

$$g(t) = \frac{d}{dt}\left\{B(t)\,F(t) - \frac{B(t)}{\pi(a^2-t^2)^{\frac{1}{2}}}\int_{-a}^{a}R(t,\xi)\Gamma(\xi)d\xi\right\}. \tag{10.191a)'}$$

From (10.185) and (10.191) we derive

$$B(t)\frac{d}{dt}[B(t)\Gamma^{(1)}(t)] + \Gamma(t) = B(t)[f(t) + g(t)]. \tag{10.192}$$

If we consider the right-hand side of this expression to be a known function, this expression is a linear differential equation of the second order in $\Gamma(t)$. Then the solution must be

$$\Gamma(t) = C_1\cos\omega(t) + C_2\sin\omega(t) + \int_0^t\{f(\xi) + g(\xi)\}\sin\{\omega(t) - \omega(\xi)\}d\xi. \tag{10.193}$$

Here, put

$$\omega(t) = \int_0^t\frac{d\xi}{B(\xi)}.$$

If $t = 0$, then $C_1 = \Gamma(0)$. Then, $C_2 = 0$, with $\Gamma(t)$ being an even function.

Substitute (10.191)′ for $g(t)$ in (10.193) and finally we obtain [F_2]:

$$\Gamma(t) + \frac{1}{\pi}\int_{-a}^{a} K(t, \xi)\Gamma(\xi)d\xi = h(t), \tag{10.194}$$

$$K(t, \xi) = \int_0^t \frac{R(\eta, \xi)}{(a^2 - \eta^2)^{\frac{1}{2}}} \cos[\omega(t) - \omega(\eta)]d\eta, \tag{10.194$'$}$$

$$h(t) = \Gamma(0)\cos\omega(t) + \int_0^t \Big\{ \sin[\omega(t) - \omega(\eta)]f(\eta)$$

$$+ \frac{\cos[\omega(t) - \omega(\eta)]}{\pi(a^2 - \eta^2)^{\frac{1}{2}}} \int_{-a}^{a} \frac{(a^2 - \xi^2)^{\frac{1}{2}} f(\xi)}{\eta - \xi}\, d\xi \Big\} d\eta. \tag{10.194$''$}$$

The function $\Gamma(t)$ changes with the change of the plane shape of the wing; for example, when the wing shape is elliptic we have only to put

$$b(t) = b_0\left(1 - \frac{t^2}{a^2}\right)^{\frac{1}{2}}.$$

In this case, we clearly have $R(t, \xi) \equiv 0$, $K(t, \xi) = 0$, and (10.194)″ directly serves as the solution.

I.N. Vekua reached this result by a somewhat different method under the assumption that (10.190) is analytic. (This assumption is necessary when we use Vekua's method.) Before long, this result was extended by L.G. Magnaradze. It is chiefly his method that we have explained above. (Since he supposed that $\Gamma(-t) = \Gamma(t)$, $\Gamma(-a) = \Gamma(a) = 0$, his result has a few arbitrary constants.)

The expression (10.194) looks more complicated than (10.188) at first glance, but from the point of view of applications the former is much easier to work with. To begin with, the kernel is normal. Secondly, we can solve many important applications in quite a simple way. In particalar, when $P(t)$ in (10.190) is a rational function and can take the form

$$b(t) = \left(1 - \frac{t^2}{a^2}\right)^{\frac{1}{2}} P(t),$$

we can solve (10.194) quite easily. In the theses of Vekua and Magnaradze, we can find examples of calculations of some special cases. When we have

$$b(t) = b_0\left(1 - \frac{t^2}{a^2}\right)^{\frac{1}{2}} \frac{1 + \nu t^2/a^2}{1 + \mu t^2/a^2},$$

the solution is arrived at quite easily by putting $b_0 > 0$, $\mu > -1$, $\nu > -1$. This equation can also include many practically important cases merely by changing the constants μ, ν.

For example, put $\mu = 0$, $\nu = 0.9$, and we can get an almost rectangular wing.

t/a	0.1	0.2	0.3	0.4	0.5	0.6	0.7	0.8	0.9
$b(t)/b_0$	1.00	1.02	1.03	1.05	1.06	1.06	1.03	0.95	0.75

In the same way, we can easily solve

$$b(t) = b_0\left(1 - \frac{t^2}{a^2}\right)^{\frac{1}{2}} \frac{1 + \nu_1/t^2/a^2 + \cdots + \nu_n t^{2n}/a^{2n}}{1 + \mu_1 t^2/a^2 + \cdots + \mu_n t^{2n}/a^{2n}}.$$

A special case when all μ_j vanish is treated by H. Schmidt (1937), but the result is very complex.

Exercises

1. For a uniform elastic string fixed at the two end points (0, 1), take the x-axis along the direction of the string. Let $G(x, \xi)$ be the variation in $x = \xi$ when an external force of strength 1 is applied to $x = x$, then show that $G(t, \xi)$ is the Green function of this problem and prove that the variation caused by the distributed weght $f(t)$ can be expressed as

$$\int_0^1 G(t, \xi) f(\xi) d\xi.$$

2. Prove Lemma 10.3 and Lemma 10.9.

3. When h is real in the Sturm–Liouville system $x^{(2)} - (\lambda - q)x = 0$, $x(0) = x(1) = 0$, prove that the eigenvalue is $-h - n^2\pi^2$ and the eigenfunction is $2 \sin n\pi x$. [Churchill]

4. When $x^{(2)} - (\lambda + h)x = 0$, $x(0) = x(1) = 0$, find the eigenfunction φ_n for the eigenvalue λ_n.

5. Study the motion of an elastic string of length 1 when we make it vibrate freely by giving an initial velocity $g(x)$ to every point on it from an equilibrium position. [Churchill] [Form the integral equation and find the eigenvalue and eigenfunction. The solution is

$$y(x, t) = \frac{2}{\pi c}\sum_1^\infty \frac{1}{n} \sin n\pi x \sin n\pi ct \int_0^1 g(\xi) \sin n\pi\xi \, d\xi.]$$

6. For a thin elastic screen spread over a circle of radius 1, discuss the forced vibration of the screen when a force $A \sin \omega t$ is applied evenly to every part. [Churchill]

[The differential equation of variation is given by

$$u_{tt} = b^2\left(u_{rr} + \frac{1}{r}u_r\right) + A\sin\omega t.$$

The eigenvalue is given in terms of $J_0(\lambda_n/b) = 0$.]

7. Given a string of length 1, one end of which is fixed at a point on the rotational axis of a flat board rotating at a constant angular velocity. Suppose the string is always on the flat board, then discuss the free vibration of the string when it is set in motion with a certain initial velocity in a small deflection from the balance position. Assume that the rotational velocity is large enough to be able to neglect the weight of the string. [Lovitt]

[The differential equation of the variation $u(x, t)$ is given by

$$u_{tt} = c^2\frac{\partial}{\partial x}[(1 - x^2)u_x]; \qquad u(0, t) = 0, \qquad u(1, t) = \text{finite};$$

$$u(x, 0) = f(x), \qquad u_t(x, 0) = g(x),$$

the Green function is

$$G(x, \xi) = \frac{1}{2}\log\frac{1+x}{1-x}\ (0 \leqq x < \xi);$$

$$\frac{1}{2}\log\frac{1+t}{1-t}, \quad \xi \leqq x \leqq 1, \quad \lambda_n = 2n(2n-1),$$

$$f(x) = \sum C_r P_{2n-1}(x), \quad C_r = (4n-1)\, f \,.\, P_{2n-1}$$

where P is Legendre's polynomial.]

8. In the theory of inner friction of plastic deformation,

$$\sigma = E\varepsilon + A\frac{d\varepsilon}{dt}$$

is assumed to be the relation between stress and strain. Find the relation between this theory and hysteresis theory.

[Hysteresis theory corresponds to the case in which all derivatives of $\varepsilon(t)$ are taken into consideration.]

9. Form an electric circuit similar to the mechanical model shown in Fig. 10.2.

10. Show that the mechanical model shown in Fig. 10.2 has hysteresis of the second order, and study the forced vibration of this system.

11. When the effect period is 1, discuss the forced vibration of a plastic material with hysteresis of the first order.

12. With regard to the free vibration of a uniform beam with hysteresis, discuss the following cases:

(a) one end is fixed, the other end is free;

(b) one end is fixed, the other end is supported;

(c) both ends are fixed.

13. Given a uniform beam whose ends are both fixed. Discuss the solution when a forced external force $F_0 \sin \omega t$ is applied to it at a point which is at the distance $x = l_0$ from one end.

14. Calculate the free vibration of a uniform beam which has hysteresis of the first order and is fixed at both ends. Assume that the hysteresis function is $K(t) = A\, e_1^{-\alpha_1 t}$, and the initial condition is $u(x, 0) = f(x)$, $u_t(x, 0) = 0$. [$mu_{tt} + EIu_{xxxx} = 0$]

15. Discuss the deformation of a plastic screen of radius R whose circumference is fixed.

[Place polar coordinates (r, θ) and make the fundamental equation with $u(r, \theta, t)$ for the variation at time t. Separate the variables as $u = v(r)w(\theta)z(t)$ and the fundamental equation becomes

$$r^2 v^{(2)}(r) + r v^{(1)}(r) + (k^2 r^2 - n^2) v(r) = 0,$$

$$w^{(2)}(\theta) + n^2 w(\theta) = 0,$$

$$\rho z^{(2)}(t) + k^2 \left[T z(t) + \int_0^t K(t - \xi) z^{(1)}(\xi) d\xi \right] = 0.$$

Here ρ denotes surface density, T tension, and k^2, n^2 are parameters.] [Volterra]

16. Study the problems of the beam and the screen when the effect period is 1.

17. Assume purchase and sale are both continuous and stored goods are sold off uniformly in time T. When we neglect the time between buying and opening for sale, how do we buy in goods to keep stored goods constant? [Goldstein]

[Assume a unit quantity stored at time $t = 0$ leaves $K(t)$ at $t = T$. Under this assumption we have

$$K(t) = \begin{cases} 1 - t/T, & t < T, \\ 0, & t \geqq T. \end{cases}$$

Let $x(t)$ be the stocked-in quantity at t, and we obtain the integral equation

$$K(t)+\int_0^t K(t-\xi)x(\xi)d\xi=1.$$

We have only to solve it by operational calculus; see § 6.8].

Miscellaneous problems (II)

1. Arrange various types and solutions of the singular integral equations into a table.
2. When the kernel takes the form $K(t, \xi)=P(t)k(t, \xi)$, $P(t)\geqq 0$ in I and $k(t, \xi)=k(\xi, t)$ in S, prove that the integral equation can lead to the case with a symmetric kernel. [Put the unknown function as $(F(t))^{\frac{1}{2}}x(t)\equiv y(t)$, and the kernel is expressed as $(P(t)P(\xi))^{\frac{1}{2}}k(t, \xi)$.]
3. Solve the following dual integral equation:

$$\left.\begin{aligned}&\int_0^\infty \xi^\alpha x(\xi)J_\nu(t\xi)d\xi=f(t), \qquad 0<t<1,\\ &\int_0^\infty x(\xi)J_\nu(t\xi)d\xi=0, \qquad t>1.\end{aligned}\right\}$$

[Titchmarsh, Sneddon]
[The Mellin transform of $\xi^\alpha J_\nu(t\xi)$ becomes

$$\begin{aligned}J_\alpha(s)&=\int_0^\infty \xi^{\alpha+s-1}J_\nu(t\xi)d\xi\\ &=\frac{2^{\alpha+s-1}}{t^{\alpha+s}}\Gamma\left(\frac{(\alpha+\nu+s)}{2}\right)/\Gamma\left(\frac{2-\alpha-s+\nu}{2}\right)\end{aligned}$$

By the convolution theorem of the Mellin transform, let $X(s)$ be the Mellin transform of $x(t)$, so that

$$\int_0^\infty \xi^\alpha x(\xi)J_\nu(t\xi)d\xi=\frac{1}{2\pi i}\int_{c-i\infty}^{c+i\infty}X(s)J_\alpha(1-s)ds,$$

$$\int_0^\infty x(\xi)J_\nu(t\xi)d\zeta=\frac{1}{2\pi i}\int_{c-i\infty}^{c+i\infty}X(s)J_\nu(1-s)ds.$$

The solution reads

$$x(t)=\frac{2}{\Gamma\left(\frac{1}{2}\alpha\right)}\int_0^1 \eta^\alpha d\eta\int_0^1 g(\eta\zeta)\zeta^{\nu+1}(1-\zeta^2)\frac{1}{2}^{\alpha-1}d\zeta$$

$$\times\frac{1}{2\pi i}\int_{c-i\infty}^{c+i\infty}2^{s-\alpha}(t\eta)^{-s}\Gamma\left(\frac{1+\nu-s}{2}\right)/\Gamma\left(\frac{2+\nu+\alpha-s}{2}\right)ds$$

$$= \frac{(2t)^{1-\frac{1}{2}\alpha}}{\Gamma\left(\frac{1}{2}\alpha\right)} \int_0^1 \eta^{\,1+\frac{1}{2}\alpha} j_{\nu+\frac{1}{2}\alpha}(\eta t) d\eta \int_0^1 g(\eta\zeta)\zeta^{\nu+1}(1-\zeta^2)^{\frac{1}{2}\alpha-1} d\zeta.$$

4. Using an integral equation, find the torsional vibration of a uniform column with one end fixed and the other end free.

[The kernel is $K(t, \xi) = \text{Min}(t, \xi)$, the eigenvalue is

$$\lambda_n = \left(\frac{2n-1}{2}\pi\right)^2,$$

and the eigenfunction is $\sin\sqrt{\lambda_n}t$.]

5. Describe the characteristics of integral equations from the viewpoint of applications.

References

Bairstow, L. (1939) *Applied Aerodynamics* (London) 533–550.

Bairstow, L. and Berry, A. (1919) Two dimensional solutions of Poisson's and Laplace's equations. *Proceedings of the Royal Society* A **55**, 457–475.

Bolza, O. (1904) Lectures on the Calculus of variations (University Press, Chicago)

Churchill, R.V. (1944) *Modern Operational Mathematics in Engineering* (McGraw-Hill, New York).

Doetch, G. (1936) Zerlegung einer Funktion in Gaussche Fehlerkurven und zeitiliche Zurückverfolgung eines Temperaturzustandes. *Mathematisches Zeitschrift* **41**, 283–318.

Glauert, H. (1926) *Elements of Aerofoil and Airscrew Theory* (University Press, Cambridge).

Goldstein, S. (1932) Operational solution of an integral equation. *Journal of the London Mathematical Society* **6**, 262–268.

Kneser, A. (1924) (14) *Die Integralgleichungen und ihre Anwendungen in der mathematischen Physik* (Braunschweig).

Kondo, J. (1954) Some notes on Wiener's prediction theory. *Bulletin of Statistical Research* **6** (Statistical Bureau, Prime Minister's Office) 1–10 (in Japanese).

Lee, E.H. (1940) The impact of a mass striking a beam. *Transactions of the American Society of Mechanical* Euqineers **62**, 66.

Lennertz, J. (1937) Beitrag zur Frage nach der Wirkung eines Querstosses auf einen Stab. *Ingenieur-Archiv*. **8**.

Lotz, I. (1931a) Berrechnung der Auftriebsverteilung beliebig geformter. *Flügel* **22**, 189–195

Lotz, I. (1931b) Mémoire sur le calcul des differentielles à indices quelconques. *Flügel* **22**

Lotz, I. (1933) *Zeitschrift Flugtechnik und Motorluftschiffahrt* **24**, 664.

Lovitt, W.V. (1924) *Linear Integral Equations* (Dover, New York).

Moriya, T. (1936) Method to calculate pressure distribution of an arbitrary profile. *Journal of the Japan Aeronautical Society*, **4**, pp. 365–366 (in Japanese).

Muskhelishvili, N.I. (1946) *Singular Integral Equations* (tr. by Radok, J.R., Stechert, New York).

Riemann, G.F. and Weber, W. (1935) *Differentialgleichungen der Physik*, Vol. 1,

Part 3 (Berlin).
Schmidt, E. (1906) Entwicklung willkürlicher Funktionen nach systemen Vorgeschriebener. *Mathematisches Annalen* **63**, 433–476.
Schmidt, H. (1937) Strenge Lösungen zur Prandtlschen Theorie der tragenden Linie. *Zeitschrift für Angewandte Mathematik und Mechanik* **17**, 101–116.
Tani, I. (1940) A simple method of calculation of aerodynamic characteristics of a monoplane. *Aero-Insitute, Tokyo University Report* **197** (in Japanese).
Timoshenko, S. (1913) Zur Frage nach der Wirkung einer Stostze auf einer Balken. *Zeitschrift für Mathematik und Physik* **62**, 198–203.
Titchmarsh, E.C. (1937) Introduction to the Theory of Fourier Integrals (University Press, Oxford) 128–129.
Trefftz, E. (1921) Prandtlsche Tragflächen und Propeller-Theorie. *Zeitschrift für Angewandte Mathematik und Mechanik* **1**, 206–218.
Volterra, E. (1950) Vibration of elastic systems having hereditary characteristics. *Journal of Applied Mechanics* **10**, 363–371.
Volterra, E. (1951) On elastic continua with hereditary charasteristics. *Journal of Applied Mechanics* **11**, 273–279.
Volterra, V. (1929) Alcune osservazioni sui fenomeni ereditari. *Rendiconti della Reale Accademia Nazionale dei Lincei* **9**.
Wiener, N. (1949) *The Extrapolation, Interpolation and Smoothing of Stationary Time Series* (Wiley, New York).
Whittaker E.T. and Watson G.N (1953) *A Course of Modern Analysis* (Cambridge).

Bibliography

Books

1. Bocher, M. (1914) *An Introduction to the Study of Integral Equations*. Cambridge Tracts in Mathematics 10 (Cambridge University Press).
2. Bückner, H. (1952) *Die Praktische Behandlung von Integralgleichungen* (Berlin).
3. Courant, R. and Hilbert, D. (1931) *Methoden der Mathematischen Physik*, Chapters 3, 5 (Berlin).
4. Davis, H. (1936) *The Theory of Linear Operators*. Monograph of the Waterman Institute (Indiana University) (especially pp. 468–514).
5. Hayashi, T. and Takenaka, A. (1929) *Elementary Integral Equations* (Okura Tokyo) (in Japanese).
6. Heywood-Fréchet, ? (1923) *L'équation de Fredholm et ses Applications à la Physique Mathématique* (Paris).
7. Hidaka, K. (1941) *Theory of Integral Equations* (Kawade-Shobo, Tokyo) (in Japanese).
8. Hidaka, K. (1943) *Theory and Application of Integral Equations* (Kawade-Shobo, Tokyo) (in Japanese).
9. Hilbert, D, (1912) *Grundzüge einer allgemeinen Theorie der lincaren Integralgleichurgen* (Leipzig)
10. Hoheisel, G. (1936) *Integralgleichungen* (Sammlung Göschen).
11. Ikeda, Y. (1928) *Theory of Integral Equations* (Iwanami-shoten, Tokyo) (in Japanese).
12. Inui, T. (1939) *Applied Partial Differential Equations*, Chapters 4, 6 (Iwanami-shoten, Tokyo) (in Japanese).
13. Kakeya, S. (1934) *Theory of Integral Equations* (Iwanami-shoten, Tokyo) (in Japanese).
14. Kneser, A. (1924) *Die Integralgleichungen und ihre Anwendungen in der mathematischen Physik* (Braunschweig).
15. Kowalewski, G. (1930) *Integralgleichungen* (Berlin).
16. Lalesco, T. (1912) *Introduction á la Théorie des Equations Intégrales* (Paris).
17. Lichtenstein, L. (1931) *Vorlesungen über einige Klassen nichtlinearen Integralgleichungen und Integro-Differentialgleichungen* (Berlin).
18. Lovitt, W.V. (1924) *Linear Integral Equations* (New York).
19. Morse, P.M. and Feshbach, H. (1953) *Methods of Theoretical Physics*, Chapter 8 (McGraw-Hill, New York).
20. Muskhelishvili, N.I. (1946) *Singular Integral Equations* (tr. by Radok, J.R., Stechert, New York).
21. Parodi, M. (1950) *Equations Integrales et Transformations de Laplace* (Paris).
22. Riemann, G.F. and Weber, W. (1935) *Differentialgleichungen der Physik*, Vol. 1, part 3 (Berlin).
23. Sato, T. (1946) *Integral Equations and Green Functions* (Kashiwa noha Shoin, Tokyo) (in Japanese)
24. Schwank, V. (1929) *Elemente der Theorie der linearen Integralgleichungen* (Hannover).
25. Takeuchi, T. (1933) *Integral Equations* (Kyoritsu-sha, Tokyo) (in Japanese).
26. Terazawa, K. (1931) *Outlines of Mathematics for Natural Scientists*, Chapter 13 (Iwanami-shoten, Tokyo) (in Japanese).

27. Volterra, V. (1913) *Leçons sur les Equations intégrales et les Equations Intégro-differentielles* (Paris).
28. —— (1930) *Theory of Functionals and of Integral and Integro-Differential Equations* (London).
29. —— and Pérès, J. (1936) *Théorie Générale des Fonctionnelles* (Paris).
30. Whittaker, E.T. and Watson, G.N. (1953) *A Course of Modern Analysis*, Chapter 11 (Cambridge).
31. Wiarda, G. (1930) *Integralgleichungen* (Leipzig).
32. Yoshida, K. (1950) *Theory of Integral Equations* (Iwanami-shoten, Tokyo) (in Japanese).

Among these, (5, 11, 13, 24) treat almost the same subject as chapters 2 and 6 of this book; the central theme is the introduction of integral-equation theories by Volterra, Fredholm and Hilbert. Takeuchi (24) includes exercises and methods of solution. In (7, 8) we find detailed explanations of the numerical method of solution and its applications which appear in chapters 4 and 5. Sato (23) gives many examples of applications. Yoshida (32) gives lectures on the eigenvalue problems of differential equations which are fully developed in chapter 10.

Of the reference books in languages other than Japanese, (1, 9, 10, 15, 16, 26, 27) mainly treat the theory of integral equations, and (1), although a small volume of only 72 pages, gives precise explanations. Primarily it deals with applications of the Sturm–Liouville boundary-value problem.

Problems that can be considered special are dealt with in (2, 17, 20, 21). Lichtenstein (17) describes the theory and applications of nonlinear integral equations; Parodi (21) gives precise comments on integral equations that can be solved by way of the Laplace transform with the form of the kernel restricted. Bückner (2) explains the method of finding eigenvalues, and especially their minimum values, numerically. Muskheilishvili (20) systematically treats integral equations that take Cauchy's principal value (referred to in chapter 8 in this volume) and also one-variable nonlinear integral equations, giving explanations of the boundary-value problems of potential that appear in elasticity and hydrodynamics. Radok's English version gives us the opportunity to study Russian documents with which we rarely have contact. It is necessary to know the theory of integral equations to the extent presented in part I of this volume to go through these books, and, especially, (20) requires knowledge of the theory of functions.

Some reference books of a relatively high level are (4, 28, 29). On the whole the present book is written chiefly in the light of (29). In addition, for the existence theorem of precise solutions, we may mention references on linear operators and functional spaces, but the present book does not comment on these kinds of abstract theory.

The books (3, 12, 19, 22, 25, 30) give considerable comments on integral equations. The present book is in debt to (3, 12) for the section on the Sturm–Liouville system.

In the course of preparation of the present volume, the author utilized following two books as the basis of the manuscript.

33. Kondo, J. (1954) *Integral Equations* (Baifukan, Tokyo) (in Japanese).
34. Kondo, J. (1959) *Integral Equations and their Applications* (Corona Pub., Tokyo) (in Japanese).

The book 33 had been used as a textbook of Applied Analysis at the Engineering School, University of Tokyo, Tokyo, Japan.

Papers

The following papers on integral equations, except for the historical ones, are confined to those original papers to which the author directly referred. We can find full lists of the papers on integral equations in [11], [21], [23], [24], and [27], which were mentioned above, and in (7) that follows hereafter.

1. Abel, N.H.: *Résolution d'un probléme de Mécanique*, (Euvres) **1**, 97.
2. ——: *Solutions de quelques problèmes à l'aide d'intégrales définies. ibid.* p. 113.
3. Aprile, G.: Un integrafo per la valutazione delle espressioni simboliche del calcolo operatorio funzionale, *Comment. Pontificia Acad. Sci.* **8**, 1944. pp. 31–44.
4. Bairstow, L., Berry, A.: *Two dimensional solutions of* Poisson's and Laplace's equations. *Proc. Roy. Soc. A.* **55**, 1919. pp. 457–475.
5. Batemann, H.: On the numerical solution of linear integral equations. *Proc. Roy. Soc. A.* **100**, 1921. pp. 441–449.
6. ——: Some integral equations of potential theory. *Journ. App. Phys.* **17**, 1921. pp. 91–102.
7. Bertrand, G.: Équation de Fredholm à intégrales principales au sens de Cauchy. *Comptes Rendus.* **172**, 1921. pp. 1458–1461.
8. Browne, P.J.: Sur quelques cas singuliers de l'équation de Volterra. *Comptes Rendus* **154**, 1912.
9. Carleman, T.: Sur la résolution de certaines équations intégrales, *Arkiv fur Mat., Astron. och Fysik* **16**, 1922, Bd. 16, No. 266, pp. 1–19.
10. Carmichael, R.D.: Algebraic guides to transcendental problems. *Bull. Amer. Math. Soc.*, **28**, 1922. pp. 179–210.
11. Carrier, G.F.: Integral equation boundary layer problems, *50 Jahre Grenzschichtforschung*, Braunschweig. 1955, pp. 13–20.
12. Carrier, G.F.: On the determination of the eigenfunctions of Fredholm equations. *Journ. Math. Phys.* **27**, 1948. pp. 82–83.
13. Davis, H.T.: The present status of integral equations. *Indiana Univ. Stud.* **13**, No. 70, 1926.
14. Doetch, G.: Zerlegung einer Funktion in Gausssche Fehlerkurven und zeitliche Zurückverfolgung eines Temperaturzustandes. *Math. Zeitschrift* **41**, 1936. pp. 283–318.
15. Elliot, J.: On some singular integral equations of the Cauchy type. *Ann. Math.* **54**, 1951. pp. 349–370.
16. Evans, G.C.: The integral equation of the second kind of Volterra, with singular kernel. *Bull. Amer. Math. Soc., 2nd series*, **16**, 1909.
17. ——: Volterra's integral equation of the second kind with discontinuous kernel. *Trans. Amer. Math. Soc.* **11**, 1910; **20**, 1911.
18. Feller, W.: On the integral equation of renewal theory, *Ann. Math. Stat.* **12**, 1941, pp. 243–267.
19. Fredholm, I.: Sur une classe d'équations fonctionnelles. *Acta Mat.* **27**, 1903.
20. Giraud, G.: Equations à intégrales principes, étude suivie d'une application. *Ann. de l'Ecole Normale Superieur* **51**, 1934. pp. 251–372.
21. Goldstein, S.: Operational solution of an integral equation. *Journ. London Math. Soc.* **6**, 1932. pp. 262–268.
22. Hammerstein, A.: Nichtlineare Integralgleichungen nebst Anwendungen, *Acta Math.*, **54**, 1930. pp. 117–176.
23. Havelock, T.H.: The solution of an integral equation occurring in certain problems of viscous fluid motion. *Phil. Mag.* **42**, 1921. pp. 620–628.
24. ——: On the decay of oscillation of a solid body in a viscous fluid. *Phil. Mag.* **42**, 1921. pp. 628–634.

25. Hille, E., Tamarkin, J.D.: On the theory of integral equations. *Ann. Math.* **31**, 1930.
26. Hilbert, D.: Grundzüge einer allgemeine Theorie der linearen Integralgleichungen. *Gött. Nachr., math. phys.* Klasse, 1904–1908.
27. Ingram, W.H.: On the integral equations of continuous dynamical systems. *Proc. Nat. Acad. Sci.* **30**, 1940. pp. 370–376.
28. Kakeya, S.: On an infinite number of linear integral equations. *Scientific Rep. Tohoku University* **5**, 1916.
29. Kameda, T.: A general method for solving linear integral equations, Part I. *Jour. of Math-Phys Soc. Japan, 3rd series* **10**, 1927. pp. 231–235. Part II., *do., 3rd series,* **11**, 1928. pp. 17–27. *Part III., do., 3rd series,* **11**, 1928. pp. 169–180.
30. Kanazawa, T.: Oscillation of a rectangular plate under an arbitrary load. *Journ. Soc. Naval Arch. Jap.* **82**, 1940. pp. 31–45. (In Japanese.)
31. Kitagawa, T.: On the approximate solution of Volterra's integral equation of the 2nd type. *Functional Equations and Applied Analysis* (Osaka Univ.) Nos. 19, 20, 21. 1930. (In Japanese.)
32. Keune, F., Oswatitsch, K.: An integral equation theory for the transonic flow around slender bodies of revolution at zero incidence, *KTH AERO TN* 37.
33. Kondo, J.: Notes on the numerical solution of integral equations. *Seishin Studies.* **1**, 1954. pp. 10–27. (In Japanese.)
34. Kondo, J. Some notes on Wiener's prediction theory. *Bulletin of Statistical Research*, Statistical Bureau, Prime Minister's Office, **6**, 1954. pp. 1–10. (in Japanese.)
35. ——: Hiraki, H.: The critical Mach number as a characteristic value. *Appl. Math.* **1**, 1946. pp. 26–34. (In Japanese.)
36. ——, J.: Renewal problem of active carbon, *Journ. Op. Res. Soc. Japan* **2**, 1959 (in printing).
37. Kondo, J.: An estimation of mortality, *Management Science* (*Journal of the Japan Operations Research Society.*) **2**, 1957. pp. 66–71. (in Japanese).
38. ——: A numerical solution of non-linear ocillation, Read to the *Annual Meeting of the Japan Mathematical Society,* May 26 1949 (unpublished report).
39. Lee, E.H.: The impact of a mass striking a beam, *Trans. A. S. M. E.*, Vol. **62**, 1940. p. 66.
40. Lennertz, J.: Beitrag zur Frage nach der Wirkung eines Querstosses auf einen Stab. *Ing. Archiv.* **8**, 1937.
41. Levi-Civita, T.: Sull' inverzione degli integrali definiti nel campr reale. *Atti. Ac. Torino.* **31**, 1895.
42. Liouville, J.: Mémoire sur quelques questions de géométrie et de mécanique et sur un nouveau genre de calcul pour résoudre ces questions. *Journ. École Polytechn.* **21**, 1882.
43. Lotz, I.: Berrechnung der Auftriebsverteilung beliebig geformten. Flügel, **22**, 1931. pp. 189–195. *Zeit. Flugtechnik u. Motorluftschiffahrt.* **24**, 1933. p. 664.
44. ——: Mémoire sur le calcul des differentielles à indices quelconques. Flügel, **22**, 1931.
45. Lotkin, M.: On a certain type of nonlinear integral equation. *Bull. Amer. Math. Soc.* **50**, 1944. pp. 833–841.
46. Mikhlin, S.: Integral equations and their application to certain problems in mechanics, mathematical physics and technology Leningrad, 1944. (transl. Armstrong. A.H. 2nd ed., New York, 1957); International Series of Monographs on Pure and Applied Mathematics Vol. 4, Pergamon Press.
47. Nickel, K.: Lösung einer Integralgleichungs system aus der Tragflügeltheorie. *Math. Zeit.* **20**, 1951. pp. 81–96.
48. Noble, B.: The Wiener–Hopf Technique. (New York, 1958, Vol. 7, Pergamon Press).

49. Picard, E.: Sur les équations intégrales de troisième espèce. *Ann. École supér. 3rd series*, **28**, 1911.
50. Poincaré, E.: Remarque diverses sur l'équation de Fredholm. *Acta Math.* **33**, 1910. pp. 57–89.
51. Prager, W.: Druckverteilungen au Körpern in ebene Potentialströmung. *Physikalische Zeitschrift* **29**, 1928. pp. 865–869.
52. Prasad, G.: On the numerical solutions of integral equations. *Abst. Intern. Congress of Toronto.* 1924.
53. Radenkovoć, D.: Bending of a curved bar in its own plane, *Quart. Journ. Mech. and Applied Math.*, **7**, 1954, pp. 385–398.
54. Rothe, R., Schmeidler, W.: Höhere Mathematik. 1956, (Stuttgart) Teil VII, Chapter 2.
55. Schmidt, E.: Entwicklung willkürlicher Funktionen nach systemen Vorgeschriebener. *Math. Ann.*, **63**, 1906. pp. 433–476.
56. Schmidt, H.: Strenge Lösungen zur Prandtlschen Theorie der tragenden Linie. *Z. A. M. M.* **17**, 1937. pp. 101–116.
57. Sears, W.R.: A new treatment of the lifting-line wing theory, with application to rigid and elastic wings, *Quart. Applied Math.*, **6**, 1948. pp. 239–255.
58. Spence, D.A.: The lift coefficient of a thin, jet-flapped wing, *Proc. Roy. Soc. London, A.* **238**, 1957, pp. 46–68.
59. Tani, I.: A simple method of calculation of aerodynamic characteristics of a monoplane. *Aero. Inst. Tokyo Univ. Rep.* No. **197**, 1940. (In Japanese)
60. Tea, P.L.: A menchanical integraph for the numerical solution of integral equations. *Journ. Franklin Inst.* **245**, 1948. pp. 403–419.
61. Timoshenko, S.: Zur Frage nach der Wirkung einer Stostze auf einer Balken. *Zeit. für Math. u. Phys.* **62**, 1913. pp. 198–203.
62. Tranter, C.J.: A further note on dual integral equations and an application to the diffraction of electromagnetic waves, *Quart. Journ. Mech. Applied Math.* **7**, 1954, pp. 317–325.
63. Trefftz, E.: Prandtlsche Tragflächen und Propeller-Theorie. *Z. A. M. M.* **1**, 1921. pp. 206–218.
64. Tricomi, F.: On the finite Hilbert transform. *Quart. Journ. of Math.* **10**, 1951. pp. 199–211.
65. Tricomi, F.G.: Integral Equation. (New York, 1957, Series of texts and monographs, Pure and Applied Mathematics Vol. 5).
66. Vekua, I.: On Prandtl's integro-differential equation, *Priklandnaya mat. i mech.* **9**, 1945. pp. 143–150.
67. Villat, H.: Sur la résolution de certaines équations intégrales. *Acta Math.* **40**, 1916. pp. 101–178.
68. Volterra, E.: Vibration of elastic systems having hereditary characteristics. *Journ. Applied Mech.* **10**, 1950. pp. 363–371.
69. ——: On elastic continua with hereditary characteristics. *Journ. Applied Mech.* **11**, 1951. pp. 273–279.
70. Volterra, V.: Alcune osservazioni sui fenomeni ereditari. *Rend. della. Reale Accademia Nazionale dei Lincei.* **9**, 1929.
71. ——: Sulla inversione degli integrali definiti. *Atti Ac. Torino.* **31**, 1895–1896.
72. ——: Sulle equazioni integro-differenziali della teoria dell' elastica *Rend. della Reale Accademia Nazionale dei Lincei, 5 ser.* **18**, 1909. pp. 295–301.
73. Watkins, C.E. and Berman, J.H.: On the kernel function of the integral equation relating lift and downwash distributions of oscillating wings in supersonic flow. *NACA Rep.* 1257, 1956.
74. Watkins, C.E., Runyan, H.L., Woolston, D.S.: On the kernel function of the integral equation relating the lift and downwash distribution of oscillating finite wings in subsonic flow, *NACA Rep.* 1234, 1955.
75. Whittaker, E.G.: On the numerical solution of integral equations. *Proc. Roy.*

Soc. London. A **94**, 1918. pp. 367–383.
76. Woolston, D.S., Sewall, J.L.: Use of the kernel function in a three-dimensional flutter analysis with application to a flutter-tested delta-wing model, *NACA TN* 4395, 1958.

Historical Notes

Short history of the theory of integral equations

The first appearance of integral equations in the history of mathematics was in the thesis of N.H. Abel (1802–1829) on tautochrones, which was published in 1823 and 1826. J. Liouville (1809–1882) introduced integral equations for himself in the problem of the gravity of a stick of infinite length (§ 10.8) in 1832 without ever knowing of Abel's work. Then in the year 1837, he published a discussion on the relation between integral equations and differential equations in which he showed that the particular solution of a certain differential equation is given by an integral equation and originated the method of solution by successive substitution.

Studies on the vibration of a continuous material and its associated potential boundary-value problem started from the remote age of Daniel Bernoulli (1700–1782) and a number of other mathematicians who had focused their attention on these, while in 1861 Joachinstahl discussed Liouville's problem, and E. Beltrami (1835–1900) found the symmetrical potential which has a given value on a circular plane and whose normal line derivative vanishes. This is a special case of Dirichlet's problem. In 1885, J.T. Schwarz (1930–) proved the existence of the eigenvibration of a screen. Then, at the end of the nineteenth century, H. Poincaré (1854–1912), C.G. Neumann (1832–1925), V. Volterra (1860–1940), and others each developed his own original method. Around the same time, H. von Koch (1870–1924) introduced the idea of the infinite-dimensional determinant which is associated with simultaneous equations having infinitely many unknowns. It may well be said that Fredholm's masterly theory on integral equations that we are studying came into being because of circumstances. E.I. Fredholm (1866–1927) studied the method for solving the integral equation of the second kind [F_2], associating it with Dirichlet's problem of potential theories, and completed the theory in close analogy with the case of simultaneous integral equations of the first order (Chapter 2). Then he expressed the solution as the ratio of the determinant of von Koch's form, and showed that each can be expressed as a power series of λ, introducing the parameter λ.

In the winter of 1900–01, a Swedish mathematician, E. Holmgren

(1872–), attended Hilbert's seminar held in Göttingen and there introduced Fredholm's theories. These gave Hilbert no small stimulus and he gave himself up entirely to this problem for more than ten years, publishing many studies in the Göttingen report, beginning in 1904, and latter summing them up in a famous work *Grundzüge einer allgemeinen Theorie der Linearen Integralgleichungen*, Leipzig (1912 edition) (see (9) in the bibliography). He proved that the differential equation of the vibration of a screen can lead to a Fredholm-type homogeneous integral equation with symmetric kernel by a Green function, and studied this problem, converting it to the eigenvalue problem of an integral equation, and established a general theory on orthogonal system expansion, paying attention to the fact that this problem can be considered to be the limit of the problem of transforming a quadratic form of n variables to the standard form with regard to a principal axis. In this process, Hilbertian space was born as its condition.

The fundamental study of functionals begins with a series of theses by Volterra, which were published from 1887 in *Rendiconti delle R. Academia dei Lincei*. He established the method for solving integral equations by iterated kernels (§ 2.1) using the idea of the convolution of functions (§ 1.4), and extended the theory of integral equations, to include integro-differential equations (§ 9.5), and so on (Chapters 7 and 9).

In this way, the study of integral equations marked the central problem in the mathematics field at the beginning of the twentieth century, and we see studies by F. Riesz (1880–1956), Toeplitz (1881–1940), and E.D. Hellinger (1883–1960) carried out over long periods, many reports of studies on the continuous spectra of integral equations by Hellinger and T. Carleman (1892–1949), and many contributions to integral equations by C.E. Picard (1856–1941) and E. Goursat (1858–1936).

We may arrange the boundary-value problems of linear ordinary differential equations into a well-organized system by applying the theory of integral equations. As for the Sturm–Liouville type of boundary–value problem, the explanation of which is given in § 10.1, we have the expansion theory by H. Weyl (1885–1955) and M.H. Stone (1903–) for the case when the boundary points $t = a$, $t = b$ become singular points, and this theory was completed by E.C. Titchmarsh (1996–1963) and K. Kodaira (1915–). In September 1954, Kunihiko Kodaira was given the Field Prize for this study at the twelfth general assembly of the International Mathematicians Conference held in Amsterdam.

The integral equation with a singular kernel (Chapter 8) holds an important place in applications, yet its theory leaves room for further investigation, and strong demands for practical methods of solution are being met, especially with the development of computers. The finite element method is introduced to obtain a numerical solution for a boundary-value problem when the fundamental equation is expressed in the form of integral equations. In 1918, L. Prandtl (1875–1953) published the theory of lifting lines of a three-dimensional wing and expressed lift distribution along the wing breadth in an integro-differential equation (§ 10.9). The nonlinear integral equation (Chapter 9) is studied by the theory of the linear operators in functional space, which has developed into the central problem of modern analysis; yet to give the survey here is beyond the scope of the present volume.

Although the above is generally accepted, there is an opinion that the first appearance of an integral equation was marked by P.S. Laplace (1749–1827) when he treated

$$\int e^{t\xi}x(\xi)d\xi = f(t), \qquad \int \xi^{t-1}x(\xi)d\xi = f(t)$$

in solving differential equations in 1782; but we take Abel's case as the one where the form of an integral equation was first explicitly considered.

From the point of view of applications the development of the theory in the case of nonlinearity is important, as in the case of differential equations. Several examples are included in the volume.

Index